POWDER METALLURGY

POWDER METALLURGY

Processing for
AUTOMOTIVE, ELECTRICAL/ELECTRONIC AND ENGINEERING INDUSTRY

P. Ramakrishnan
Professor Emeritus, I.I.T., Bombay

NEW AGE INTERNATIONAL (P) LIMITED, PUBLISHERS
New Delhi • Bangalore • Chennai • Cochin • Guwahati • Hyderabad
Jalandhar • Kolkata • Lucknow • Mumbai • Ranchi
Visit us at **www.newagepublishers.com**

Published by New Age International (P) Ltd., Publishers
First Edition: 2007
Reprint: 2008

Branches:

- 36, Malikarjuna Temple Street, Opp. ICWA, Basavanagudi, **Bangalore**. ✆ (080) 26677815
- 26, Damodaran Street, T. Nagar, **Chennai**. ✆ (044) 24353401
- Hemsen Complex, Mohd. Shah Road, Paltan Bazar, Near Starline Hotel, **Guwahati**. ✆ (0361) 2543669
- No. 105, 1st Floor, Madhiray Kaveri Tower, 3-2-19, Azam Jahi Road, Nimboliadda, **Hyderabad**. ✆ (040) 24652456
- RDB Chambers (Formerly Lotus Cinema) 106A, Ist Floor, S.N. Banerjee Road, **Kolkata**. ✆ (033) 22275247
- 18, Madan Mohan Malviya Marg, **Lucknow**. ✆ (0522) 2209578
- 142C, Victor House, Ground Floor, N.M. Joshi Marg, Lower Parel, **Mumbai**. ✆ (022) 24927869
- 22, Golden House, Daryaganj, **New Delhi**. ✆ (011) 23262370, 23262368

ISBN (10): 81-224-2030-3

ISBN (13): 978-81-224-2030-2

Rs. 495.00

C-08-05-2579

2 3 4 5 6 7 8 9 10

Printed in India at Saras Graphics, Rai, Haryana.
Typeset at Akriti Graphics, Delhi.

PUBLISHING FOR ONE WORLD
NEW AGE INTERNATIONAL (P) LIMITED, PUBLISHERS
4835/24, Ansari Road, Daryaganj, New Delhi-110002
Visit us at **www.newagepublishers.com**

PREFACE

Powder metallurgy processing conserves energy, raw materials; minimize pollution and machining in relation to other manufacturing technologies resulting in reduced cost. Another attraction of the process is the ability to fabricate high quality complex parts to close tolerances from high performance and specialty materials in an economical manner. As a material processing technology, rapid developments are taking place including in the upcoming area of nanotechnology and the field of applications are steadily expanding in automotive, electrical, electronic, machine tools and other engineering industries. Automotive industry is the major customer of powder metallurgy. In India there is a substantial growth in automotive sector consisting of passenger cars, commercial vehicles, multi-utility vehicles, tractors, two and three wheelers. Several Indian and international companies in this industry have operations in India and using this facility for local and export oriented production. Powder Metallurgy Association of India, the premier organization in our country, has organized a very successful International Conference on Powder Metallurgy for automotive and engineering industry with Trade Exhibition at the Renaissance Mumbai Hotel and Convention Center during Feb. 3–6, 2005. I am thankful to the members of the Organizing, National and International Committees, authors, speakers, chair persons, sponsors, co-sponsors, advertisers, exhibitors and all those who have contributed to the success of the conference and publication of the proceeding.

The book contains selected 36 papers in 5 sections. These are trends in powder metallurgy, powder metallurgy for automotive industry, properties of powder metallurgy products, powder metallurgy for electrical/electronic industry and powders, processing and applications. A variety of powder metallurgy topics are covered ranging from status of powder metallurgy in Europe, technology and markets for copper/bronze/brass powders, high density multilevel components by high velocity compaction, metal injection molding of low alloy steels, sinter hardening, fatigue, friction and wear of steels, brake materials, nano and ultra fine titanium, nano and fine ceramic processing by self propagating high temperature synthesis, nano materials and powder metallurgy to particulate materials processing of high performance cements for construction industry. The articles presented in the book have been contributed by leading experts from educational institutions, industries and research and development laboratories from India and abroad. The book will serve as a valuable source for current state of developments and future trends in the expanding area of powder metallurgy and particulate materials.

P. Ramakrishnan
Convener, Intl. PM-05
Professor Emeritus, I.I.T., Bombay.

CONTENTS

Unit–I

Trends in Powder Metallurgy

1

Status of PM World in Europe

Dr. Cesar Molins, Jr.
AMES, Director General
EPMA President, Spain

ABSTRACT

This paper is an update on the evolution of the different areas of Powder Metallurgy in Europe, from the viewpoints of business volumes, corporate strategies and research institutions. It also presents the evolution of usage of PM products in the automobile industry, main customer of PM parts. The main areas of PM industry considered are raw materials, machinery and equipment, parts making and hard materials. Special consideration is given to the development of the East-European countries.

It shows the positive evolution of the PM industry in Europe in spite of the rather flat situation of car production and the trend of moving industry from west European countries towards the east European ones.

POWDER SHIPMENT STATISTICS AND FORECASTS

The latest data gathered from the main PM associations worldwide for a whole year refer to 2003, as 2004 data is not yet fully compiled. The worldwide shipment of powders for 2003 ammounted to an estimated 880,000 metric Tons, about half of which were shipped to the USA, Europe and Japan following with 18% and 13% respectively. It is to be noted, however, that the rest of the world, which includes a number of developing countries such as China, India, Brazil, Iran and so on accounted already for a total sum equaling that of Europe and that these countries' share is very likely to increase in the years to come, considering the fast development of some of those areas. Figure 1 is a pie chart displaying the percentual distribution of powder shipments to the different areas of the world.

On a worldwide scale, manufacturing of parts by the PM route enjoys continued growth after the reduction experienced at the end of the economic expansion in 2001. Its present growth rate is about 8% per annum, mostly due to the usage of PM in new applications and to an increased acceptance by designers, which is increasingly turning into PM the manufacture of parts that were previously designed for other metal forming technologies. We can see as new PM applications the VVT technology in cars, which is a heavy user of PM due to the complexity and precision required in its components and an example of increased market penetration can be seen in applications for gearbox components. The car sales trends of increasing volumes of Sports Utility Vehicles, which are larger, heavier and with more powerful engines than standard cars is also a factor influencing the growth of sales of PM parts. Figure 2 shows the evolution of metal powder sales for PM applications since year 2000, including the forecast for 2004.

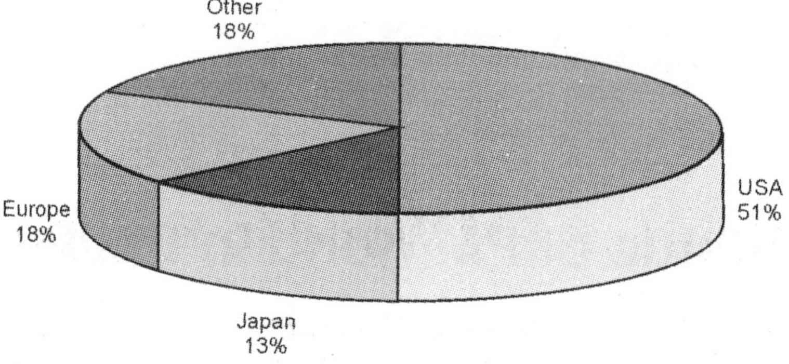

Fig. 1. Powder Metal shipments in the world in %. 2003 data.

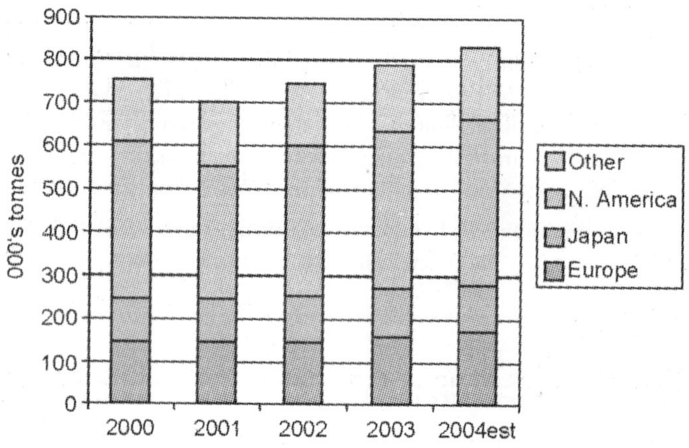

Fig. 2. Shipments of ferrous powders for PM parts applications.

SOME MAJOR FACTORS INFLUENCING EUROPEAN PM INDUSTRY

Taking a closer look at the European situation (Fig. 3) we can see that the dip of production experienced in the worldwide scale in 2001 did not affect Europe at that time (the main worldwide dip was caused by the slowdown in the USA) but it did affect Europe to a lesser degree, in 2002. Since that time powder shipments in Europe have developed in a sound manner in spite of the essentially flat and even slightly downwards trend of car sales in western Europe, as can be seen in Fig. 4. This further proves the increased usage of PM in European cars, as was previously explained.

The growth of the contents of PM parts in European cars becomes apparent from the two above figures and it can be seen in the graph in Fig. 5. We should note, however, that European (and Japanese as well) cars contain significantly less weight of PM parts than their American counterparts, which are today in the 18 Kg range, the main reasons being the smaller size of European cars and the smaller penetration of automatic transmissions in European markets.

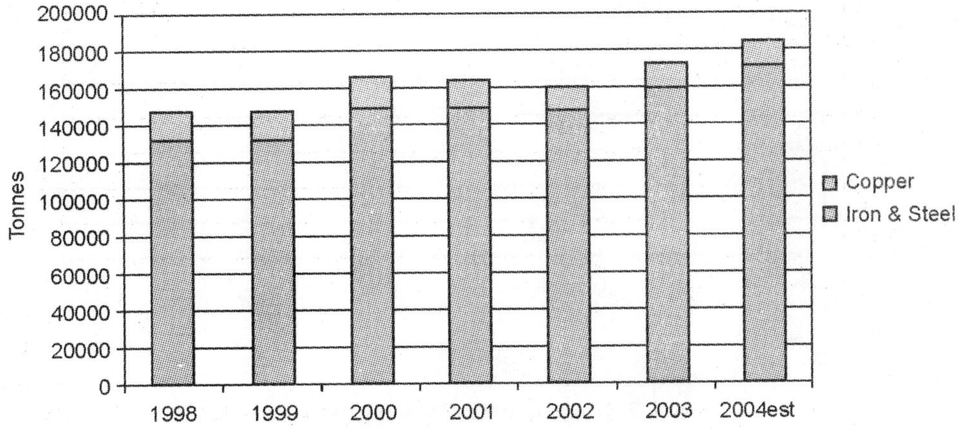

Fig. 3. European power metal shipments.

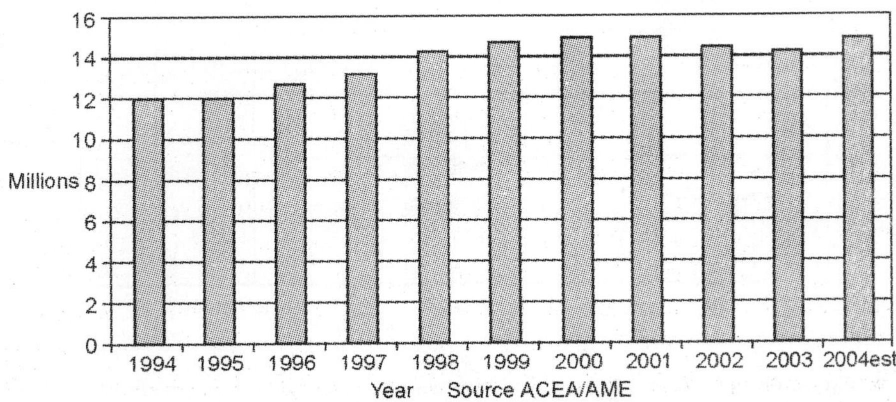

Fig. 4. Western Europe cars sales.

The smaller market share of automatic transmissions in Europe, which can have different explanations ranging from simply cultural likes to morphology of roads and relative abundance of twisty mountain roads in comparison to the USA or even sheer cost is expected to move in the sense of a wider acceptance of automatic transmissions in the future, as can be seen in the rather optimistic outlook graph in Fig. 6. Another reason to expect further increase of the PM parts contents per car in Europe is the increasing acceptance of modern diesel engines, which already account for a large share of cars manufactured in Europe and which is still hoped to grow in the future, as shown in Fig. 7.

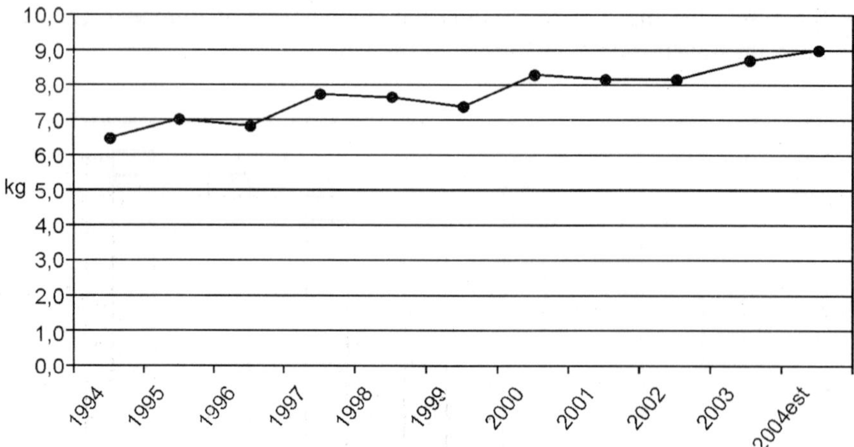

Fig. 5. Usage of PM parts in European cars.

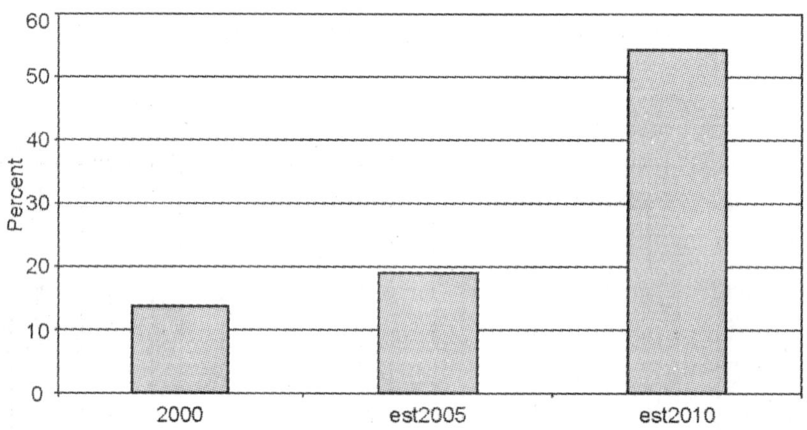

Fig. 6. Forecast growth of usage of automatic transmissions in European cars.

The expectations of better business environment in Europe for the PM industry described so far have however some threats as well, such as the dramatic surge in raw materials, lubricants and energy prices occurred between end of 2003 and beginning of 2005 which in the case of powders have ranged from 10 to over 25% and the development of a very stringent piece of environmental legislation, previously called NCP (New Chemicals Policy) and now labelled as REACH, which aims at requiring proof of being "harmless to humans and to the environment" for all "substances" purchased, sold or used in processes in the EU. This legislation is scheduled to come into effect in 2006. There are strong lobbying battles trying to limit the scope of what is considered as "substances" and the treatment of alloys and their different mixing ratios as just mixes of elements or as different substances. The threat to European industry of this over-restrictive draft legislation goes beyond PM industry as it affects most heavily the chemical industry, and all industries in general, as very few industries could be considered to use, buy or sell absolutely "no substances". The latest news from the lobbying activities seem to indicate that several governments of EU countries are aware of the big threat posed by this draft legislation for the competitiveness of the European industry and that there is activity in the sense to try and make the requirements less stringent than originally drafted.

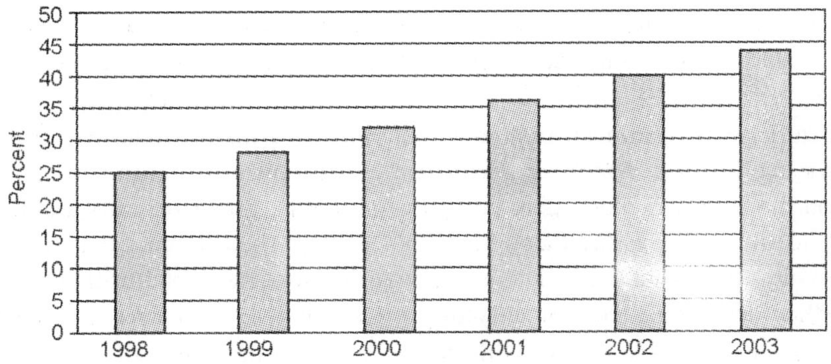

Fig. 7. Percentage of cars made in Europe with diesel engine.

Again a factor on the positive side for the outlook of PM industry in Europe is the development of central and eastern European countries, many of which just recently became members of the Union, while others are yet working their way into the EU. All those countries have a number of cars per inhabitant well below the average for west Europe and this is a situation clearly bound to converge in the future. This means a sizeable market, while on the other hand, and also with a positive effect, there is the contribution of lower labour costs by the new EU member, states which will make European industry more competitive. This will happen, however, at the expense of some efforts and sacrifices on the side of the wealthier EU member states, which are already suffering a clear migration of industry towards the lower cost areas. This migration is likely to continue for a number of years and will definitely bring headaches to some of the western European countries. The imbalance of cars and wealth can be seen in the table of Fig. 8.

GDP

Name	*Population (m)*	*GDPI capita $000*	*Cars/00 people*
EU15	377,9	23,2	50
New EU 10	74,9	10,7	28
Bulgaria	7,9	5,7	26
Romania	22,4	5,5	14
Turkey	68,7	5,2	7
Total	173,9	7,6	19

Fig. 8. New and projected EU members compared with EU-15.

EUROPEAN FERROUS PARTS SECTOR

As can be deduced from the previous explanations, the European PM parts sector is in general healthy and expanding. The frentic activity of purchases and mergers of the late 90's is today reduced, although some merging activity still exists. However, in addition to consolidation movements, it is possible to find new startups, both in eastern and in western Europe and both as subsidiaries of large existing groups or as completely new starts as well. In general, there is a marked trend of migration towards lower cost countries, driven by the fierce competition surrounding all activities related to the automobile industry.

On the technical aspects, European customers continue to demand higher performing materials for more demanding applications, which is driving the efforts towards higher density materials and new alloys, while on the geometry side, more complex and higher precision parts are demanded on a constant basis. Applications such as the mentioned VVT require tight tolerances and geometries that were all but impossible a few years ago. There is a non-declared race for fulfilling the demands of gearbox gears, with their demanding requirements for strength, fatigue resistance and geometry demands.

Quality continues to play a key role while it is coming to the limits of what is conceptually understandable from an engineering viewpoint. It is common today for customers to demand and accept quality levels in the area of the low-ppm defects, as it is hardly understandable to demand absolute zero defect. However, it takes only one defective part detected to trigger vigorous activity in search of its cause and possible root remedy. More worrisome in this respect is the increasing pressure from the market for liability related to quality problems. While it seems logical that one faulty part priced at a few cents of an Euro should not bear liability in the range of millions of Euro in costs of recalls, replacements, plant stoppages, etc., there is no clear policy as to what is the limit to the liability demands on the part of the market. So far, most known problems are settled in agreements between customer and supplier, but the mounting pressure and volume of the claims may represent a problem for the future handling of such situations.

In the same way as we mentioned the technological advancements of PM technology in its various fronts as the way to new markets and applications for PM, the same thing can be said of competing technologies which, in their progress, conquer some of PM's traditional applications. Such is the case, for example, of ABS sensor rings, which just a few years ago represented a nice and increasing field for PM while today they are being increasingly transferred to other technologies.

In addition to competition from different technologies, European PM parts makers are facing an increasing presence of competitors from Asia and South America.

EUROPEAN HARD MATERIALS SECTOR

Europe is the main manufacturer worldwide of hard materials and, like the PM parts sector, the hard materials sector is experiencing steady growth. Consolidation activity also continues in small scale but more importantly, European hard materials manufacturers are increasing their sales by presenting new materials and applications for niche markets. Functionally graded materials and hard tips shaped by MIM technologies are some examples. Fig. 9 shows the evolution of hard materials shipments in Europe.

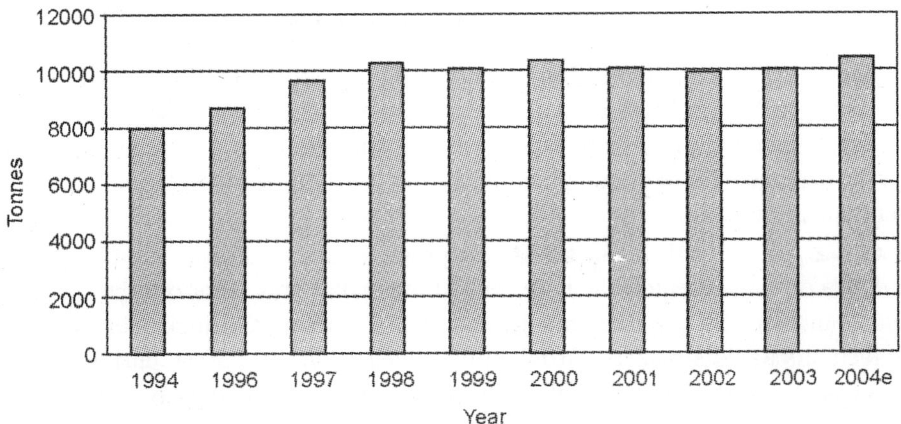

Fig. 9. Estimated European hard materials shipments.

EUROPEAN EQUIPMENT SECTOR

The European PM equipment sector is now recovering from a very slow period associated to the downturn years 2001-2002. Its present sales come mainly from developing countries and in particular from Asia.

There are continuous and remarkable improvements in all areas of equipment. On the presses side, there is the improved high-speed multiplaten hydraulic and mechanical-hydraulic presses or the development of the High-Velocity Compaction concept. On the furnaces side, improved atmosphere controls, better rapid-cooling furnaces for sinter-hardening applications or medium-high temperature furnaces using mesh belt are some examples of a technologically very active sector.

EUROPEAN ACADEMIA AND RESEARCH CENTRES

Europe enjoys a very extense and dynamic academia in PM related fields. Over 400 researchers in some 40 universities and research institutions, spread across more than 12 countries, form an active research network with frequent multi-national research projects, which combine research institutions and industry as well. This networking activity is fostered from the EU authorities as a means to develop a more united Europe and in order to better take advantage of all expertise within the EU, wherever this may be found.

European academia is heavily committed to EPMA and is the motor behind the very successful PM Summer School which has been held for a number of years, in varying countries and counting with high experts both from academia and industry as teachers.

Research is carried on new materials as well as improved processes, materials testing and characterization, process modelling, etc.

Academic institutions with PM activities, whether research or teaching, are spread through a large number of European countries, some of the most active being Germany, Spain, Italy, France, Sweden, the U.K., Austria, etc.

EPMA MAIN ACTIVITIES

EPMA as the main association in Europe gathering and representing the PM-related industries, academia and research has amongst its goals the promotion of PM, coordination and promotion of research efforts and the representation of industry before the European governmental bodies. EPMA's members currently include 175 corporate members from 27 countries worldwide, over 50% of which are components manufacturers.

EPMA was, along with MPIF and JPMA, the promoter and sponsor of the Global PM Database, the first worldwide on-line accepted database for PM materials. This notorious data set is the result of gathering data available from the three main member associations and their members, ranging from powder manufacturers to parts makers. Officially launched in October 2004, during the PM'04 World Congress in Vienna, the World Database is the most comprehensive set of data available for PM materials and it is prepared to be directly accessible from CAD packages in addition to direct on-line consultation. It represents a major step forward to clarify to the designers of the world the characteristics of PM materials, backed by the most respected organizations in the world.

Other important activities and services of EPMA to the PM community is the organization of benchmarking exercises, needed for the ISO 16949 certifications or the elaboration of industry statistics. The yearly series of PM conferences, including the rotating biannual series of World PM Conferences are some of the most visible activities of EPMA.

When organizing its conferences and seminars, EPMA has a clear policy of choosing varying countries in its largest possible scope in an effort to make it clear that EPMA's goal is that of promoting the whole of European PM and to help PM worldwide.

In organizing and coordinating multinational research efforts, EPMA represents all PM industry before the EU agencies wherever there is a possibility of obtaining governmental support for research or networking activities.

EPMA is also very active representing PM industry in lobbying activities and in Health, Safety and Environmentally related matters before the EU. Such is the case when presenting facts about health and environmental matters related to the REACH policy which is in preparation.

CONCLUSIONS

European PM industry is evolving in a positive manner, in spite of a rather weak European economical situation.

Use of PM parts continues to increase in European cars.

East-European countries playing an increasing role in industry in general, and in the PM industry as well.

Raw materials and energy cost increases to have a strong impact on the financial results of European PM industry.

Continuous technical improvements both in parts making and in new equipment.

Increased pressure on quality and liability issues.

Need to monitor and lobby EU legislation.
EPMA very active promoting and uniting PM industry.
Major steps forward in world PM promotion through the Global PM Database.
New applications appearing in the market for PM.

ACKNOWLEDGEMENT

The author wants to thank especially the staff of EPMA for supplying most of the data contained in this paper and for the help in preparing it.

2

Technology and Markets for Copper/Bronze/ Brass Powders–Present and Future

Edul M. Daver
ACuPowder International, LLC, U.S.A.

ABSTRACT

The paper provides a brief background about manufacturing technology evolution from "Reduction" to "Electrolytic" to "Atomization" and application evolution from bronze bearings and filters to a host of other application. The paper will then discuss specific **Copper/Bronze/Brass powders** which have been developed by technology customization for relatively new applications such as admixing, infiltration, brazing, micro-bearings, injection molding, bi-metallic split bearings, composite materials, diamond tools, water filtration rapid prototyping, bullets, and artistic applications. It will finally provide a glimpse into future trends and opportunities.

THE BIRTH OF COPPER POWDER

No story of copper powder is complete without the story of Professor Everett J. Hall - assayer, teacher, inventor, pioneer, and humanitarian. [1] Everett Hall attended Columbia University in New York City in the late 1890's focusing on chemistry and assaying of metals. He graduated in 1900, and around this time John Randolph, a representative of Tata Iron Company, India showed up at Columbia University looking for an "assayer" for their Manganese mines in Hyderabad, India. Everett Hall, who was now 23 years old, took the job, and his diary of Tuesday, September 18, 1900, reads as follows, "I sailed on the SS Maria Theresa N.G.L. in the employ of John Randolph engaged as assayer and general utility man with salary of $100/month and expenses. Destination Hyderabad, India".

Sunday, Nov. 4: "From Madras to Raichur. I have arrived at the scene of my labors for the next several months. Visited assay labs at Wondalli. Looked over some of the dumps on the Tata Iron Company's property at Topaldoddi and settled in with Mr. Harris. Between assaying I played tennis and billiards and did some riding".

Everett Hall completed his contract, continued sailing around the world and returned to Columbia University as Assistant Professor of Metallurgy. His interest started evolving toward metal powders, especially zinc powder for the chemical industry and aluminium powder for the explosives, flares and military pyrotechnic applications.

He decided to commercialize his inventions, and with the help of William Prime, an investor, Metals Atomizing Company was incorporated in New York on July 5, 1916. The name was changed to Metals Disintegrating Co. (MD) on October 10, 1916, and manufacturing started in Bound Brook, NJ. MD produced Zinc Powder/Dust in 1916, Tin and Aluminium powders in 1917. In 1919 a new plant was commissioned in Elizabeth, NJ (later to become part of Union) which remains the current site of ACuPowder International, LLC. Copper powder production by the chemical reduction of mill scale (copper oxide) from rod/wire/sheet mills started in 1920. In 1921, Harry Williams of General Motors Research Corp. in Dayton, Ohio started testing MD copper and tin powders for bearing applications. By 1922 Williams & Hall had developed a commercial porous self-lubricating bearing.

Within a few years, because of mill scales purity constraints and limited supply, work started to produce copper powders by other techniques such as electrolytic deposition and atomization. Like all new developments, commercialization was slow and required innovation, hard work and persistence. MD's process of air atomization, followed by roasting and reduction was quite successful, and by 1941 a very respectable 7,000 tons was being used for self-lubricating bearings. The Times of September 29, 1941 called "Bronze Self-Lubricating Bearings, the most radical innovation in metallurgy since the Bronze Age". During this period MD also developed bronze powders for Bronze Filter for use as a gasoline filter and, most amazingly, Professor Hall continued to teach and do research at Columbia University. He frequently said, "I teach because Youth are so interesting and so unpredictable". Professor Everett Hall died on September 3, 1931 at age 54.

With demand and competition increasing, in the 1950's MD developed water atomization of copper powders. Between the 1950's to today, three major processes compete for the wide range of applications that have developed. These manufacturing processes include:

1. Electrolytic Deposition
2. Oxide Reduction
3. Water Atomization.

STATE OF CURRENT TECHNOLOGY

A. Copper Powders

There are three major production processes that continue to be used for a range of applications depending on specific properties required and economics.

1. Electrolytic Copper Powders

The electrolytic deposition process is relatively expensive and environmentally unfriendly. However, its particle shape is inherently dendritic and irregular with apparent densities from 0.7 g/cc to 2.8 g/cc, with resulting good green strength characteristics. Because of its higher cost, it is generally used when other types of powders cannot be used effectively and its higher price is justifiable.

2. Oxide Reduced Copper Powders

In this process, copper oxide is chemically reduced to copper powder. The source of copper oxide could vary but it needs to be clean, with controlled particle size. The most effective commercial process is one in which copper powder is atomized by air or water and then roasted to form copper oxide. The copper oxide is then ground and chemically reduced to form copper powder.

The process is tedious and relatively expensive, though less than Electrolytic powders. The powder shape is relatively irregular with apparent densities in the range of 1.0 g/cc to 3.2 g/cc, with resulting good green strengths. It competes with electrolytic copper at the low apparent density range and with water atomized copper at the higher apparent density range.

3. Atomized Copper Powders

Atomization is a flexible and cost effective process. Powders may be atomized horizontally, vertically down or vertically up utilizing gas, air or water depending on particle shape and size required.

Gas/Air atomization is used to produce spherical copper powder with an apparent density range of 4.0-5.0 g/cc. Water atomization is used to produce irregular and fine powders with apparent densities in the range of 2.3-4.0 g/cc with moderate green strengths.

The processes are efficient and cost effective, and are the preferred choice if they can be utilized effectively for a specific application.

Table I. Summary – Copper Powders

Process	*Shape*	*Size*	*Apparent Density*	*Green Strength*	*Cost*
	Spherical ↑	Coarse ↑	High ↑	Low ↓	High ↑
Gas Atomization					
Air Atomization					
Water Atomization					Low
Oxide Reduction					
Electrolytic					
	Irregular	Fine	Low	High	↓ High

B. Pre-alloyed Bronze Powders

The preferred for pre-alloyed Bronzes is atomization – gas, air or water. For special applications they are also produced by diffusion alloying a tin coating into copper powders.

Table II. Pre-Alloyed Bronze Powders

Process	*Shape*	*Size*	*Apparent Density*	*Green Strength*	*Cost*
	Spherical ↑	Coarse ↑	High ↑	Low ↓	High ↑
Gas Atomization					
Air Atomization					
Water Atomization					Low
Diffusion Alloying					
	Irregular	Fine	Low	High	↓ High

C. Brass Powders

The preferred method for pre-alloyed Brass powders is atomization - air or water. Air atomization yields irregular shaped powders because of zinc's tendency to oxidize quickly, not allowing the particle to spherodize. As such, there is only marginal difference between air and water atomized brass powders.

Water atomized powders tend to be slightly finer but also more expensive because of the need to de-water and dry.

MARKETS AND TECHNOLOGY — PRESENT AND FUTURE

1. Bronze Self–Lubricating Bearings

(a) Press and Sinter Bearings

Bronze premixes, containing elemental copper, tin, (graphite), and lubricant are pressed and sintered. Shifting trends include:

(*i*) Utilizing partially alloyed powders to reduce segregation and improve dimensional change stability.

(*ii*) Utilizing diffusion coated and/or fine alloyed powders to produce fine channeled porosity for micro-bearings.

(*iii*) Dimensional change stability to accommodate the shift in reducing gases from Dissociated Ammonia or Endothermic to N_2/H_2 mixes. The reduced H_2 content reduces the growth which then has to be compensated.

(*iv*) Higher Green strengths for increased mechanical handling of green parts.

(b) Steel-Backed Split Bearings

Pre-alloyed bronze powder of relatively narrow size distribution is bonded onto steel strip and porosity filled with a solid lubricant such as PTFE and lead, or overlaid with acetal resin. Strips are then slit, cut and bent to form a split bearing. The steel strip is pre-plated with a copper coating to allow the bronze powder to be sintered onto the strip.

Future trend is to allow the bonding to take place without the need for the expensive copper pre-plating on the steel strip.

(c) Bronze-Plastic Composite Bearings

Pre-alloyed bronze powder is mixed with plastics such as Nylon (PTFE), pressed and cured (sintered) in air or nitrogen. One of the key needs is to retain color without oxidation/decomposition discoloration. This is being achieved with the use of zinc in the bronze powder. A popular composition is 89 Cu/9 Sn/2 Zn. The bronze filler reduces the wear rate of Teflon bearing many fold.

2. Highly Porous Applications

(a) Bronze Filters

Narrowly graded pre-alloyed bronze powders are gravity sintered in graphite or stainless molds. Powders for these applications are produced by

(*i*) screening narrow cuts from pre-alloyed bronze powders;

(*ii*) diffusion coated, narrowly sized spherical copper powders or mono-sized cut copper wire coated with tin. This is a mature market facing competition from other materials such as plastic filters and pressed wire/fibers. As such, the challenge here is to compete effectively by reducing powder cost and increasing processing efficiency.

(*b*) *Copper Heat Pipes*

In a computer the Heat Pipe takes the heat from the silicon chip to the Heat Sink. It consists of copper tubing with copper powder loose sintered to the tubing's inside wall. The tube is filled with water and sealed under vacuum. The heat from the chip vaporizes the water to steam which travels to the cool end and condenses. The porous copper layer wicks the condensed water and acts as a pump to transport it back to the hot side. Although each heat pipe uses a very small amount of powder the millions of computers being produced makes it an attractive new and growing application. The application demands very exacting chemical and physical powder properties.

(*c*) *Brass for Water Filtration*

Relatively coarse brass powder (85/15 Brass and 50/50 Brass) are loosely packed in a container. The medium removes chlorine, heavy metals and bacteria from the water passing through it. Such a filter bed is utilized by water treatment plants and consumer products such as shower heads. The application requires exacting chemical purity and powder size distribution.

3. Structural Components

(*a*) *Copper*

High density parts are pressed and sintered for maximum sintered density and accompanying maximum electrical (>90% IACS) and thermal conductivity; e.g., electrical contacts. The trend is towards the utilization of higher purity copper powder and special clean burning scavenging lubricant.

Heat sinks with complex high surface area are a popular growing application which are being produced by metal injection molding (MIM). MIM requires very fine spheroidal copper powder with a specially binder system. Very fine spherical copper powder is very expensive which required development of a suitable compromise powder with less fines but also less expensive.

(*b*) *Bronze and Brass*

High density brass parts, especially for lock components is a mature market while high density bronze parts is growing consistently as a replacement for currently cast or machined parts. Specialty pre-alloyed bronze compositions develop >40,000 psi tensile strength cost effectively.

(*c*) *Green Ammunition*

Ever since the US Government declared its "Green Ammunition" initiative, which requires the elimination of lead from ammunition, it has become a huge opportunity for powders. Besides being eco-friendly, various other advantages are also being realized. Ammunitions being made for target practice are frangible and break-up at target without ricochet and therefore are not only safer, but also less expensive. Various material systems have been developed and are being commercially utilized for niche markets.

The military is interested in tungsten powder based ammunition held together with tin or a polymer such as nylon. Recent estimates are that the army plans to buy 400 million cartridges in 2005.

The law enforcement agencies and individual hunters are using lower density, lower cost bullets made from all metal copper or copper/tin press and sinter systems or combined with polymers and then cured by hot pressing.

4. Ferrous P/M Applications

(a) Copper as Alloying Admix

Copper powder is mixed into Iron premixes (2-20%) as an alloying element to impart strength and toughness. The trend is towards finer copper powders to improve diffusion and alloying efficiency and also dimensional control.

(b) Copper Infiltrant

Copper premix and alloys are used to melt and infiltrate Iron parts during sintering to provide higher densities and strength. Copper infiltrants are of two types – residue type and non-residue type. Copper infiltrants' demise was predicted at least 10 years ago with the advent of high density technology, but it keeps on going and going.

5. Composites

I am using composites as a very general term for material mixtures containing metals and non-metals. Metals in powder form is the natural choice for composites. Under this category, the following markets will be discussed:

(*a*) Carbon Brushes
(*b*) Friction Materials
(*c*) Cutting and Grinding Tools
(*d*) Fillers For Plastics
(*e*) Cold Casting.

(a) Copper-Base Carbon Brushes [2]

Copper powder for carbon brushes was one of the first applications, even before self-lubricating bearings. Brushes are components that transfer electrical current between the stationary and rotating elements in electrical motors and generators. Most common brushes are composites of graphite and a conductive metal such as copper or silver. Graphite provides the lubrication while the metal provides current-carrying capability. Typical copper contents is in the range of 20- 75%. Typically lower apparent density coppers such as Electrolytic or oxide reduced are utilized, but atomized powders are also used because of cost effectiveness. Graphite and copper powders are blended and molded and then sintered in a protective atmosphere between 500-800°C. Copper-graphite brushes are widely used in battery-powered tools and automotive applications that require high power output in small, lightweight packages.

(b) Copper-Base Friction Materials [3]

Sintered-metal-base friction materials are used in applications involving the transmission of motion through friction (clutches) and for deceleration and stopping (brakes). In these applications, mechanical energy is converted into frictional beat, which is absorbed and dissipated by the friction material. Copper-base materials are preferred because of their high thermal conductivity, however lower cost alternate systems are being used for less demanding applications.

Most friction systems containing copper powder are blended with other metal powders (brass, Sn, Ni Zn, Pb), solid lubricants, oxides and other non-metallic compounds. These are blended, compacted and sintered with simultaneous bonding to a backing plate. Metal powders with high surface area are preferred to provide a strong, thermally conductive matrix. Typically lower apparent density powders such as electrolytic or oxide reduced are utilized but atomized powders are also used because of their cost effectiveness.

(*c*) *Cutting and Grinding Tools*

Diamonds and other refractory ceramics are held in a matrix of metal and then processed by either hot pressing or cold pressing and sintering. The size of the abrasive particles, the method of molding and material to be cut, ground or polished influence the selection of metals powders. Copper, tin, bronze, nickel, cobalt and Iron are used to achieve the required melting point, toughness, brittleness and glazing properties required. The actual recipes are highly guarded proprietary information.

In general, goal is to match the wear rate to the material being cut/polished so as to expose fresh diamonds at the appropriate rate. First generation systems used elemental copper and tin to develop bronze matrix compositions from 90 Cu/10 Sn to 50 Cu/50 Sn. Higher tin contents lower the melting point and increase the hardness. Elemental premixes provide good green strength for cold pressing and sintering. However, segregation of tin could form a soft area and rejection of entire grinding wheel. Second generation systems utilized pre-alloyed bronze powders. This eliminated segregation issues but, because of the spherical shape, green strength was insufficient and hot pressing had to be utilized.

Third generation systems used either tin coated copper or irregular bronze powders (water atomized or CuSn/Zn) to eliminate segregation issues and also provide sufficient green strength such that both cold pressing and sintering or hot pressing could be utilized for processing.

(*d*) *Fillers for Plastics*

Metal powder fillers are added to various types of plastics to increase density and hardness, thermal and electrical conductivity, wear resistance, frictional characteristics, magnetic properties and radiation shielding. Metal-powder-filled resin composites find important applications as bearings, seals/gaskets, adhesives, magnetic tapes, and radiation shielding protection. Both copper and pre-alloyed bronzes are popular fillers. The process consists of mixing metal powder with resin matrix material and modifiers. Mixture is then molded or poured and resin matrix cured. Oxidation and/or decomposition is a common appearance problem with bronze filled systems which is resolved by the use of a special alloy containing zinc; *i.e.*, 89 Cu/9 Sn/2 Zn. Bronze filled self-lubricating bearing, discussed in an earlier section, is an important application, as is industrial seals and gaskets. Copper powder is used in numerous commercial adhesives where good electrical conductivity is required. Copper in anti-fouling paints and bronze filled maintenance build-up compounds are other applications.

(*e*) *Cold Casting*

Copper, bronze, and brass powders are added to polyester resin, urethane, epoxy, plaster, cement, etc. to simulate solid metal castings or introduce metallic flecks. It is widely used by artists, sculptors, and commercially for decorative floor/wall tiles and inserts. Metal powder is mixed with medium and added to a mold where it is cured. The finished artifact can be further burnished or given a "patina" finish.

6. Joining, Coating and Filling

Copper and copper alloys are used as filler material for joining, coatings, and fillings. The material either melts or softens to allow adherence or filling of the base materials. The specific applications discussed in this section include:

(*a*) Brazing and Soldering
(*b*) Thermal Spraying and Cold Spraying
(*c*) Rapid Prototyping.

(a) Brazing and Soldering

Metal powders are used extensively as filler metals for brazing. When heated to the proper temperature with a suitable flux or under protective atmosphere, the powder melts and flows between closely fitted surfaces of the joint by capillary action. Copper, Phos-Cu, and Bronze alloys (94/6, 90/10, 80/20) with and without Phos are all utilized. The powders are generally air atomized as spherical morphology is important in paste formulations. The material being brazed and melting temperature desired determines the alloy composition.

One big worldwide ongoing effort is by the CuproBraze Alliance. CuproBraze is a revolutionary technology for manufacturing radiators and heat exchangers from brass and copper for tubes and fins respectively using a special brazing alloy of Copper/15 tin/4 Nickel/5 Phos. CuproBraze has been developed by the International Copper Association (ICA) and is being implemented by members of the CuproBraze alliance. ACuPowder International is licensed to manufacture and supply this brazing alloy. Technology employs flux-free, lead-free alloy and brazed copper heat exchanger, has better performance than thick-walled aluminum heat exchangers.

Historically, solders are low melting Tin-Lead alloys. Numerous alloys are being developed to eliminate the presence of toxic lead. Some commercial alloys which are gaining acceptance are 97 Sn/3 Cu, and 96 Sn/3 Cu/l Bi.

(b) Thermal Spraying and Cold Spraying

In Thermal Spraying, metal powders are heated and accelerated by an energy source, such as chemical combustion (flame) or electrical plasma/arc to melt or soften them. Their high temperature and velocity result in significant droplet formation on impact at a surface, producing thin layers that conform and adhere to the surface and build up rapidly.

Narrowly sized spherical copper and bronze, which flow freely through the spray gun are commonly used to provide conductance and/or corrosion resistance to other surfaces. The powders also need to have low oxygen content for good bonding and conductivity and no organics that will carburize with heat.

Cold Spraying is a new developing technology in which ductile metal powders (copper, tin, zinc, etc.) are gas dynamically accelerated to supersonic velocities, with no to minimal heat. The powders plastically deform on impact fusing to the substrate without melting. Original development was funded by US auto companies consortium for new manufacturing technologies, with the objective to replace chemical plating.

(c) Rapid Prototyping

Rapid prototyping is the technology to produce prototype components quickly and at a relatively low cost. For example, stainless steel powder is deposited by computer modeling as a porous structure to create a mold to injection-mold a plastic prototype part. The powder is laser sintered very carefully to avoid distortion and infiltrated with bronze to achieve a solid mold. Bronze is a popular infitrant with a low enough melting point to infiltrate but high enough to handle the heat of plastic injection molding.

7. Chemical and Metallurgical Applications

(a) Catalyst

Copper powder is used as a catalyst in the production of silicones and also synthetic diamonds.

(*b*) *Chemical Compounds*

Fine copper powder is used as an effective form for efficient dissolution for generation of copper compounds; e.g. copper naphtanate, a wood preservative, and other medicinal compounds.

(*c*) *Injection Alloying*

Copper powder is injected with high pressure nitrogen into molten metal for fast and complete alloying; e.g. copper into aluminium alloys.

REFERENCES

[1] Gene Romig, The Professor's Powder Works, A Story of Everett Joel Hall, 1942.

[2] Pierre W. Taubenblat, Importance of Copper in Powder Metallurgy, International Journal of Powder Metallurgy-Volume 39, No. 4, p. 869, 2003.

[3] Pierre W. Taubenblat, Importance of Copper in Powder Metallurgy, International Journal of Powder Metallurgy-Volume 39, No. 4, p. 867, 2003.

3

Thermal Spray Powders–The Indian Happening in Global Perspective

Dr. R. Chattopadhyay
Wear Prognosis Technology, Mumbai-400053, India

ABSTRACT

In the present article, an attempt has been made to discuss the developments in manufacturing, applications and marketing of thermal spray powders in India in the global perspective. In thermal spraying, the metallic, ceramic, polymeric and composite powders are used to form coatings on the substrates of similar or dissimilar materials. Metallic and metal-based composite powders are also used in associated processes, such as, 'spray and fuse' and plasma transferred arc (PTA) fusing. Atomization process is used for commercial production of the main bulk of metallic powders. Ceramic powders are produced by fusing or sintering followed by milling. Polymeric powders are normally produced by cryogenic grinding and also by atomization. The sustained development of custom products, processes and procedures followed by application oriented market development, by the author and his teams has resulted in substantial growth in thermal spray powder business during the last two and half decades.

INTRODUCTION

Wear, including corrosion of materials occur by interaction of surface with the environment[1]. Surfaces can be engineered to minimize wear [1,2]. An unique surface engineering process for coating virtually any material surface with similar or dissimilar materials powders is termed as thermal spraying. The immense possibilities of minimizing wear in the widest spectrum of materials have led to the development and use of equally broad range of powder materials during the last two and half decades. The list of available thermal spray powders includes metals, ceramics, polymers and composites. It is also possible to produce newer compositions in the sprayed and fused powders. RSP techniques are used to produce both amorphous metallic powders and deposits. Amongst the various processes used to produce metallic powders and alloys [3], atomization is widely used to produce most of the Fe-, Ni-, Co-, and Cu- base alloys and metallic zinc. Ceramic powders are produced by fusing or sintering, followed by milling. Polymers are produced in powder form by sub-zero grinding below glass transition temperature (Tag) or by atomization. The development of advanced thermal spray processes, such as, plasma, D-Gun, HVOF, and HVAF along with newer applications has given the extra thrust to produce powders to suit the specific equipment and application.

Powders for Thermal Spray and Related Applications

Metallic, ceramic, polymeric and composite powders are used for thermal spraying. Largest group of powders belong to metallic and metal-matrix composites, followed by ceramics plus composites. Polymeric materials and polymer-composites are late additions.

Metals, Alloys and Metal Matrix Composites

Bulk of the common metallic materials, such as, metallic zinc, Fe-, Ni-, Co-, Cu-base alloys are commercially produced by atomization processes. The processes, such as, rotating electrode and ultra-rapid solidification (melt spinning and extraction) are used for producing specialized materials [3] like titanium [4] or met-glass powders [5] of unique compositions.

Atomization Process: In this process pure metal and metallic alloy powders are produced by atomization of molten material by high pressure fluid. The basic advantages of atomization process include, greater latitude for alloying or making newer alloys, uniform composition across the section, control of particle shape and size by varying parameters and high purity compared to mechanical attrition.

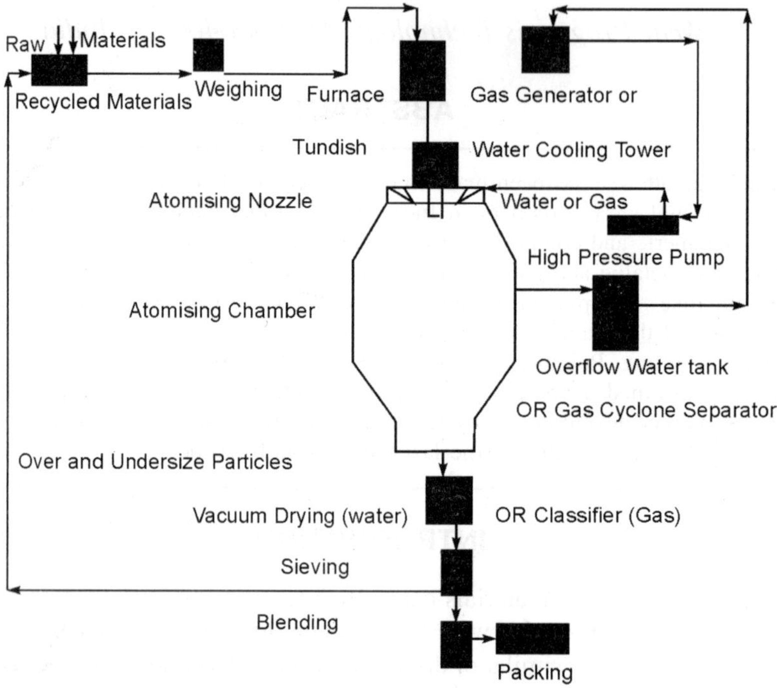

Fig. 1. Atomization process for powder production flowsheet.

The two types of fluids are used for atomization, viz, water and gas. Air is used for production of powders from low melting point metal such as zinc. Inert gas such as, nitrogen is used extensively for gas atomization of most of the specially powders, excepting few superalloys, where argon is used as atomizing medium.The water atomization is extensively used for the manufacture of iron and steel powders including stainless and high speed tool steels. The gas atomization is used for the production of superalloys of nickel and cobalt as well as super stainless steels and other alloys prone to oxidation during water atomization. The production flow sheet of atomized powder [6] is shown in Fig. 1.

The powder quality should conform to stipulations made in the specification with respect to chemistry, sieve size analysis, Hall flow rate and the hardness of the deposit.

Atomized Powders: The bulk of the atomized powders belong to Ni-, Co- and Fe-base alloys meant for spray, spray and fuse and PTA applications.

A wide range of nickel base self-fluxing alloys with fusion temperatures in the range of 950 – 1050°C has been produced by water atomization and used for both spray and spray-fuse applications. These are basically Ni-Cr-B-Si-alloys, where B and Si are used to lower the fusion temperature and to control the hardness from 90HRB to 62 HRC.

Table 1. Metallic Bond Coat Materials

Materials	*Maxm. Service Temp (°C)*	*Bonding Interface*
Molybdenum	315	Metal/metal
80 Ni – 20Al	620	Metal/metal/ceramic
89 Ni/6 AI/5 Mo	–	–ditto–
94 (Ni + Cr) + 6AI	980	–ditto–
95 Ni + 5 AI	1010	–ditto–
80 Ni + 20 Cr	1260	–ditto–
NiCrALY, e.g. NiCrALY/CoCrAIY	1260 – 1316	For ceramic coat on metal
Zn/AI	–	For plastic substrate
AI – Si	–	AI-substrate/ceramic top

For sprayed coatings, a bond coat material is used for adhesive bonding of the top coat material to the prepared substrate. The list of bond coat materials for all types of coating is included in Tab.1.Some of the Ni-base bond coat materials can be manufactured by atomization. Most of the bond coat material is also used for protective coatings, such as Mo for MG cylinder and piston ring [2]. A low to moderate thermal capacity oxyacetylene torch is used for spraying on jobs, such as, bearings, spindles, shafts etc. Even big jobs like coating of cast iron cylinders for paper (MG) and carbide drying have been successfully carried out using Ni-base self-fluxing powder (37-40HRC) using moderate thermal capacity torch [2].

With the introduction of high thermal capacity torches having supersonic flame speed, such as plasma and HVOF in 80s, it has been possible to spray high melting point materials, such as SS, Stellites and ceramics for critical applications, such as, jet engine components and also bigger jobs, such as MG cylinder [2]. Powders for use with plasma and HVOF torches are of smaller size range (15 to 30 micron), due to shorter dwell time in the flame compared to those used for low or moderate thermal capacity torches(1 00 to 325 ASTM mesh size). Single stage powders incorporating the bond coat materials are developed to take care of thicker coating and to improve productivity.

Some of the highlights of these developments are indicated in Tab.1. Two single stage iron base powders (Fe-Cr-C-Al) were introduced in our production range to take care of erosive wear in boiler tubes(Tab.2).ln order to cater to the needs of challenging big applications, such as, Osepa air separator in Cement plants, single stage wear resistant powders were devolved for use with high capacity torch like plasma.

Similarly spray fused powder range got a boost due to introduction of higher capacity torches and automation leading to development of Ni-base self fluxing custom powders (CP) for OEM applications like compressor cylinder (replaced electroplating by mirror finish spray fuse coat of CP), conveyor rolls in steel plants' light merchant mill (high temperature adhesive wear resistant coating of CP), wear ring, wheel burn in railroad rails (deposit of same hardness as rail steel), glass mould etc. Two cobalt base self fluxing powder with properties similar to Stellite 6 and 12 were also developed.

Table 2. Selected Metallic and Metal-base Suitable for Different Processes

Metals/Alloys	*Production Process*	*Application Process*	*Major Applications*
Metals			
Mo	S	Sp	Bond coat/piston ring/ MG cylinder
Zn	G	Sp	Bridges/structures
Ni-base Alloys			
Self-fluxing Ni-Cr-B-Si-Fe-C series Hardness-90RB-60RC	W	Sp Sp+F	Bearings, shafts, CI MG cylinder Off-shore splash zones Compressor plungers/wear ring, conveyor rolls in merchant mill
Inconel 625 Ni-22Cr-9 Mo-3. 7Cb	G	PTA	Chemical/marine pollution control equipments
Hastelloy 276 Ni-16Cr-16Mo-4W	G	PTA	Heat exchanger, hot shear blade
Nimonic (Udimet 520)	G	PTA	Bar-forging hammer, dies, engine valve
Ni-60 (Ni-Cr-Fe-Si)	G/W	PTA	Engine valve
Cobalt base alloys			
Stellites Co-Cr-W-C (St306, St6, St12, St190, St32) Co-Cr-Mo-C (St21)	G	PTA Sp	Engine/process control valves, hydro turbine components Erosive wear applications.

[Special (S), Gas and Water atomized (G and W), Spray (S), Fuse (F), Plasma Transferred Arc (PTA)]

PTA powders: In the plasma transferred arc (PTA) process, powder particles carried in argon gas stream are fed in the plasma arc resulting in the deposition of molten material on the substrate [2].

All the grades of spray fused powders can be used in PTA process. A range of water atomized Stellite powders developed for use with PTA system has been used successfully for applications like engine valves, process control valves and critical applications like spindles and shafts for hydroturbines (Table 2).This unique development was possible through sustained studies on batches of Stellite powders with respect to the oxygen content, weldability(PTA), microstructure, composition, hot hardness, size analysis and Hall flow and subsequently fixing the process parameters for optimum product properties. The welding crack problem in PTA deposit using water atomized powder was found to be caused by excessive oxygen and thus the maximum limit for oxygen was incorporated in product specification [7].

Similarly nickel base superalloys, such as, Inconel 625, Hastalloy C 276, and Nimonics 80 were developed and used for few applications. Selected batches were tested for hot hardness variation over a range of temperatures in order to establish credentials for high temperature applications, such as bar forging hammer, hot shear blades etc [8]. An improved version of Ni-60 giving porosity free PTA deposit was developed and sold in tons for engine valve applications [9] Metal matrix composites such as, NiCrBSi + WC, HSS + TIC powders are produced by mechanical mixing of atomized metal powders with respective carbides. The composite with self fluxing alloy matrix has been used as spray fuse coats for applications like ID fan blades, and also as PTA coat for tricone bits. The composites with self fluxing alloy base has also been produced in the form of paste (2). The dry paste coat on the surface is subsequently fused by fusion torch. Titanium carbide plus HSS composite has been used as PTA coat for plastic extrusion screw (2).

Ceramic Powders

Four basic types of ceramic oxide powders used in thermal spraying include Al_2O_3, TiO_2,Cr_2O_3 and ZrO_2 (stabilized by Ca, Mg or Y).

The list of processes for ceramic powder production (Fig. 2) includes fused-milled, sintered-milled, agglomerated (sintered), atomized, spereodised, and sol-gel [10]. Almost all the requirements of ceramic powders are imported. Attempts to indigenously produce thermal spray grade oxide powders were made by some leading ceramic material manufacturers met with limited success. WC used for thermal spraying also falls into a special type which is not manufactured in India. Typical applications, properties of some of the ceramic powders are indicated in the Table 3.

Polymer Powders

Thermoplastic and thermosetting resins can be deposited on clean metallic surface by thermal spraying. Thermoplastic polymers flow under the action of heat to form continuous coating by fusion coalescence above fusion temperature. On reheating, the coating will remelt, since thermoplastic never reach a cured or cross-linked state. Most thermoplastic materials requires bond coat. Thermosetting are lower molecular weight resins, which on melting undergo polymerization and form higher molecular weight polymers. The fully heat cured material cannot be remelted. Thermosetting resins have excellent adhesive bonding properties with metals and thus do not need bond coating. Examples of thermosetting and thermoplastic resins, important properties and applications are indicated in Table 4 [1].

The polymeric coatings are ideally suited to coat large structures, such as bridges, shafts, ship decks, pumps rudders and impellers, for long time corrosion and weathering protection. These are normally produced by cryogenic grinding below glass transition temperature [1] of respective resins. Few commonly used polymer powders were developed and the coated samples were successfully tested for corrosion resistant in a leading chemical industry. No further progress was made due to lack of demands. However there is a good potential market for dual coating of polymer and zinc in pipelines, bridges and structures for long time protection against atmosphere and marine environment.

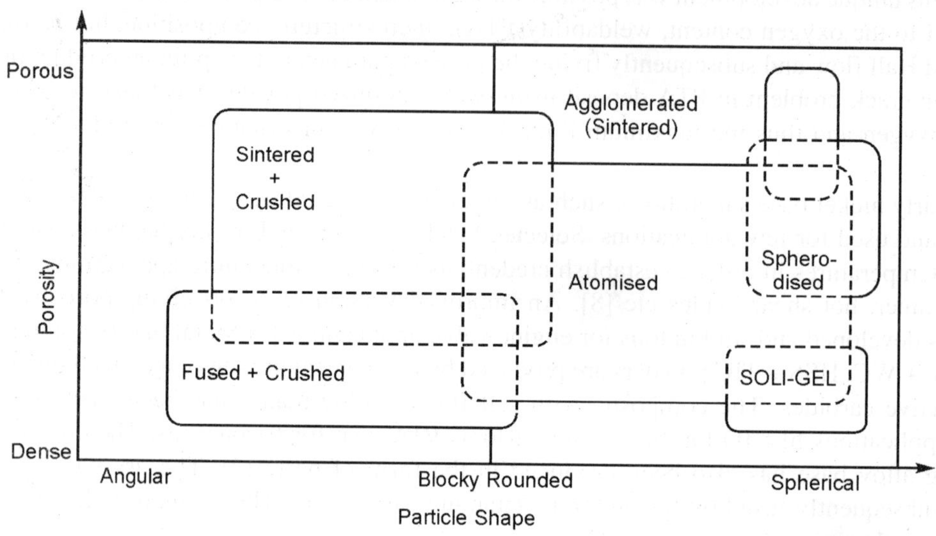

Fig. 2. Powder characteristic—porosity and shape due to production method

Table 3. Selected Ceramic Powders, Properties and Applications

Ceramics	*HK/HV(GPa)*	*Therm. Expn.* 10^{-6}*/K*	*Major Applications*
Alumina	18 – 23	7.2 – 8.6	Seals, engine valve diffusion barrier
Titania	7 – 11	9.4	Used with Al_2O_3 in textile industries
Cr_2O_3	29	7.5	Hard, resists oxidation and corrosion
Zirconia	10 – 15 (cubic)	13.5 (cubic)	Stabilized for thermal barrier
TiC	28 – 35	7.4 – 8.6	With metallic matrix for spray + PTA
Cr_2C_3	10 – 18	9.8	With Ni – Cr matrix for boiler tubes
WC	8 – 10	4-8.3	Sink rolls in Galvanizing

Composites and Multiple Metal Coatings

In addition to composites, coating comprising of multiple metals have been used to cater for specific application requirements. Selected composite materials and applications are listed in tab5.Most of the composite materials are made by mechanical mixing. In addition to those listed in Tab.5, number of composite materials are made as per requirements for application oriented graded coatings.

Table 4. Properties of Selected Polymers for Coating

Polymer	*Primer*	*Processing Temp °C*		*Tg °C*	*Hardness*
		Softening	*Curing*		
Thermosetting					
Epoxy	No	180	200		...
Polyester	No	140	200	75	...
Thermoplastic					
Nylon 6/6	No		>270	50	109 – 110 HRR
Polypropylene	No	167	>176	– 16	105 – 106 HRR
Polyvinyl chloride	Yes		>212	87	52 HRM
Polyethylene					
Low density	No		>115	– 120	23 – 25 HRR
High density	No		>137	– 120	59 – 62 HRR

Table 5.Composites and Multiple Metal Coatings for Thermal Spraying

Coating	*Process*	*Applications*
Metal and composites		
MCRALY (M = Ni/CO/Fe)	HVOF/plasma	Bond coat/high temperature
$SiO_2 + ZrO_2$ + CO-Cr-Al-Y + Al_2O_3	HVOF/plasma	Hearth rolls in continuous annealing
COCRALY-$Y_2O_3 + CrB_2$	HVOF/plasma	Hearth rolls in continuous annealing
Cr_3C_2- NiCr	HVOF/plasma	Piston ring/boiler tube
Ni-Cr-B-Si-C + Cr_3C_2	HVOF/plasma	Continuous casting molds
5.5 BN + $Ni_{14}Cr_8Fe$ + 3.5 A1	HVOF/plasma	Self cleaning of rolls in stainless steel annealing furnace
BN/Graphite + Ni	Plasma	Aerospace engine-abradable coating
WC-CO + Stellite	HVOF/plasma	Fan blades
TiC + HSS	Plasma tungsten arc	Extrusion screws in plastic molds
WC + Ni Cr B SiC	Plasma tungsten arc	Mining tricone bits
Plastic and composites		
Al-bronze + polyester	Oxyfuel	Corrosion resist coating
Stainless steel + low-density plastic	Oxyfuel	Prime coat for stainless steel on plastic
HVOF, high velocity oxygen fuel		

Atomized Powders for Related Applications

The list of non-thermal spray applications, where there is a large requirement of atomized powders, includes HSS (for manufacturing cutting tools through P/M route), AISI 316L (for polymer filter) and metals and prealloyed powders for diamond tool industries. HSS powders were produced by us to cater for in-house requirements of producing sintered tools [11]. Polymer filter were made and supplied to leading petrochemical industries. Few prealloyed powders for diamond tool industry were developed but failed to get anchored in the range between highly priced cobalt and cheap tin bronze [12].

Market in India

The market survey of metallic and metal matrix composite powders suitable for thermal spray (includes spray-fuse, PTA) and related applications (HSS, 316L and prealloyed diamond tool powders) indicated a requirement of around 300 tons in 1998 worth Rs.55 crore. With a 7% normal growth rate to-day's requirement is around 500 tons worth 100 crores. If one takes into account of the high industrial growth rates in all sectors during this interim period, specially in automobile, power, infrastructures like highways and construction industries, the present requirement can be anywhere between 900 to 1200 tons. In a manner similar to that of the market developed by us for metallic powders, it is necessary to develop a good number of large applications using ceramic and polymeric powders to ensure sustainable growth in these areas.

REFERENCES

1. R. Chattopadhyay, author of the book 'Surface Wear-Analysis, Treatment, and Prevention',ASM International, Materials Park, OH,USA.2001.
2. R. Chattopadhyay, author of the book 'Advanced Thermally Assisted Surface Engineering Processes', Kluwer Academic Publishers, MA, USA, 2004.
3. R. Chattopadhyay: Special powders for wear and corrosion resistant coatings, Nat. Sem. 'Powd.Met.Opportunities for Eng. Industries, IIT, Mumbai, Dec.14, 1985. pp. 80-87.
4. N.J. Grant, Powder Metallurgy for High Performance Application, ed., J.J.Burke and V. Weiss, Syracuse Univ. Press, 1972, p. 85.
5. C. Suryanarayana & others,' Development of Light Alloys by Rapid Solidification Processing, The Intl. Jove Powder Met., vol 26, no. 2, 1990, pp 117-129.
6. R. Chattopadhyay, Project Report on 'Special Powder Production by Atomization and the Domestic Market' 11 August, 1998, submitted to Mapin Industries, India.
7. R. Chattopadhyay, 'Study on Crack Formation in a Cobalt Alloy PTA Deposit', Int. Thermal Spray Conference, 18-22 May, 1998, Nice, France.
8. R. Chattopadhyay, Study on Microstructure and Wear of Plasma (PTA) Deposited Superalloys, Proc.Int. Conf. on 'Structure-Property Relationships with the Environmental Degradation of Materials' Int. Metallography Society, Monterey, CA, 29 July-1 August, 1991, pub. 'Microstructural Science' vol 19, ASM Intl, pp. 477-504.
9. R. Chattopadhyay, 'Wear Resistant Plasma Transferred Arc Coating on Engine Valve', Proc. of 14th Int. Thermal Spray Conf., Seminar on Automobile Industry, High Temp. Soc. of Japan, Osaka held at Kobe, 22-26 May, 1995, pp. 31-37.
10. R. Chattopadhyay, Wear-resistant Powder Materials, Proc. Nat. Workshop. on 'Metal, Ceramic and Composite Powders', Ed. P. Ramakrishnan, IIT, Bombay, 18th Jan, 1989. pp 41-48, Oxford & IBI Publishing Co. Pvt. Ltd
11. L. Albano Muller, R Waehling, & R Chattopadhyay, 'Manuf. & Propts. of Sintered HSS Products', Int. Conf. Powd. Met., IIT, Bombay, 7-10 Dec.1987.
12. R. Chattopadhyay, Diamond Tools-in Retrospect and Prospect, Keynote Address, Workshop on Diamond Tools, ASM Int., India Chapter, Bombay, 29th Sept. 1994.

4

P/M Alloys for Aerospace Applications

Anish Upadhyaya
Materials and Metallurgical Engineering
Indian Institute of Technology, Kanpur, India.

ABSTRACT

This article reviews the consolidation of various components for aerospace application through powder metallurgy (P/M) with emphasis on aero-engine and spacecraft materials. Aerospace materials are moving out of the established world of conventional alloy design strategies through relatively simple chemical modification into a more complex environment whereby materials are 'engineered' to meet specific component requirement. P/M provides one such net-shape fabrication technique.

INTRODUCTION

The increased performance requirement for aerospace systems applies to both airframes and aeroengines. For example, a fighter aircraft must operate over broader combat envelopes with supersonic speeds (increasing Mach numbers), at greater turn rates (g's), and at higher load factors, as shown in Fig. 1.[1] The same requirements are also being translated to the design of civilian aircrafts as well. To enhance the performance, there are ongoing efforts to improve and innovate through newer structural concepts, automated design, active load control and flutter suppression techniques.[2] . One area, which has contributed significantly to performance enhancement, is advancement in materials both from novel processing strategies viewpoint and newer compositions.

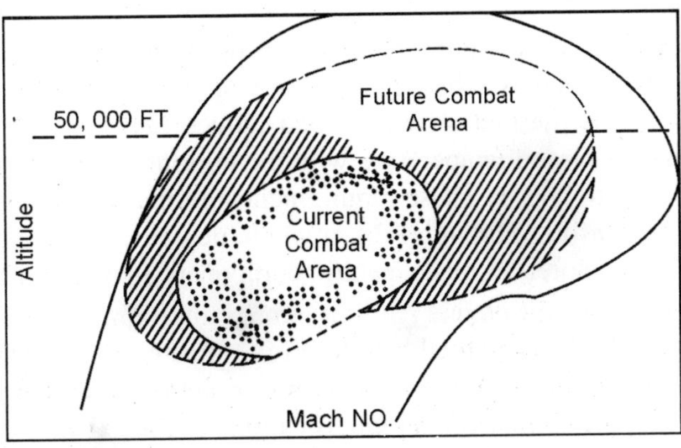

Fig. 1. Combat envelope of future fighter aircrafts.[1]

Among the various materials processing technologies, powder metallurgy (P/M) is the most diverse manufacturing approach. One attraction of powder metallurgy (P/M) is the ability to fabricate high quality, complex parts to close tolerances in an economical manner. Key steps in P/M processing include the shaping or compaction of the powder and the subsequent thermal bonding of the particles by sintering. The process effectively uses automated operations with low relative energy consumption, high materials utilization and low capital cost.[3] Fig. 2 shows schematically the section of the production-run-part complexity diagram comparing powder injection molding (P/M)–which is one variant of P/M–with the conventional materials fabrication processes including die casting, investment casting, and machining. Clearly, P/M scores in applications requiring complex parts with higher production rates. These characteristics make P/M well aligned with current concerns about productivity, energy, and raw materials. Furthermore, it is a flexible manufacturing process capable of delivering a wide range of new materials, microstructure and properties. That creates several unique niche applications for powder metallurgy components. [3,4] One such emerging area is the application of P/M materials for aerospace applications.

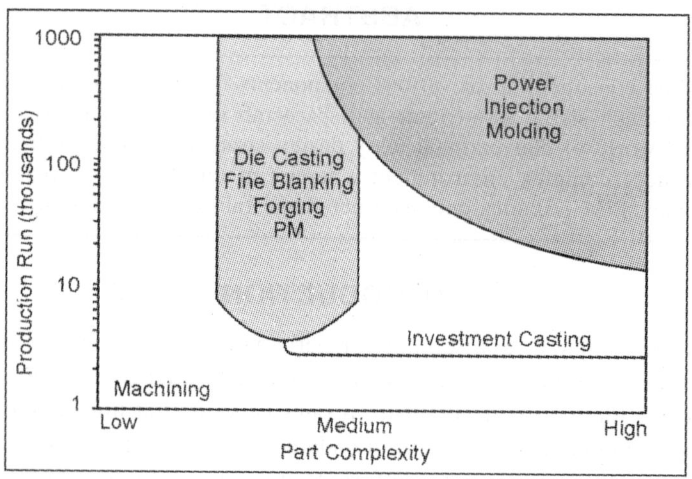

Fig. 2. Comparison of various manufacturing technique *vis a vis* part complexity and production rate.

P/M Aircraft Components

In aerospace applications the engine requires strong, lightweight materials. Since the early jet engines, pressure ratios have increased by a factor of seven while turbine temperatures have more than doubled. In order to increase thermal efficiency and reduce fuel consumption, aeroengines need to operate with high gas pressure and temperature with the current objective being to increase the thrust to weight ratio from 10 : 1 to 20 : 1. These requirements are making the use of sintered metallic and ceramic materials very extensive. Through P/M, a whole new array of materials are being researched wherein the properties are tailored to meet specific component requirement.

In aerospace applications, the engine requires strong, ultralight materials. In modern aircrafts, the engine accounts for about 10% of the total weight. Fig. 3 illustrates the development in the aerospace engine materials *vis a vis* the strength and temperature of application. The trend suggests that one of the key developments include compositional modifications of the existing superalloys, Al- alloys, and titanium-based alloys by various composite reinforcements. Owing to the density differences and wide differences in the fusion temperatures, the reinforcements are not homogeneously distributed through conventional casting techniques. In addition, the poor machinability of these advanced composites requires a near-net shape manufacturing technique. Both these limitations are well addressed by P/M.

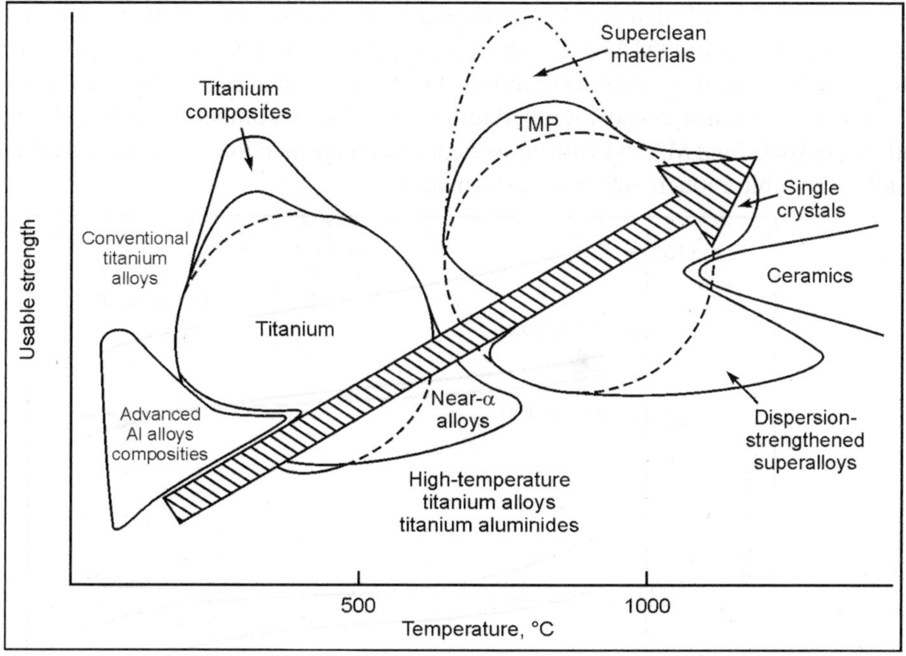

Fig. 3. Trend in the major aeroengine materials.

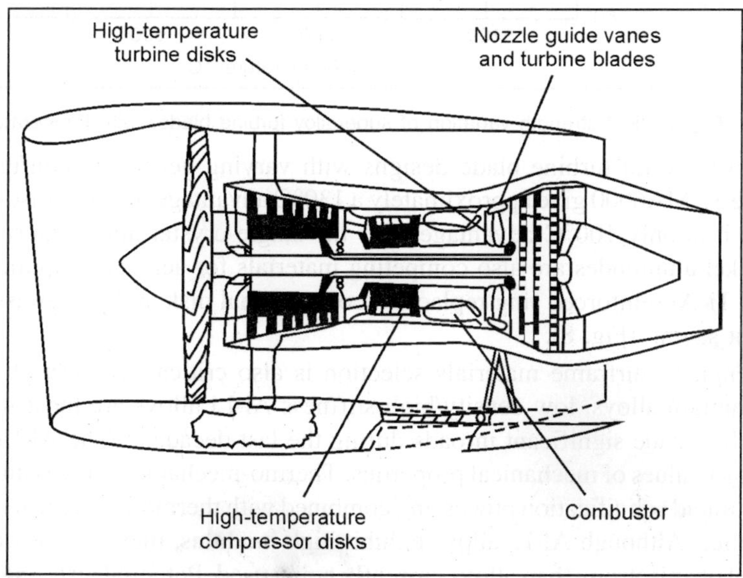

Fig. 4. Principal engine components using superalloys.

Fig. 4 shows the principal areas of superalloys in engine components. One of the key property requirements for the turbine disk is their low cycle fatigue strength. Use of P/M alloys and controlled thermo-mechanical processing (TMP) has significantly increased the fatigue properties. As shown in Fig. 5, PIM disk superalloys have higher static strength than the conventional disk materials. Because of the susceptibility of failure initiation due to defects, it is very important to use ultraclean superalloy powders consolidated to full-density. A major trend in the material development for the turbine blade is the use mechanically alloyed (MA) oxide-dispersion strengthened (ODS) superalloys by the P/M route. One such ODS P/M superalloy MA6000 exhibits superior performance over the conventionally fabricated (as-cast) alloys in the high stress-low temperature regime. Fig. 6 compares the thermal fatigue resistance of P/M alloys-MA6000 and MA754 with equivalent cast and dispersion-strengthened (DS) alloys.[6] Clearly the P/M alloys exhibit a much superior performance.

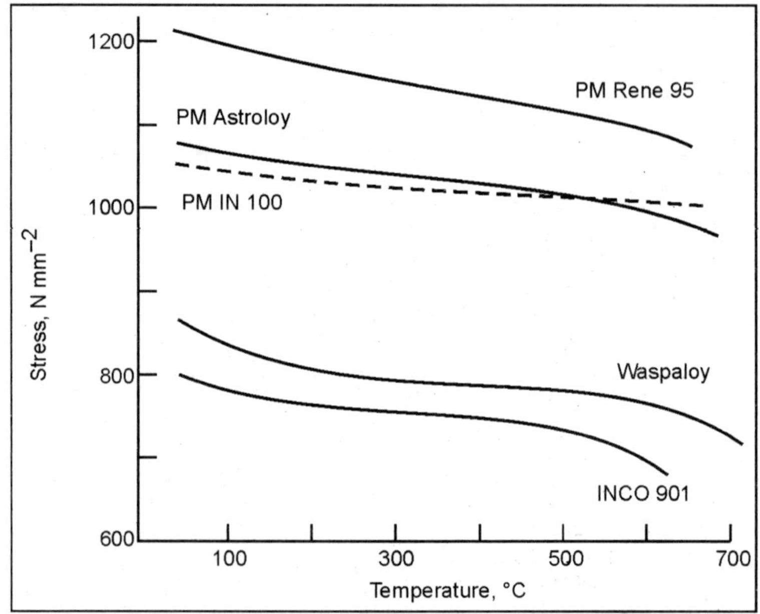

Fig. 5. Yield strength variation of superalloy turbine blades with temperature.

Fig. 7 shows several turbine blade designs with varying geometric configuration. For *tapered shroudless* (blade 3), MA6000 give approximately a 130°C advantage in comparison to *tapered shrouded* (blade 2) where it is only 100°C advantage over the single-crystal alloy. Intermetallics, particularly titanium and nickel aluminides are also competing materials for aerospace applications. Froes *et al.* [6] have shown that Ti_3Al reinforced ring replaces a conventional disk and spacer assemble and results in upto 75% weight saving (Fig. 8).

Like aeroengines, airframe materials selection is also critical for aircraft operating efficiency. Among the aluminium alloys, low density/high-stiffness Al-Li alloys are most widely used, in which P/M processing has made significant inroads during the last decade.[7,8] The Al-Li-Cu-Mg alloy fulfils the minimum target values of mechanical properties. Thermo-mechanical processing can further enhance these properties. Rapid solidification processing combined with thermomechanical operation have further enhances properties. Although Al-Li alloys exhibit high modulus, they have lower fracture toughness and a higher level of anisotropy than alloys currently being used. Potential airframe applications include: fuselage skins, fuselage stingers, lower wing skines, access covers, slats, fixed leading edges, spoilers, trailing edges, floor beams, seat frames and seat tracks.

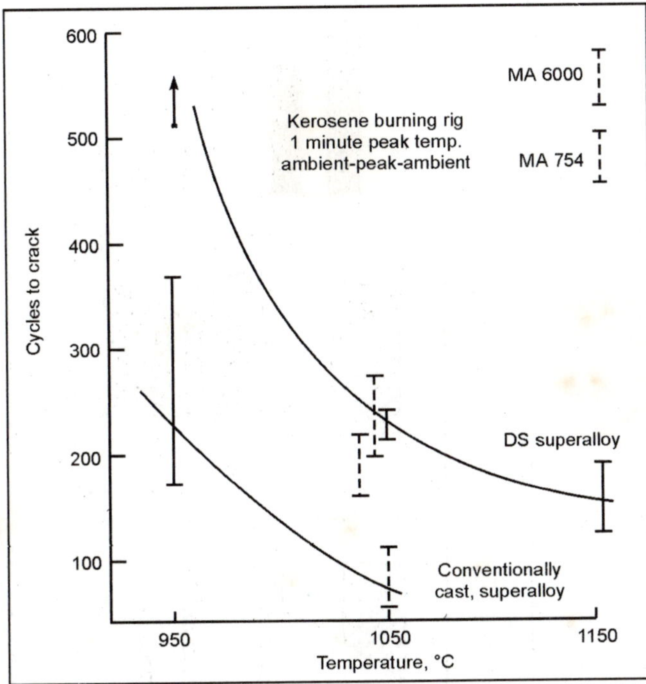

Fig. 6. Thermal fatigue of as-cast and P/M superalloys.

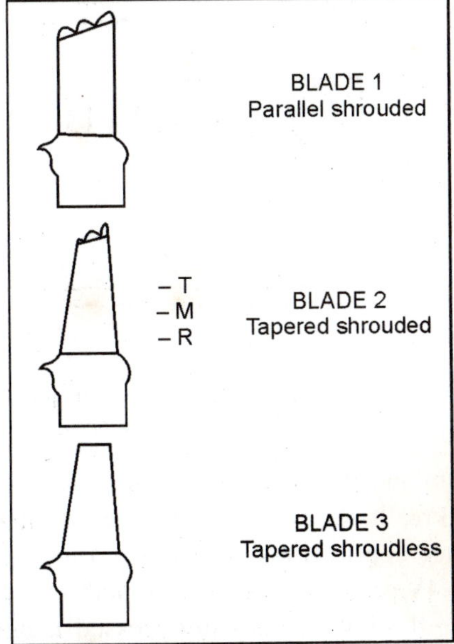

Fig. 7. Aircraft turbine blade design.

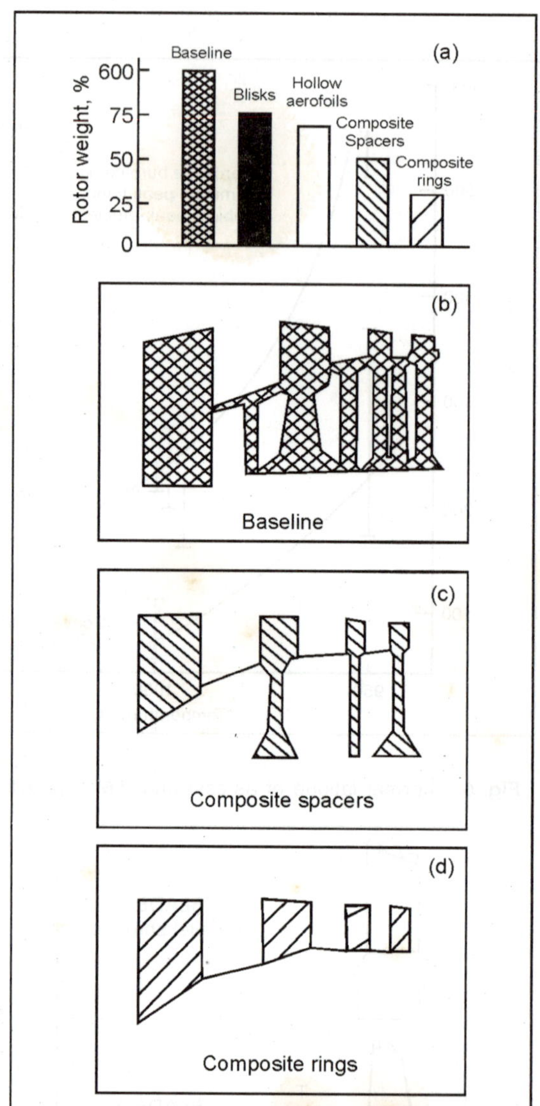

Fig. 8. Saving in the rotor blade weight by replacing conventional disk and spacer assembly with a P/M Ti_3Al composite ring.

The high strength of titanium and its moderate density make it attractive on strength to weight basis for aerospace applications. Prealloyed centrifugal or gas atomized powders are in common use for high density, high performance applications. The most common alloy is 90Ti-6A1-4V. Table 1 contrasts the properties of Ti-6A1-4V processed by several methods.[3] In terms of strength and ductility, the cast and forged materials and the hot isostatically pressed materials are competitive. However, better materials utilization makes hot isostatic compaction the most favored route.

Table 1. Mechanical Property Comparison for Ti-6A1-4V alloys.[3]

(P + S: press and sinter; HIP: hot isostatically pressed)

Process	*Porosity %*	*Yield strength MPa*	*Tensile strength, MPa*	*Elongation %*	*Reduction in area, %*
as – cast	0	840	930	7	15
cast + forge	0	875	965	14	40
*P + S	2	786	875	8	14
* HIP	<1	805	875	9	17
**HIP	0	880	975	14	26

* blended elemental powder mix; **prealloyed powder

One of the main applications for P/M titanium is aerospace components where the fatigue properties are a main concern. The allowed stress cycle without fatigue crack growth depends on the inverse square root of the defect size. Consequently, powder cleanliness and minimization of defects are important to those applications. The consolidation of the powder is relatively straightforward. A hot-pressing cycle of 100 MPa for 3 h at 900°C is sufficient to give full density. Alternatively, cold pressing at 400 MPa and sintering for 16h at 1200°C is a traditional method for consolidating blended elemental powders. Sintering dissolves the surface oxides into titanium, resulting in a clean surface for easy diffusion bonding. The processing must be performed under inter or vacuum conditions due to reactivity of titanium. In certain applications, the interstitial elements like oxygen and nitrogen provide significant strength and toughness at levels below 1500 ppm.[3]

One of the recent successes of P/M Ti-alloys is a sintered Ti-6A1-4V aircraft fastener. These are fabricated from elemental powder mixture with is pressed in a doughnut preform, hen mechanically hot coined to the twelve-point nut shape. The threads are tapped in the nut after P/M forming. Cost savings by producing this component from wrought titanium are substantial.[9]

Counterbalance weights and rotational gyroscope members in inertial guidance systems are integral part of any aircraft or spacecraft. To minimize the volume requirement, these components are made from high-density materials. Tungsten (density: 19.3 g/cm^3) is widely used metal for this. However, tungsten is also a refractory metal with a fusion point of 3420°C. Hence, these components can only be fabricated through P/M processing. Typically, for such an application an alloy of tungsten is used. A typical tungsten heavy alloy (WHA) contains 80 to 95 wt.% tungsten. The balance is generally a mixture of relatively low melting transition elements, such as nickel, iron, copper, and cobalt. These alloys are processed *via* sintering to temperatures ranging between 1300 to 1500°C where at least one of the constituents melts.[10] Liquid phase sintering offers the advantage of relatively low sintering temperature, enhanced densification and microstructural homogenization, and near theoretical sintered density. Sintered WHAs offer a unique combination of properties such as high density (16-18 g/cm^3), high strength (1000–1700MPa), high ductility (10-30%), good corrosion resistance, and easy machinability. In addition, these alloys also offer a favorable combination of high thermal conductivity and low thermal expansion coefficient. Fig. 9 shows a typical liquid phase sintered microstructure of a WHA.

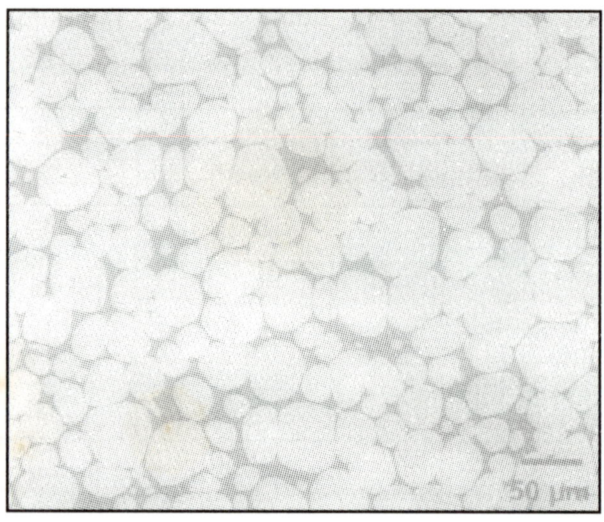

Fig. 9. Microstructure of liquid phase sintered W-Ni-Fe Alloy.

P/M in Space Vehicles

Spacecraft experiences a series of varying thermal and mechanical stresses (especially) during takeoff and re-entry. Fig. 10 shows the approximate temperature contours that a space shuttle must withstand during takeoff and re-entry and to be re-usable for 100 missions.[11] Table 2 shows the typical environment experienced by a spacecraft. The severity of the conditions is dependent on the fuel type, payload size and weight, and its configuration. A manned spacecraft has additional restrictions on flammability and toxicity. Ceramic tiles on the heat shield of space shuttle and ceramic composites at the nozzle are typical examples of ablative materials. Raman *et al.*[12] have described the one-step synthesis of dense titanium carbide tiles of 15 cm × 15 cm × 2.5 cm by self-propagating high temperature synthesis (SHS) combined with a 'CERACON' quasi-isostatic consolidation technology. The innovative P/M processed composites have resulted in 30-50% cost saving over the prevailing 'conventional' manufacturing processes. Fig. 11 shows various stages involved in preparation of consolidated Ni_3Al through SHS or reaction sintering.[13]

Table 2. Hyper-environment Encountered by a Space Vehicle

Temperature	–185° to 2760°C
Heating rate	23×10^4 to 57×10^7 $Jm^{-2}s$
Total heating	12×10^7 to 170×10^7 jm^{-2}
Heating time	1 to 3600 s
Pressures	0.01 to 400 atm
Loading	1 to 30g
Shear stress	48×10^{-2} to 7×10^2 kgm^{-2}

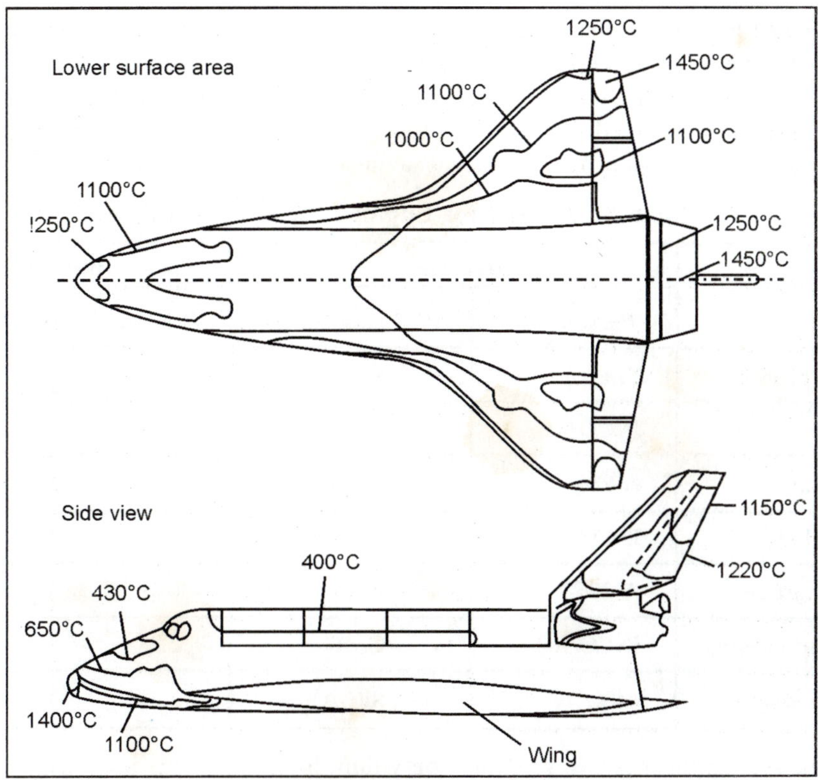

Fig 10. Approximate temperature distribution that the space shuttle must withstand during take-off and re-entry. [11]

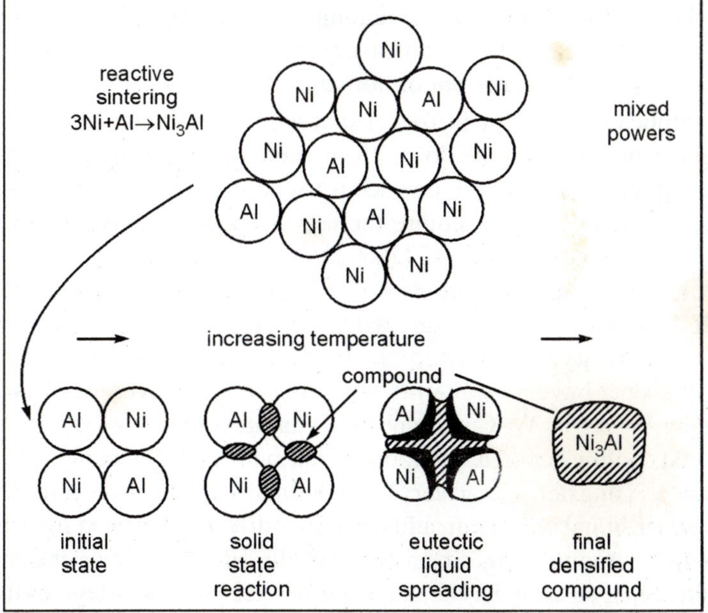

Fig. 11. Schematic diagram showing reaction sintering process for a mixture of Ni and Al powders. During heating, interdiffusion between the particles results information of Ni_3Al.

As indicated in Fig. 10, the highest temperatures that a reusable craft experiences ranges from 1425 to 1650°C and are on the nose and the heading edges of the wings. A reinforced carbon-carbon composite with a surface SiC based coating is used which prevents oxidation. For low temperature applications, ceramic-reinforced metals and alloys are finding wide application particularly in liquid engine components. As shown in Table 3 weight savings upto 30% are achieved by such P/M composites.

Table 3: Material for Engines for Space Vehicle. [14]

Components	*Materials*		*Weight, kg*	
	Present	*P/M composites*	*Present*	*P/M composites*
Gimbal bearings	Ti alloy	SiC-A1	70	50
Ducts	INCO 903	SiC-A1	579	389
Nut retainer	A-286	SiC-A1	1.1	0.4
Pump-shaft	INCO 718	SiC-A1	2.3	1.1
Wheel turbine	A-286	B-A1	0.45	0.25
H_2 pump housing	INCO 903	Gr-A1	2.88	2.25
Bearing support	440C	SiC-A1	1.73	1.15

For structural applications in satellites, beryllium is finding wide use owing to its high thermal conductivity and high heat capacity, lower thermal expansion coefficient, lightweight, and excellent specific stiffness. Some of the components include, antenna supports, solar panel booms, and structural supports. Owing to its toxicity, automated handling is a must during the beryllium-component fabrication units. In this respect, P/M is more amenable to automation. Furthermore, P/M beryllium alloys have much finer grain size than the as-cast alloys, and have therefore better structural integrity and reliability. Hence, P/M processing is the only viable option for such alloys.

Another recent application of P/M in rocket/spacecraft include rocket nozzle throat fabricated from sintered tungsten and tungsten-based alloys. The rocket nozzle experiences very high temperatures (>2200°C). A critical requirement for these nozzles is to maintain its dimensional stability during the period of flight. Such high temperatures require the use of transition refractory metals like tungsten. Owing to the high melting point of tungsten (3420°C) P/M is the only route to fabricate near net shape nozzle components. Ultrafine tungsten (particle size < 2 mm) is pressed to 200 MPa. The as-pressed (green) part has about 40% porosity and is sintered to about 1700°C for 1-5 h. The sintered tungsten compact has about 20% porosity by volume, which is subsequently infiltrated by silver or copper. Though both silver and copper have no solubility with tungsten, however they have very good wetting behavior. Thus, resulting W–Ag or W–Cu composite is an ideal nozzle material because the infiltrate provides a transpirational cooling action as it vaporizes during firing. The composite also as an excellent thermal shock resistance. Tungsten has a very low thermal expansion coefficient (<2 ppm/K) and therefore restricts the dimensional change. In addition, the solid-state sintered tungsten forms a 'skeletal structure' which provides structural rigidity to the nozzle. Fig. 12 shows the microstructure of an infiltrated W-composite. Such an alloy has been recently used in the first stage of the A3 Polaris rocket wherein the pressure at the rocket throat reaches to about 2.8 MPa and temperature close to 3000°C.

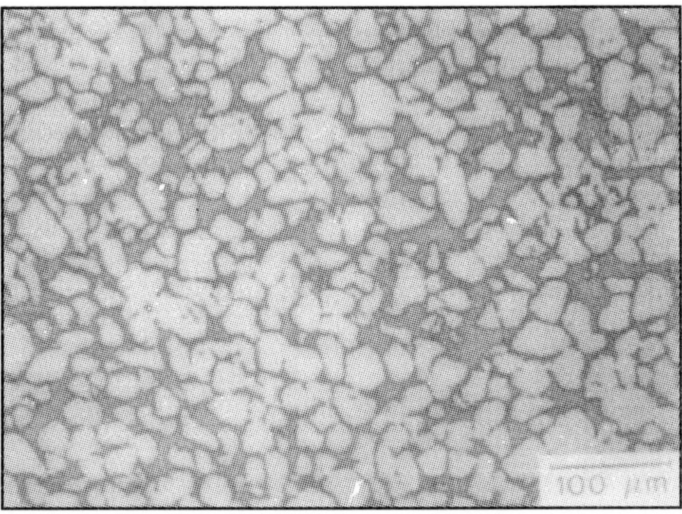

Fig. 12. Microstructure of an infiltrated W-alloy used for nozzles.

SUMMARY

P/M processing provides an important fabrication route to the creation of unique materials, novel microstructures and properties, and net-shaped complicated component geometries. The interest in P/M for aerospace application will further evolve and expand in future for processing newer compositions with properties hitherto not realized by the conventional manufacturing processes.

REFERENCES

1. Froes F H, Ward-Close C M, Taylor P R & Baeslack W A, P/M in Aerospace, Defense and Demanding Applications-1995, MPIF, Prineton, NJ, USA (1995) 3-40.
2. Marsh G, *Materials Today,* 3 (3) (2001) 21.
3. German R M, Powder Mettallurgy Science, 2nd ed., MPIF, Princeton, NJ, USA (1994).
4. Upadhyaya G S, Powder Metallurgy Technology, Cambridge lnt. Sci. Pub., Cambridge, England (1997).
5. Froes F H, *JOM* 43 (12) (1991) 20.
6. Froes F H, Suryanarayana C, Quist W, Lavernia E, Bondarev B I, Anoshkin N F, Polkin I S, Fatkulin O K, Sararov V & Notkin A B, Synthesis, Processing, Modeling of Advanced Materials, Froes F H & Khan T (eds.), Trans Tech Pub., Aedermannsdorf, Switzerland (1993) 1-10.
7. Chellman D J & Bahya T D, Advances in Powder Metallurgy, v. 7, MPIF, Princeton, NJ, USA (1992) 257.
8. Wang W, *JOM,* 45 (9) (1993) 16.
9. Garriott R E & Thellmann E L, Modern Developments in Powder Metallurgy, v. 11, MPIF, Princeton, NJ, USA, 1976.
10. Upadhyaya A, *J Mater. Chem. Phy.* 67 (2001) 101.
11. Korb L J, Morant C A, Calland R M & Thatcher C S, *Ceram. Bull.,* 11 (1981) 1181.

12. Raman R V, Rele S V, Pland S, Lasalvia L, Meyers M A & Niiler R R, *JOM,* 47 (3) (1995) 23.
13. Munir Z A, Reviews Particulate Materials, v. 1, Bose A, German R M & Lawley A (eds.), MPIF, Princeton, NJ, USA (1993) 41.
14. Sarkar B K, Advanced Ceramlics, Ganguly C, Roy S K & Roy P R (eds.), Trans Tech Pub., Zurich, Switzerland (1991) 135.

5

Nanomaterials and PM Industry

Bijoy Sarma
Scientist 'G', Defence Metallurgical Research Laboratory, Hyderabad 500058.India

ABSTRACT

The persistent demand for miniaturization led to extensive effort in the use of thin films wherein nanostructures manifest. Nanostructured materials display a host of new properties different from their bulk materials. Particulate materials technologists have realized the distinct benefits obtainable from finer particle sizes. The challenges for nanoparticulate use, particularly metallic particulates(powders) has to be appreciated in the context of their extreme reactivity, and this opens up new vistas for diversification in the PM industry. Some information on market trends and the challenges ahead are discussed.

Preamble

I wish to thank the Governing Council of Powder Metallurgy Association of India for this opportunity to share some of my thoughts with you on a subject of current relevance.

At the outset I wish to pay my homage to the Late Dr.R.V.Tamhankar, who gave this association its birth three decades ago. Let me once again recapitulate those days when Powder Metallurgy was in its infancy in the country. DMRL was one of the pioneering institutions to recognize and nurture the potential of particulate materials technology. Since then DMRL has furthered the vision of Dr. Tamhankar to raise the laboratory to international recognition and today we are riding the nanowave. Dr. Tamhankar had the true attributes of a leader whereby he brought so many of us together to share his dream of technological excellence. The Powder Metallurgy Association today has over 500 members covering every aspect of particulate science and associated manufacturing technology. Possibly, I am an appropriate choice for today's memorial lecture as I am amongst the few available PM activists initiated by Dr. Tamhankar. Dr. Tamhankar came from a background of Physics at the Masters level followed by a Diploma in Electrical Engineering and capped it all with a Ph.D. in Physical Metallurgy (Fatigue studies to be specific). He was imported from France around 1962 to head DMRL, then just conceived. DMRL today is a premier Metallurgical Institution, a proud landmark of Dr. Tamhankar's vision. Mishra Dhatu Nigam a PSU was the culmination of his vision and further he left us the legacy of some associations/societies as this one to nurture that vision. So much so in memory of the great person.

What I am here to share with you today is my excitement regarding the new wave called Nanotechnology which has become a commonplace buzzword.

Why This Concern About Nanomaterials

The subject of my talk today is Nanomaterials and the PM Industry. I shall not delve into technical aspects of nanoparticles, nanomaterials and nanotechnologies as I am sure many of my learned colleagues shall share their information during the course of this conference. Rather I would like to briefly discuss the driving force and market potential that excites me. Why at all should we be concerned above this new wave or hype of nano? The simple reason is survival. Every now and then a tidal wave sweeps across the technological sea throwing new challenges and leaves in its wake the debris of past glory. While we love to bask in past glory, it hardly helps to cope with tomorrow. We must understand that change being the only constant in life we have to learn how to manage change, in other words ride the tidal wave and not go under.The fairly recent wave which many fell victim to, was Globalisation and the associated ISO qualification needed. In the present context look at some figures and that will indicate the potential of this wave and also look at the changes likely to overcome us if we are not on the surfboard.

The total global expenditure for nanotechnology related R&D has been indicated as 8.6 billion USD. The USA's 21st Century Nanotechnology Research and Development Act (2003) allocated 3.7 billion USD on nanotechnology R&D during 2003-2008. The Japanese government had doubled its funding for nanotechnologies to 800 million USD from 2001 to 2003. Within Europe about 1.25 billion USD is the spending every year. The UK government has allocated nearly 82 million USD per year from 2003-2009. Very large spending ineed.

Today there are over 1500 companies involved in nanotechnology R&D world wide. Exact statistics are difficult to access, however it is reported the number has doubled only in the 1990's. Of these, roughly 320 produce nanomaterials in some form or other, of which around 200 are nanoparticle producers.

In the Table 1 below you can see the market share in today's range of applications but that will not remain as more people understand and exploit the immense potential being unfolded each day.

Table 1. Market Share of Companies Directly Involved with Nanomaterials Usage.

Type of product	*Number of cos.*	*Primary market focus*	*Percentage*
Nanoparticles	160	Medical/pharmaceutical	30%
Nanotubes	55	Chemicals and advanced materials	29%
Nanoporous materials	22	Information and communication technology (ICT)	21%
Fullerenes	21	Energy	10%
Quantum dots	19	Automotive	5%
Nanostructured materials	16	Aerospace	2%
Nanofibres	9	Textiles	2%
Nanocapsules	8	Agriculture	1%
Nanowires	6		
Dendrines	5		
Total	321		

The worldwide growth in interest in nanomaterials is clearly indicated by the increase in literature published and patents applied (see Fig. 1). There is a doubling of interest in the last decade and that is going to bear fruit now. What we have seen is an awesome amount of investment and so shall be the harvest. The question I want to ask ourself is: where are we particularly the PM Industry on this map.

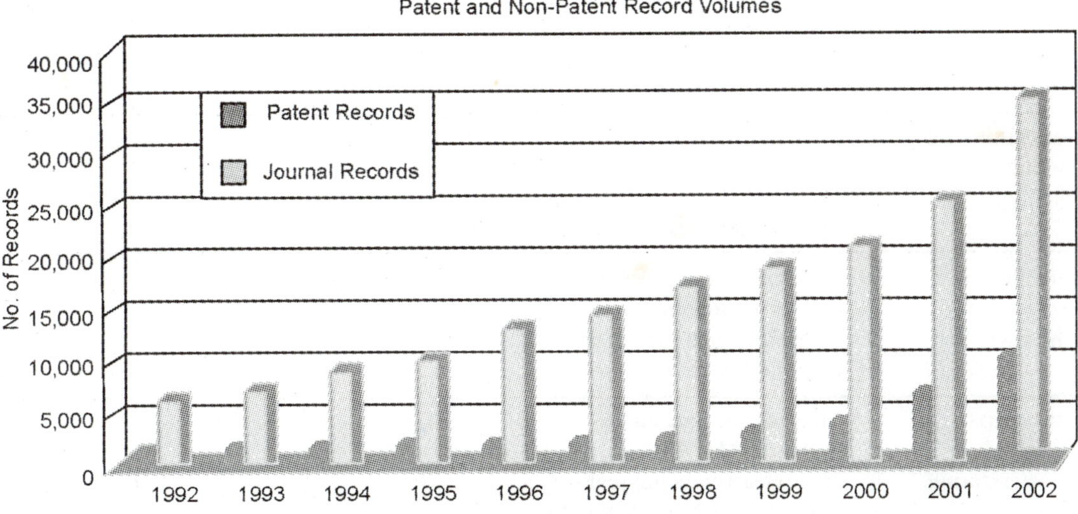

Fig. 1. Growth in number of patents and journal publication on nanotechnology.

The strongest indication that the market for nanomaterials is set to grow significantly is the fact that major materials based companies are investing heavily on R&D for modernizing and buying into existing companies.

The Market for Nano Particles

The market for nanomaterials is so vast that it has led to a rapid growth in the number of companies. The trend is governed by two philosophies.

1. The evolutionary progression into nanoscale materials.
2. The step-change in performance opening up new vistas of applications.

In the first category the market already existed. Possibility of re-engineering existing materials down to nanoscale have enabled increase performance and thus introduce new products. As example we have cermet cutting tools with improved performance and life, use of nanosize lapping and polishing compounds for chips whose track dimensions are approaching 90 nm and in magnetic recording media where higher density storage is driven by finer particle and grain sizes to enable terabyte storage capacities.

On the other side, the step-change in properties is opening up totally new applications which were only hypothesized all these years, for example targeted drug delivery, new optoelectronic devices, and miniaturization (robotics). Carbon nanotubes have demonstrated unique optoelectronic properties. Nanodispersed structural parts promise hitherto unachievable property regimes.

The wealth of potential applications is growing by the day and some estimates of the overall size of the nano technology market and of the subsidiary nanomaterials market is indicated in Table 2.

Table 2. Estimated Global Production Rates for Various Nanomaterials and Devices

Application	Materials/device			
		Present	2005 – 2010	2011 – 2020
Structural	Ceramics, catalysis, coatings, thin films, powders, metals	10	10^3	$10^4 - 10^5$
Skincare products	Metal oxides (TiO_2, ZnO_2, Fe_2O_3)	10^3	10^3	10^3 or less
Information and communication technology (ICT)	Single-walled nanotubes, nanoelectronics, optoelectronic materials (TiO^2, ZnO, Fe_2O_3), light emitting diodes (OLEDs)	10	10^2	10^3 or more
Biotechnology	Nanocapsules, targeted drug delivery, biocompatible materials, quantum dots, composites, biosensors	less than 1	1	10
Instruments, sensors, characterization	Microeletromechanical systems (MEMS), Nanoelectromechanical systems (NEMS), scanning probe microscopy (SPM), dippen lithography, direct-write tools	10	10^2	$10^2 - 10^3$
Environmental	Nanofiltration, membranes	10	10^2	$10^3 - 10^4$

The impact of nano materials will extend way beyond the immediate value of the materials themselves. One estimate puts the impact at of over 340 billion USD by 2010.

The value addition in nano materials use is very high and so the volumes are some times contradictory. Considerable effort is being put into developing advanced defence applications which are not estimated but are likely to have an even greater impact on commercial manufacture.

The growth in activity surrounding nano materials continues unabated as more R&D funds are poured into nano technology and people are looking to exploit the expanding range of novel properties being discovered each day. Advances in existing production techniques are improving quality and yields, providing encouraging prospects of commercially viable bulk manufacture. While a variety of processes are being used it is clear that these will commercially succeed only if they are developed concurrently. The drive behind all this intense activity is the continued developments in the field of nanomaterials and the constant stream of new properties and capabilities that are being discovered. These innovations appear to make possible an ever- increasing number of applications for which nanomaterials can provide improved performance and the promise of competitive advantage and value creation. In many mature industries such as automotives, commodity plastics, and power generation the slightest advantage over one's competitor can mean the difference between struggling to survive and a sustainable existence. The use of nanomaterials in these industries is increasingly seen as one way of gaining this advantage in the market place.

The question we need to ask ourselves today is: should we and if yes, then how do we relate to this new wave so we are not swept under. Let us briefly understand.

What is Nano-Technology?

There is a little consensus as to the exact domain of nano (10^{-9} m ?) And so I would go with a non-dimensional definition.

Nanoscience is the study of phenomenon and manipulation of materials at atomic, molecular, and macromolecular scales, where properties differ significantly from those at a much larger scale. Nanotechnologies involve the design, characterization, production and application of structures, devices and systems by controlling shape and size at the nanoscale.

Nano materials cross the boundary between nanoscience and nanotechnology and link the two areas together.

Industrially, we are interested in nano particles which are discrete particles having a mean diameter < 250 nm, covering a diverse range of materials; polymers, metals and ceramics. Nanomaterials come in all morphologies, viz., spheres, flakes and platelets, dendrific structures, tubes and rods. Obviously, obtaining such materials and their subsequent processing is going to demand the best of engineering capability.

The materials at the nanoscale 1 nm and 250 nm lie in between the quantum effects of atoms and molecules and bulk properties of materials. It is in this 'no-man's land' where many physical properties of materials are controlled by phenomenon that have their critical dimensions at the nanoscale. Fabrication and control of the structure of nanoparticles will give the scientist and engineer the capability to influence properties and to design materials and devices with unique characteristics. Some potential applications are listed in Table 3.

Table 3. Characteristic Lengths in Solid State Science Model

Field	*Property*	*Scale length*
Electronics	Electronic wavelength Inelastic mean free path Tunneling	10 – 100 nm 1 – 100 nm 1 – 10 nm
Magnetics	Domain wall Spin-flip scattering length	10 – 100 nm 1 – 100 nm
Optics	Quantum well Evanescent wave decay length Metallic skin depth	1 – 100 nm 10 – 100 nm 10 – 100 nm
Superconductivity	Cooper pair coherence length Meisner penetration depth	0.1 – 100 nm 1 – 100 nm
Mechanics	Dislocation interaction Grain boundaries Crack tip radii Nucleation / growth defect Surface corrugation	1 – 1000 nm 1 – 10 nm 1 – 100 nm 0.1 – 10 nm 1 – 10 nm
Catalysis	Surface topology	1 – 10 nm
Supermolecules	Kuhn length Secondary structure Tertiary structure	1 – 100 nm 1 – 10 nm 10 – 1000 nm

How Nanoparticles are Made

Nanoparticles are made by four generic routes.

(1) Wet chemical

(2) Mechanical

(3) Form in place

(4) Gas phase synthesis

It is important to note that the resultant material property is significantly affected by the route chosen to fabricate them and some of the routes are more aligned with the fabrication of certain classes of materials.

1. *Wet Chemical Processes*

These include colloidal chemistry, hydrothermal methods, sol-gels and other precipation processes. Essentially, solutions of different ions are mixed in well defined quantities and under controlled conditions of heat, temperature and pressure promote the formation of insoluble compounds and thence to dry powders. The advantage of wet chemical routes are the diversity of materials that may be handled including inorganics, organics and also some metals, with low cost equipment and in significant quantities with control over contaminants. Another advantage is the ability to control particle size closely and so produce highly monodisperse materials.

However, there are limitations too, such as bound water molecules as in sol-gels whereby yields are low.

An adjunct to these processes are those involving biological materials that provide a template into which inorganic materials can be grown. For instance, they are effectively used to manufacture materials for use in high density storage devices.

2. *Mechanical Processes*

These include grinding, milling and mechanical alloying and reaction milling, but are applicable only to non organic materials.

3. *Form in Place Processes*

These include lithography, vacuum deposition such as PVD and CVD and spray coatings. These however are not suited for producing bulk powder.

4. *Gas Phase Synthesis*

These include flame pyrolysis, electro-explosion, laser ablation, high temperature evaporation and plasma synthesis and are obviously not applicable to organic materials, because of the high temperatures involved.

While a host of methods are available, only those will be commercially viable which will have established downstream processing into user friendly products (applications).

Some potential application areas are listed in Table 4.

Gearing up for the Future

The use of nano particles and nano structured materials is set to escalate and the market has the potential to increase dramatically over the coming decade.

Table 4. Selection of Current and Future Applications Using Nanoparticles

Area	*Under development*	*Being introduced*	*Well established*
Power/Energy	Dye-sensitized solar cells using TiO_2 H_2 storage using metal hydrides improved anode and cathode materials for fuel cells Thermal control fluids using Cu	Nanocrystalline Ni and metal hydrides for batteries Environmental catalysts, ceria in diesel	Automotive catalysts.
Healthcare/medical	Nanocrystalline drugs for easier absorption Inhalable insulin Nanospheres for inhaling drugs currently injected using biocompatible Si Bone growth promoters. Virus detection using quantum dots Anticancer treatments Magnetics particles for the repair of the human body with prosthetics or artificial replacement parts Antioxidant drugs based on fullerenes	Molecular tagging using CdSe quantum dots Drug carries for drugs with low water solubility Coatings for implants such as hydroxyapatite	Ag-based antibacterial wound dressings, ZnO fungicide. Au for biolabelling and detection. Magnetic resonance imaging contrast agents using superparamagnetic Fe_2O_3. Sunscreens using ZnO and TiO_2.
Engineering	Improved thermal barrier coatings Spark plugs using nanoscale metal and ceramic powders Nanoporous silica based on aerogels for high-efficiency insulators Controlled delivery of herbicides and pesticides Improved moisture films for packaging Chemical sensors Molecular sieves	Abrasion-resistant coatings using alumina, YZr_2O_3 Nanoclay reinforced polymer composites. Lubricant/hydraulic additives: $CuMoS_2$ Pigments Improved scratch-resistant coatings Self-cleaning glass using TiO_2 Propellants using Al	Structural enhancement of polymers and composites. Thermal spray coatings based on TiO_2, TiC-Co, etc. Nanostructural Al alloys incorporating nanoparticles Inks; conducting, magnetic, etc. using metal powders Processing catalysts Cutting tool bits: WC, TaC, TIC, Co Automotive tires.

(*Contd...*)

Table 4. (*Contd...*)

Area	*Under development*	*Being introduced*	*Well established*
Consumer goods		Anticounterfeit devices Nano-starch-based adhesives for cardboard packing.	Packaging using silica. Glass coatings for antiglare, antimisting mirrors using TiO_2. White goods with easy-clean coatings Ski wax. Sports goods: tennis balls. Rackets using nanoclays. Water/stain-repellant textiles.
Environmental	More sensitive sensors Environmentally friendly antifouling paints and coatings Soil remediation using Fe	Alumina fibers for water treatment Photocatalyst water treatments using TiO_2 Pollution-destroying paints using TiO_2	Tiles coated using alumina and other sanitary ware Self-cleaning glass using nanostructured coatings based on TiO_2 Antireflection coatings
Electronics	EMI shielding using conducting and magnetic materials. Electrically conductive plastics. Light-emitting Si LED nanoparticles for displays. Electronic circuits, nonvolatile random access memory (NRAM) using Cu, Al. Display technologies including field emission devices using conducting oxides and carbon nanotubes.	Ferrofluids using magnetic materials Optoelectronic devices such as switches using rare-earth-doped ceramics Smaller multiplayer capacitors (MLCs) using Ni and Cu nanopowders Conductive coatings and fabrics having rare-earth-doped ceramics Nanoscale magnetic particles for high-density data storage	Chemical-mechanical planarization alumina, ceria. Coatings and joining materials for optical fibers based on Si.

Essentially what I want you to appreciate is that as we go into finer particulate sizes we make our materials more reactive higher surface area so much so that sintering temperatures have been noticed to drop by upto 400°C. Compressibility drastically increases with enhanced sinterability and ultrafine dispersal of constituent alloy design and component manufacture will go for significant change. If short duration sintering can be established using microwaves, we may even retain nano crystalline structures in bulk components. All these will need equipment development all along the processing cycle.

Some of the challenges Scientists and Engineers will need to address themselves to ride the nano wave include;

- Produce materials in volume commercially, at viable prices, contain contamination at scaled up level of operation.
- Supply the material in a form suitable for inclusion in a manufacturing processes-storage, transportation, and down stream processing to finished product.
- Consistency and reliability in volume production, adherence to tolerances on size, distribution and composition in bulk manufacture.
- Characterization of the nano particles and products thereof. Rapid, bulk, preferably online techniques need to he developed.
- Market focus and prioritization.
- Add and retain value - IPR issues with growing competition.
- Health, safety and environmental issues. With any high profile technology these questions will arise and suitable regulatory norms will have to be evolved to safeguard health and the environment.

CONCLUSION

It is obvious the opportunities are great. While prima facie the health and environmental hazards do not seem frightening some of the nano materials particularly of bio-technical interest could go astray as well.

The point is how does the PM industry and PM R & D benefit and come out a cropper. Some of the initiative taken by DMRL shall be elaborated by my colleagues during the course of the conference.

We need to decide what percent of that market share will be ours. There is a strong case for strategic alliances.

ACKNOWLEDGEMENTS

The author wishes to acknowledge the contributions of Director & colleagues at DMRL who have provided cause for the excitement and thence the inspiration to share the same.

REFERENCES

1. Pitkethly Michael, J., "Nanomaterials — the driving force". Nanotoday, 20, (Dec. 2004).
2. Dowling Ann P., "Development of nanotechnologies", Nanotoday, 30, (Dec. 2003)
3. International Conference on Nano-materials: Synthesis, Characterization and Application (Nov. 2004).

6

TPM Implementation in PM Industry Through Small Groups for Zero Defect Production of PM Parts

Vinod Kumar Singh
Ordnance Factory Board, Munitions House
10A, Shahid Khudiram Bose Marg, Kolkata-700001, India

ABSTRACT

Total Preventive Maintenance (TPM) is a shop floor management technique developed more precisely by JIPM and is focused about developing autonomous small groups of operators. It concentrates on increasing the robustness of production process by focusing on process plant and human resources around it. It believes in increasing the participation of workers and systematically developing there knowledge about the processes they are handling. The paper describes the methodology of implementing TPM, advantages of TPM in PM industry and continuous improvements resulting in Zero Defect Production of existing products and then developing a capability to develop new products with Zero Defect.

Key words: TPM, Manager Machine Model, Autonomous Maintenance, Zero Defect, Hygiene factors.

INTRODUCTION

Ordnance Factory Board (OFB) under the aegis of Ministry of Defence, Government of India has mandate to produce conventional weapon systems and equipment to meet the need of armed forces. Heavy Alloy Penetrator Project is a facility established under OFB. HAPP is an IS09001:2000 certified installation with Quality Control Laboratories at HAPP certified to IS017025:1999. In order to provide superior product at least cost through enhance productivity HAPP had opted for operating in TPM (Total Productive Maintenance) environment as introduced by Japan Institute of Plant Management with required modifications to suit the working environment in a typically administered departmental production unit under Government of India. The TPM approach is primarily based on the philosophy of providing equipment and process knowledge to the operators and motivating to function in autonomous manner. TPM approach as suggested by JIPM[1] was not suitable for operating in the Government sector production facilities. Modifications were necessary to suit the local conditions for suggested steps of JIPM TPM implementation methodology. HAPP has been divided into four production sections for maintaining the technology orientation. These are Powder Metallurgy (PM) Shop, Machine Shop, Surface Treatment Shop and Assembly Shop. A Divisional Officer or Head of Section commands each shop. Basic unit is controlled by a Charge-man reporting to Head of Section. Factory has 85 small groups of 5-8 workers and forms the basic unit. TPM implementation was extended to complete plant. PM shop is now successfully operating in TPM environment. PM shop in the plant is having 12 autonomous small groups.

Paper describes the journey of implementing TPM in PM plant and the transformation of working environment to a new environment full with confidence and enabling culture.

WHAT IS TPM

TPM expanded by JIPM as Total Productive Maintenance starts with concentration on Maintenance of plant and Machinery. It is a union of systematic small group activity aimed at improving the system by concentrating on hardware, equipment, people and preventive maintenance aspects in a proactive manner. It starts with increasing the understanding of operators in the equipment and systematically developing them into change agents with a clear goal in mind to reduce the losses. Literally word TPM meant Total Productive Maintenance and concentrated on increasing Original Equipment Effectiveness (OEE) through autonomous small groups using 8 pillar approach and 12 steps suggested by Japan Institute Plant Maintenance(JIPM[1]). JIPM suggests implementation of TPM through Manager Machine Model to be practiced by CEO of the plant and carrying out the seven steps of Autonomous Maintenance.

Eight Pillars started initially are Education and Training, SHE (Safety, Hygiene and Environment), Autonomous Maintenance, Planned Maintenance, Zero Defect, Initial Production & Flow Control, Continuous Improvement and Office Improvement. Technology and Energy committee were added later.

In TPM operators and other workers form small groups that carry out knowledge sharing to increase the group capability. Groups have multi group activity to understand the requirements of customer group precisely and also interact with supplier groups to precisely demand what shall improve their performance. Groups determine their Productivity, Quality, Delivery, Safety and Motivation targets.

TPM AT HAPP PM PLANT

To implement TPM, management selects a model machine and starts implementing 7 steps of autonomous maintenance (AM) activities. In PM plant Cold Isostatic Plant (CIP) was selected as manager machine model and the author being the officer in-charge of PM Plant Operations undertook the exercise of implementing seven steps. At the same time all the workers and middle management personnel were given training in the concepts of TPM and AM. Author followed a half an hour schedule once a day cleaning the plant and identifying the counter measures to be taken to reduce the cleaning requirement. Oil and powder leakage points were fixed and the plant design was modified to completely plug the areas from where the powder use to corm out in the environment while being filled in the mould bags and while being transferred from Double Cone Blender to CIP hopper. All hydraulic Oil pipes were replaced with steel pipes and white paint was applied at the joints to ensure that any leakage is visible immediately. It was noticed here that plant down time reduced and breakdowns vanished. Production Operators were initially spectator later formed a group with author and started helping improvement activities and within fifteen days they took over the actual ground execution operations.

Next step was to prepare Lubrication and Inspection Standards by identifying the points where lubrication need to be checked and the points where regular inspection be carried out to ensure that plant functioned smoothly. Thus a Cleaning, Lubrication and Inspection Standard was generated. At this stage author while working with the model group initiated group formation in the other work centers of the plant. Soon the workers seeing the model group formed small groups with specific focus on mixing and compaction, sintering, vacuum heat treatment, swaging, post swage machining and testing area. Middle management cadre gladly joined the groups along with the workers. As a first step all the group members were given extensive training about the plant they are using while implementing first step of AM activity which involves cleaning the plant thoroughly and systematically all the groups generated Cleaning Lubrication and Inspection Standards. During cleaning emphasis was laid that the person doing the cleaning understands why he was required to clean those areas, how did the imperfections, which necessitated cleaning, affected the performance of plant in terms of output quality or performance. What can affect the safety, quality, maintainability and reliability of equipment Groups then identified the Countermeasures to be taken to ensure that daily cleaning of plant and equipment is not at all required and the plant is being maintained into as cleaned level without routine daily cleaning. Operators while performing AM learnt what to inspect in a routine manner on the plant. Identify the items that should be checked regularly and fixed. Determine lubrication needs of the plant-type, frequency and amount of lubricant. Group prepared a standard schedule for cleaning, inspection, lubrication with the help of plant maintenance manual and maintenance personal. Operators followed standard schedule for each machine, plant and furnace for 3 months and again revision was done to change frequency, delete redundant items to decrease the time spent in following the standard. In the event of any breakdown or quality problem this standard was revised based upon a Why -Why Analysis of failure and eliminating the root cause.

Sequence of Goals was fixed as Zero accident, Zero Grievances, Zero Defect, Zero Breakdown, and there after constant reduction in losses in terms of material and energy by identifying the component of material and energy which was not being directly carried away by product.

After about 12 – 18 months it was seen that small groups started functioning in normal manner and breakdowns vanished. Quality of product also stabilized and batch failures and quality variations vanished. It was also noted that TPM group activities had increased workers confidence to a very high extent. Persons who were earlier resistant to any change were now suggesting changes themselves regularly and were taking active part in implementing the change. Another benefit was now they were taking more interest in analyzing the process and were experimenting to improve the same. Cost of production went down as a result of continuous improvements being undertaken by groups. It was seen that TPM implementation approach advocated by JIPM proved to be the fastest and most accepted by employees.

GROUP MANAGEMENT STRUCTURE

Management through groups is the basic philosophy of TPM. TPM aims at creating group, training the group to manage themselves in autonomous manner. Groups of an area make a group of group leaders. Divisional Officer of the shop is the leader of group of group leaders. Secretary of subcommittees are member of leader groups. CEO is the Chairman of Group at plant level. Secretaries and Chairman of all committees, administrative heads of shop level group are members of this group. Each group sends a member to participate in a sub-committee as a representative of group. This group meets once'a month to review the performance of the shop and coordinate TPM activities in the shop.

Group orientation has the advantage that any individual is not put to hardship, as he cannot be relieved to meet emergent necessities whenever they are required for official work. It has a disadvantage that certain persons may feel that they are losing their importance. This feeling, in any case, has been found to be of very temporary nature and as in a group the leader, always is a person with higher knowledge and experience the individual ego gets satisfied. It is also noted that the leader is supported by the knowledge and experience of other members in a group in a synergetic manner. In TPM we revolutionize the working so that the best part of every individual is extracted and distributed to other members of group, thus building a group-strength. Divisional Officer is chairman of all subcommittees in the shop also leads the controlling group. He maintains group focus to organizational goal. Divisional Officer sends a weekly report to TPM coordinator on a pre-designed format. He also schedules all intra-shop and inter-shop group meets. He ensures that group meetings are held regularly and the improvements identified by the group are implemented in time in a economical manner. He is required to find problems in group-growth in a proactive manner and encourage the group activity. He helps group to identify targets and helps them achieving the same. He leads and manages people, groups, and subcommittees in a networked manner. He has specific attention on pushing CI projects to conclusion. Group relationship is shown in the figure below:

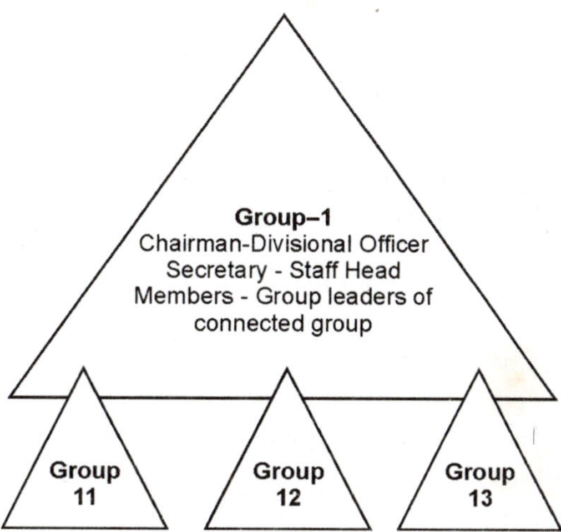

Group activities have grown with time and the customer and supplier relationship is established among the groups in the PM plant. Sintering group is customer of compaction group and powder-mixing group is supplier for compaction group. Vacuum Heat treatment, is customer of sintering group and acts as a supplier to Pre swage machining group. Swaging group is customer of pre-swage machining group and supplies to finish machining group. The ultimate customer is testing and inspection group. This group interacts with the external customers in this case machine shop.

Groups assign specific responsibility to individual members for a specific pillar activity. Generally this member represents group in the subcommittee meetings for that pillar. This arrangement offers a distinct advantage that almost every member in the group is leading a certain activity in the group and all members are interdependent for their image outside the group.

PILLAR AND COMMITTEES

Pillar is a specific area of strength to be extended to all groups in a coordinated manner. In TPM, at plant level a committee is formed for systematic furtherance of activities with reference to this strength throughout the plant. For each pillar activity there is a subcommittee in each work area. Subcommittees have their members drawn from groups in the shop. Officer in charge of the shop identifies one person from these to be the secretary of subcommittee for the shop. Secretaries are generally person who have demonstrated active interest in the particular strength and have recognition from the group. Secretaries of Subcommittees are given very specific focused training in this field and act as a referral person to all the groups in their area.

Plant level committee has all secretaries of sub committee as its member and a senior officer having committed and demonstrated capability as chairman of committee. He/She is a person who has received specific training on the strength identified by the pillar. For example, Chairman of energy committee shall be a top level executive who has experience in managing energy, understand energy audit, has knowledge about new products in the global market and has interest in constantly upgrading his knowledge in that field.

It is desirable that Chairman of subcommittee should attend some meetings of committee to keep himself aware of the developments.

Committee meetings are held once in two weeks.

Committee reviews all the activities being done in different sections and helps in identifying and providing external inputs. It also monitors that training section conducts focused program to train the operators in a particular area. They help in providing the training material and resource person. Pre determined Formats to be used by section subcommittee and groups for displaying their information are finalized by committee. It helps standardization. Committees also identify the projects/improvements to be horizontally deployed throughout the plant.

Subcommittees in a shop meet weekly and list down all the activities identified in a shop, They coordinate resources required to carry out a kaizen, condition monitoring, inter group meetings and keeping of records in a shop. They help group to identify target in the area and horizontal dissemination of knowledge and experiences across the groups. They liaison with outside agencies and arrange external resources required by the group. Every Subcommittee prepares a brief in a pre-determined format and submits it to the secretary of committee and TPM coordinator. They carry out horizontal deployment of improvement projects identified in Committee in the shop through groups.

Brief account of different committees are given below:

AUTONOMOUS MAINTENANCE COMMITTEE

Autonomous Maintenance (AM) is the most fundamental and prime activity in TPM. AM Committee systematically monitors implementation of seven steps as suggested by JIPM throughout the plant. Identifying the targets for each shop and training needs of groups for autonomous maintenance is major job of the AM committee. The committee to control the design of equipment to facilitate autonomous maintenance does interaction with new equipment suppliers. Identification of Innovative AM ideas and their horizontal deployment throughout the plant is another major task of committee.

Success of committee is reflected in achievement of Zero Breakdown and Zero Accident. Identification of hygiene factors and their uprooting is also a specific goal of this committee. Committee ensures that through Autonomous check sheets and Cleaning, Lubrication, Inspection and fixing standards all such factors that contribute to fluctuation in quality are uprooted.

CONTINUAL IMPROVEMENT COMMITTEE

Continual improvement (CI) as a result of group effort is the essence of TPM. Increasing worker participation in the improvement is mandate of this committee. To ensure that CI projects (Kaizens) progress in a smooth manner is the mandate of committee. Groups identify CI projects and register these with the committee. Horizontal deployment of the successful CI projects is a major responsibility of CI committee. CI committee arranges resource requirement of special nature.

EDUCATION AND TRAINING COMMITTEE

Education and Training Committee runs training department of Plant and arranges basic training of all employees in the basic concepts of TPM and AM. It has a mandate of raising skill level of all workers in a systematic manner to convert them from sedentary worker to knowledge worker. Identification of Master Trainers, preparation of one-point lessons, providing educational facilities to all employees to further their carriers in the organization are its major mandates.

The committee undertakes preparation of training program for new technology being introduced.

ENERGY COMMITTEE

Energy Committee analyzes the energy input into product through groups. It has a constant job of identifying potential improvements to reduce energy input in process operations. It scans improved devices and plans systematic up-gradation of plant, machinery and practices.

It has a mandate to increase energy consciousness of all employees.

Efforts of committee resulted in reduction of Maximum Demand from 3,5MW to 2.8MW. Increasing use of solar energy and other green sources is another major agenda for future. Generation of hydrogen by solar cells is identified as most promising.

INITIAL FLOW CONTROL COMMITTEE

Initial Flow Control Committee ensures that all the processes are well defined and documented. It defines the route a new product shall take. Groups to be involved in the development of a new product. Tooling, Fixtures, Gauges, Storage and Handling requirements for a new product are deliberated in Initial flow control committee. Product Design and Development group of the plant functions in close coordination with the committee. Chief Design Manager of the plant is generally designated as the chairman of the committee. It works in close coordination with all shops. Its mandate is First Time Right through process approach.

OFFICE MAINTENANCE COMMITTEE

Office Maintenance Committee keeps all outward correspondence and papers related to personnel matters, purchase, accounting and audit requirements in a structured manner. This reviews all systems and progressively plans for filing and documentation in a manner that all documents can be retrieved for reference without delay, any external query can be handled in less than ten minutes. Any reference by any worker for his requirement is made available immediately. Things are kept in a defined regimented manner using 5S. Office Management Committee has taken up systematic computerization of process data, maintenance of archives for old drawings and specifications, project management internal and external transactions. The aim is to do it in a manner that the information is available for reference by all levels through an interactive screen.

PREVENTIVE MAINTENANCE COMMITTEE

Preventive Maintenance Committee is the main stay to start TPM operations and in making TPM implementation successful. PM committee monitors that Preventive Maintenance program are implemented for all plant and machineries in the unit. With continuing training of operators about plant and process and progress of steps of AM the PM program is gradually reduced. PM Committee plans this change in a very systematic manner by providing necessary skill training and support for special tool requirements to groups.

PM Committee also monitors the expenditure on maintenance, spares and initiates kaizen for improvement of spare life thus constantly increasing mean time between failures (MTBF). Once step 6 of AM has been implemented on the equipment, it takes action to ensure that required spares are scheduled properly. Condition Monitoring program to ensure that time based PM schedules are changed to condition based proactive maintenance schedules, is introduced by PM Committee. Consumption of lubrication oil and other consumables in maintenance is monitored over a period and PM Committee ensures its availability.

SAFETY, HYGIENE AND ENVIRONMENT COMMITTEE

Safety, Hygiene and Environment Committee ensures that there is no accident in the operations being carried out. A specially trained officer in safety matters chairs it. They systematically monitor activities to remove causes of accident. Groups to ensure operational safety critically analyze all operations. CI projects are taken to modify plant design to ensure zero accident probability. Check points are also identified and introduced in the AM sheet to ensure that all critical points are monitored periodically.

Adherence to Statutory requirements and industry standards like ISO14000 is the minimum standard to which committee targets initially and goes much beyond.

TECHNOLOGY COMMITTEE

Technology Committee develops process and plant to meet specified needs. It has specific focus on different technologies available globally. Committee reviews the developments in related fields and takes strategic decision to acquire or evolve the technology for adaptation. As committee has links down to the worker groups, adaptation of any new technology is very quick and benefits can be harnessed with least lead-time. This committee decides the manner of acquisition or development of technology. Persons at all levels are pre-specified and do participate in negotiations and planning. A check on acquisition process by groups ensures that no minor detail is overlook at any time and the speed of acquisition is well adjusted.

ZERO DEFECT COMMITTEE

Zero Defect Committee has mandate that in a production process no defective is produced. Monitoring customer complaints and refining of process, material specifications, quality maintenance and assurances activities are reviewed by the committee. Identification of technology factors and hygiene factors and providing necessary inputs to keep them under check so that no defective is produced is basic job of the committee. It monitors the cost of quality and plans to reduce the same in a continual manner. To update quality matrix, root cause analysis whenever a defective is produced is major effort taken by the committee. Mistake proofing of production line to ensure that no defective moves to next stage is major exercise taken by committee members throughout the plant.

Identification of data storage and processing requirements, training of person in QC and QA techniques with emphasis on implementation in the area of work, reducing the customer complaint and process rejection to zero are the targets pursued by committee. Monitoring of quality CI projects and their horizontal integration is the prime responsibility of committee members. Organizing Supplier-Customer Group meets and resolving the issues generated there in, ensuring that development process has visible quality assurance system and any error is prevented from occurrence in a pre-defined manner are the activities undertaken by committee. Committee also pursues preparation of single point lessons on quality matters.

Committee also monitors the use of Analysis tools where Zero Rejection targets have not been achieved. Vendor education on matters of quality is also done by this committee.

MANAGING MEETINGS

At this stage one will have a feeling that executive and staff in the plant will be always be busy in organizing the meetings and many of these meetings. will be very prolonged affair. But it is not so. Initially the meetings are in a haphazard manner but soon these become very systematic. Each meeting is kept limited to 30 minutes. First five minutes are spent to review developments of previous period, next twenty minutes are devoted on specific business of the day and last five minutes to record decisions, responsibilities and targets for next period. Every participant in the meeting presents his view in Transparent Open and in Totality (TOT). It is seen meetings reach to the TOT state in six months generally. Minutes of meetings are made available in few minutes and delivered electronically, Minutes also serve as information sheet during the next period under review. The first line management uses minutes as review sheet. All developments are recoded on these sheets. In the next meeting the sheet is available with updated details and new ideas or projects for a quick reference and discussion.

Meeting is also an informal forum to teach all the formalized procedure being evolved. Development of group is the essence. If the volume of work increases in a particular area, frequency of meeting was increased to daily with invitation to other supporting group leaders to be available during the meeting. Later this was changed to supply the inputs during the first five minutes of meeting schedule telephonically or through mail to secretary/chairman prior to start of the meeting. Expectations from other groups are communicated at the end of meeting.

On the subjects of mutual interest multi group meetings are organized. CI projects often require multi-group participation. Committee and subcommittee have a distinct role to play in multi- group working.

ACTIVITY BOARD

Every group has prepared an activity board. This board displays photographs of conditions of plant and machinery and shop floor before and after implementation of seven steps of AM. Targets of the group, CI projects under implementation and Single point lessons are displayed on this board. A look at the board gives a clear picture of the road map of group for next six months. All past achievements are displayed on the activity board and constantly motivate group members and other groups. Biggest advantage is that during visits of customers and discussions no preparation is needed. It also develops group creativity and ability. Display Boards are focal center for group weekly activity. Format is predefined and flexible and is determined by group by discussions with customer groups and Divisional Officer.

ONE/SINGLE POINT LESSONS

One point lesson is a major tool for transferring the knowledge from expert to others. One point lesson is prepared in the form of a chart and teaches a single activity. Names of person who can be classified as expert are mentioned in the chart. Expert are given specific time at their disposal to teach others. More and more persons acquiring skill to master's level are developed. Once a person acquires skill equivalent to master his name is also entered in the chart. Group leader and member responsible for Education and training of group systematically monitor acquiring of skill. He ensures that all persons in the group acquire at least two skills and develop proficiency to master level. In TPM one point lesson is systematic way of disseminating skill across the plant. PM plant developed Single point lessons for Relation of Microstructure with sintering cycle variation, Effect of Annealing on quality, Effect of Die alignment on swaging, CIP cycle, density measurement etc. are prepared for penetrator bar. It is also ensured that all members are attempting to acquire higher education qualification in the field of their interest.

TPM NEWSLETTER

After reaching a stage TPM newsletter is started. Generally this stage is reached with two to three years. Frequency of newsletter in starting is monthly and then increased to weekly. It is basic mass communication device. Targets and Achievements, Plant policy, Area of Thrust, Message from CEO, Exemplary CI projects, Status of AM achieved in different shops, persons upgraded to masters, New One point lessons, Future training program are announced through TPM Newsletter. It replaces routine circulars. Committee Activity during previous period is an integral part of Newsletter. Plan is to introduce an evening channel to increase the efforts in this direction.

INFORMATION MANAGEMENT IN TPM

Success of TPM in a plant depends on speedy solutions to problems identified by small groups and thus leading the small group by example. Intra and inter group information exchange in TPM is key to success. There is a necessity to communicate the customer feedbacks to each group. Information management in TPM is done through display boards. Each group maintains a display board. In this board, targets and achievements against each pillar activity is displayed and is updated at the end of day by a designated member of the group. One-point lessons are very important piece of information on the board. Production scheduling, drawings and process capability is displayed on each plant and machine. Targets, Achievements and action plans are displayed with the road map. It will look as a very dedicated Information Management team is necessary for all these activities, but it is not so, this is a result of daily group meeting. Where on a pre-designed format the information are updated as a matter of routine in a playful relaxed environment. Divisional officers or charge-man based upon his discretion made ERP system available accessible to select groups, which have achieved a degree of autonomous working. This motivated other groups to pursue their sincere efforts for achieving a level of autonomous working. A LAN is planned to increase the speed and reduce use of paper.

RESULTS OF TPM

All out improvement was noticed in the plant as a result of implementing TPM. Cost of product dropped to 50% of original. Stage and Batch rejections became nil. Energy input per job was reduced to 60%. Maintenance Expenditure for plant and machinery came down to 10% of original. It was noticed that all workers have acquired creative skill and were more interested in developing new products in their area once they acquired Zero defect and breakdown. Some took part in developing control algorithum, while other started concentrating in increasing degree of automation. Many groups in PM plant started surfing internet and started taking keen interest in development of new products.

DIVERSIFICATION IN TAPs AND OTHER TUNGSTEN BASED ALLOY PRODUCTS

TPM Efforts made it possible to plan for diversification in the plant. Improved working permitted management to think beyond routine. Experimental facilities for Mechanical Alloying were added. A planetary mill and hydraulic press was installed with a meager investment of about US$8000. Worker-groups developed Oscillating Blanks of Watch, Balancing Weight Blanks for Chetak Helicopter, Balance Weight for Inertial Navigation System for Missiles, Thrust Plate for Trishul Missile, Boring bars for Sandvik India Ltd., Tungsten Copper Electrodes, Inserts for frangible bullets, Tungsten weight balancing granules for Advanced Light Helicopter etc.

IMPORTANCE OF MAN AND MANAGEMENT

Most important factor in PM plant is man, most vital link between various operations and subsystems. Defining training requirements, career progression, motivational schemes, psychological needs of the employees need to be studied and monitored and given careful implementation. Any problem in industrial relations can have very high cost and should be settled in a proactive manner before seeds of unrest are laid. Managing the process of HRD through groups avoids grievances. Each person has a full knowledge of the capability of others and is able to appreciate the contribution individual can make to the total effort. Group is able to select right men for the right job and is able to implement corrections almost instantaneously. They are able to groom employees in a informal environment. PM personnel need to be controlled through mind, heart and soul to make them fully engrossed in his work when he is on job. Training costs of each employee involved in PM installation is likely to be in the range of Rs. 4 to 5 Lakhs (US$10000) and its gain are made very effective using small group philosophy in TPM. Managers will have to keep a watch and use innovative methods to keep employee turnover in control. They have to plan a systematic changeover program. HRD is a very key concern in developing enabling environment and should be handled by the Top Management as a group concern. Every technical manager and Group Leader should also be trained in basic HRD principles.

One hour training to every employee per week was done through in house resources.

CONCLUSION

TPM methodology propagated by Japan Institute of Plant Management is a very effective strategy for starting group working and making them autonomous. Such Group work is key to unleash the creativity of persons and drives the cost of product to minimum. It is best to convert a line organization into matrix organization. New expansion for TPM was defined by workers in the PM plant as TOTALLY PERFECT MANAGEMENT.

ACKNOWLEDGEMENT

Author is thankful to Officers, Staff and Employees of PM Plant, HAPP, Chairman OF Board and Ministry of Defence for extending their support, all family members, and colleagues for their untiring efforts and help in implementing the TPM methods.

REFERENCES

BOOKS

1. Introduction to TPM by Seiichi Nakajima, published by Japan Institute of Plant Maintenance Reprinted in India by Productivity Press, Chennai, ISBN 81-85985-15-4.
2. Suchiro Kunio, Eliminating Minor Stoppages on Automated Lines, Productivity Press, Cambridge, Massachusetts.
3. Joiner Associate Consulting Group, The Team Hand Book Joiner Associates Inc. Madison, U.S.A., 1988.

Unit–II

Powder Metallurgy for Automotive Industry

1

High-Density Multi-Level PM Components by High Velocity Compaction

Gerd Hinzmann
Hawk Precision Components
286 Vance Drive
Oakville, Ontario, L6L 3L2
Canada

Dirk Sterkenburg
Hydropulsor AB
SE-691 80 Karlskoga
Sweden

ABSTRACT

Today the High Velocity Compaction (HVC) process and equipment enable the production of high-density, single-level PM components. Large parts in particular can be formed very cost efficiently. The compaction of multi-level components utilizing HVC is the next essential step to broaden the range of applications for this new technology. This paper describes process options for multi-level compaction and discusses their technological merits as well as risks and limitations. All of these approaches involve the compaction of a multi-level pre-form prior to densification by impact. Conventional multi-level tool adapter designs do not meet the rigidity requirements of the HVC process. A novel approach to tool rig design can overcome this constraint and will also be discussed in the paper.

INTRODUCTION

In traditional compacting the main ram motion of a press is either force- or position- controlled through hydraulic cylinders, crank drives or knuckle drives. The top tooling is constantly connected with the ram and upon entering the die cavity exerts a continuously increasing pressure on the powder charge.

High Velocity Compaction employs a ram with a discreet mass, which is accelerated to a predetermined speed and than impacts with the top tooling which has previously been brought into contact with the powder charge in the die cavity. This energy-controlled motion compacts powder by a shock wave. It reaches impact speeds of up to 10m/s.

Powders typically start to precompact, that is, particles start to mechanically bond, at a density far below the desired green density. To avoid inhomogeneous density distribution and/or cracks between sections of a multi-level part independently movable punches for each level are required as soon as the desired height difference between levels represents a substantial percentile of the overall length of the part. The multiple punches are actuated either force- or position-controlled and their motions relative to each other accommodate a uniform and simultaneous compaction of all sections of the part.

Approaches to Multi-Level HVC

The subject of this paper is to describe how multi-level parts can be made in conjunction with HVC. It appears to be extremely difficult to synchronize the auxiliary motions of multiple punches with a top tooling that is accelerated by an impact and decelerates while the impact energy is released at a speed that is two orders of magnitude higher than conventional, continuous compaction.

It has been suggested to compact the various sections of a multi-level part by impact in sequence rather than simultaneously [1]. This would not only require additional means to prevent radial powder transfer into not yet compacted sections, but it will certainly lead to cracks between the sequentially compacted sections. The more successful approach is the introduction of a dual operating mode for the main ram. With the first operating mode a multi-level pre-form is compacted to a density that provides suitable green strength, utilizing force or preferably position-controlled drives for the main ram and the multiple punch motions. In the second operating mode the main ram utilizes an energy-controlled drive to further compact the pre- form to high density, while no or little auxiliary punch motions are required. A dual operating mode was first published in conjunction with the short-stroked electro- dynamic shock compaction [2]. The adaption of the dual operating mode to High Velocity Compaction [3], as preferred by the authors, enables single-sided multi-level shapes, using multiple bottom punches.

Dual Operating Mode for HVC

The HVC dual operating mode works with closed loop position-controlled hydraulic drives for all tooling members to compact the pre-form. Preferably the pre-form features length to density ratios of its various sections, which allow further compaction to a uniform high density by impact with no auxiliary punch movements. Thus the bottom punches can rest on rigid mechanical stops to maximize impact energy discharge within the green part.

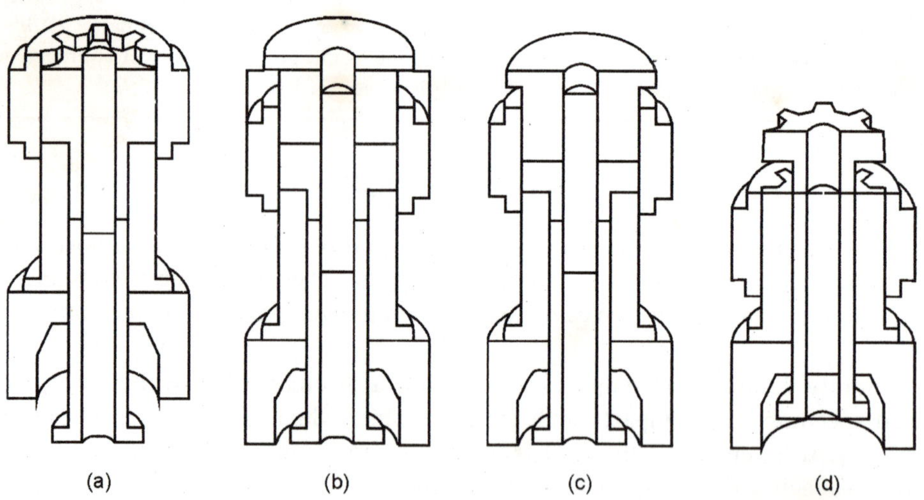

Fig. 1. Dual operating mode cycle. (a) Under-fill position (b) Position-controlled pre- compaction position (c) Impact compaction position (d) Ejection position.

Alternatively, if a highly responsive pressure relief system, which enables floating of pre-lifted punches, is feasible, the pre-form can also be of uniform density. In this case a pre-lifted punch will be driven onto its mechanical stop by impact, see Figure 1b. In the absence of a pressure relief system and with a pre-form of uniform density, a pre-lifted punch would have to be retracted onto its mechanical stop prior to impact, leaving a gap between punch and pre-form.

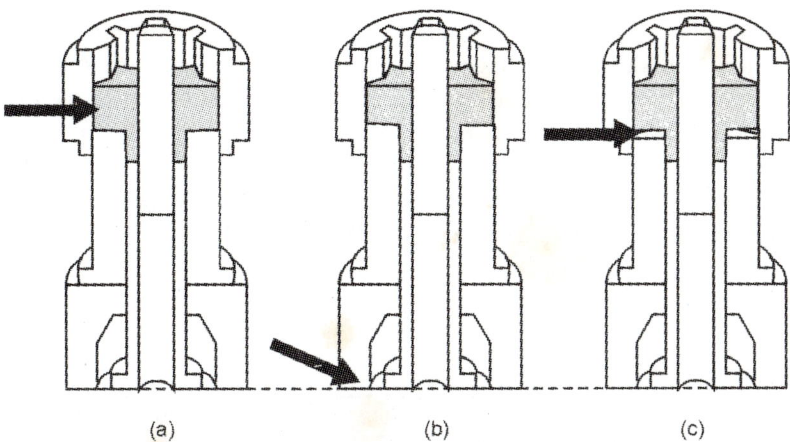

Fig. 2. Pre-form stage alternatives prior to impact. (a) Density gradient in pre-form (b) Outer punch in float mode (c) Retracted outer punch.

To extend the principle to multiple top punches for two-sided multi-level shapes appears not feasible, because the tooling with adaption and drive systems represent a large dampening mass detrimental to shock wave transmission.

Levels with moderate drop heights compared to the overall length of a part however can be formed with a stepped upper punch. The limits are similar to those of conventional compaction, with shape distortion at high densities adding another factor to consider.

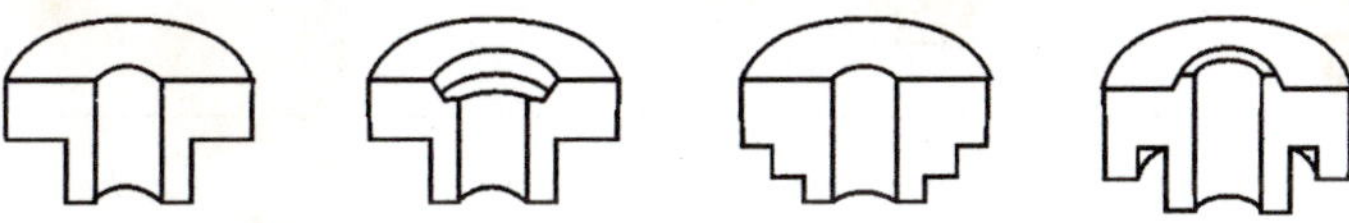

Fig. 3. Multi-level shapes requiring a bottom punch per level and a single (stepped) top punch.

In the absence of a suitable multi-level HVC press, the concept of single-sided multi- level compaction can be suitably simulated, by producing the pre-form in a conventional powder press and than inserting the pre-form into stepped tooling in a single-level HVC press.

Tool Rig for Dual Operating Mode

To minimize impact energy losses, the whole HVC press system including tools has to be as rigid as possible. While the shape to be formed inherently limits the optimization of tools and to a lesser degree tool adaption, the tool rig design is more open to improvements toward high rigidity. A new design concept has been developed, which is substantially more compact and therefore rigid compared to existing solutions [4].

The main characteristic of the new design is, that a number of concentric cylinders operate from essentially the same elevation rather than stacked on top of each other. This is made possible by supplying the pressure medium for the internal pistons through the base plate of the cylinder block and the internal cylinder housing(s). Linear encoders for position control are connected from the base plate. The concentric cylinders drive punches and cores. An additional, stationary punch can be placed on the internal cylinder housing. Lateral cylinders located in the corners of the cylinder block actuate the die plate, the top plate, which drives the top punch in the first, positional-controlled, mode can be operated the same way. To avoid spacers for punch length adjustments, adjustable mechanical stops can be integrated into the base plate. It has yet to be determined, whether adjustable stops compromise the desired tool rig rigidity to an unacceptable degree, in which case they would be replaced by hard stops.

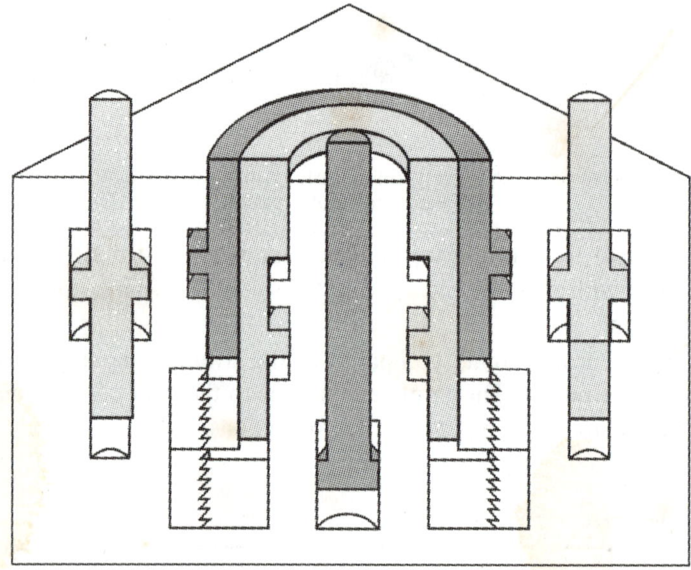

Fig. 4. Tool rig with three concentric cylinders, adjustable mechanical stops and lateral cylinders to drive not shown die platen.

The new tool rig concept and the proven HVC press frame form a very compact and robust unit. The tooling stack up, with punch cross-sections and aspect ratios dictated by the part geometry, is now the limiting factor for rigidity of the whole system. This holds particularly true with each added level, increasing substantially the length of the innermost punch.

SUMMARY

Multi-level parts with a substantial height difference on one side can be made utilizing the dual operating mode, which employs HVC after a pre-form has been generated. Levels of moderate height difference can be formed on the opposing side using stepped punches. The requirement for a robust and rigid tooling stack up limits the range of feasible shapes.

REFERENCES

[1] E. Cacolorni, J. Jerklind, Powder Compacting Tool, European Patent Application EP 1 228 856 A2 2002

[2] R. Fedrigo, B. Borrini, V. Garattini, G. Pino, Method of and Press for Compacting Materials in Powder Form to Produce Pieces to be Sintered, UK Patent 1 395 180 1975

[3] G. Hinzmann, Apparatus and Method for High-Velocity Compaction of Multiple-Level Parts P.C.T, Patent Application WO 03/008131 A2 2003

[4] G. Hinzmann, Tool Rig for the Compaction of Particulate Material, P.C.T. Patent Application WO 03/059608 Al 2003

2

Super-High Strengthening of the Low Alloy Steels Through Conventional and Advanced P/M Processing (MIM) Techniques

Hideshi Miura
Dept. of Intelligent Machinery and Systems, Faculty of Engineering Kyushu University
6-10-1 Hakozaki, Higashi-ku, Fukuoka, 812-8581 Japan

ABSTRACT

The properties of powder metallurgy products are strongly influenced by the density, chemical composition and microstructure. This study has been performed to clarify the relationship between the mechanical properties and the homogeneity of microstructure for a wide density range of low alloy steels (4600 grade, 0.4mass%C) by die pressing and metal injection molding (MIM) processes, using prealloyed and mixed elemental powders. Mechanical properties of sintered and heat treated steels for both powders are investigated, and some factors affecting the properties are discussed taking the distribution of the resultant phases and transformation products into account. Finally, it was found that the heterogeneous microstructure is quite useful for strengthening the steel in the high density range, at which all of the pores are closed and spheroidized, but the distribution of the phases should be fine and uniform.

INTRODUCTION

Sintered ferrous materials are considerably utilized as machine structural parts, and the demand has been expanded favorably in wide variety of industrial fields. Especially, the study on ultrahigh strengthening the sintered steels has been accelerated by the requirement of high performance properties for such as an automobile parts. The creation of high quality steel powders and the development of powder metallurgy (P/M) techniques have greatly contributed to the improvements in mechanical properties of sintered ferrous materials.

Metal injection molding (MIM) process is one of the advanced powder processing technique for producing the high density and complex metal components [1]. Through this MIM process, high performance sintered low alloy steel (4600 grade, Fe-2 mass % Ni-0.5 mass % Mo-0.4 mass % C) which showed the tensile strength of 1.4GPa and the elongation of 9% by heat treatment was obtained in our previous studies [2, 3]. This was due to the fine heterogeneous microstructure consisted of tempered martensite, retained martensite and austenite by the segregation of Ni added as alloying element.

This study has been performed to clarify the relationship between the mechanical properties and the microstructures for strengthening the sintered low alloy steels through the conventional P/M (die pressing) and new (MIM) techniques. Mechanical properties of the steels using prealloyed and mixed powders were investigated for a wide range of density. The effect of heat treatment was also investigated to inquire the contribution of transformation products to the properties. Finally, the contribution of each phase in the heterogeneous microstructure to the strength was discussed quantitatively by the image analysis and thin film diffusion theory.

EXPERIMENTAL PROCEDURE

A prealloyed 4600 H powder produced by water atomization as well as 4600 grade mixture from a water-atomized iron powder and several additive powders were used for this study. The chemical compositions and mean particle sizes of the powders are shown in Table 1. A 0.75 mass % zinc stearate as lubricant was added to both powders, which were further mixed with 0.50 ± 0.03 mass % graphite powder in order to attain 0.4 mass % C as the final carbon level in the sintered compacts.

Table 1. Chemical compositions and mean particle sizes of the powders used in this study

Prealloyed powder		*Chemical composition (mass%)*						*Particle size (μ m)*
		C	Mn	Cr	Mo	Ni	Fe	
		0.01	0.2	0.01	0.48	1.85	Bal.	56
Mixed	Fe	—	0.21	0.01	—	—	Bal.	48
elemental	Ni	<0.1	—	—	—	Bal.	<0.01	5
powder	Mo	—	—	—	Bal.	—	0.01	4

The blended powders were compacted into tensile test bars by single or double pressing and sintering to obtain the compacts under the wide density range. The compacts were covered with the mixture of alumina and 5 mass % graphite powders to prevent the decarburization, and sintered at 1523~1623 K for 3.6 ks in 10~30 vol. % hydrogen and nitrogen gas mixtures.

For MIM process, at first the powders were kneaded at 418 K for 1.8 ks with 40 vol % binder composed of 69 mass % paraffine wax, 20 mass % polypropylene, 10 mass % carnauba wax and 1 mass % stearic acid. The mixtures were injection molded into tensile and rotating bending fatigue test pieces. After molding, solvent debinding with n-heptane of 343 K for 18.8 ks followed by thermal debinding in 20 vol % H_2 and 80 vol % N_2 gas mixtures was applied. Sintering was performed at 1573 ~1623 K for 3.6 ks in H_2 and N_2 gas mixtures to control the carbon content to 0.4 mass %.

The heat treatment procedure involved reheating the sintered compacts to 1173 K for 1.8 ks in argon, followed by a oil quench (set at 323 K) and temper for 3.6 ks at the temperatures of 473 K and 823 K in argon. The density of the compacts was measured using water immersion. Tensile tests were performed on unmachined specimens in an Instron Testing machine at a cross-head speed of 3×10^{-6} ms^{-1}. Average tensile strength and elongation were based on three specimens in each condition, and also average hardness was determined from five measurement on each specimen. Also fatigue strength were determined by the rotating bending fatigue test. The microstructure of the specimens was examined by photomicroscopy. The distribution of alloying elements in some of the microscopic specimens was determined by scanning electron microscope (SEM) and electron probe micro analysis (EPMA). In addition, an image scanner (EPSON, GT-9000ART) was used for quantitative analyzing the route of fatigue crack and the fraction of resultant phases and transformation products.

RESULTS AND DISCUSSION

1. Die Pressing

The carbon strongly influences on the mechanical properties as well as the microstructures for the steels. Therefore, the amount of carbon must be carefully controlled to compare the properties of prealloyed and mixed elemental powder compacts. In this experiment, nearly constant carbon level was attained on the wide range of density, and the carbon content of all compacts were controlled to be within 0.4 ± 0.01 mass %.

Figure 1 shows the effects of sintering temperature on the tensile properties of both powder compacts. At the sintering temperature of 1523 and 1573 K, the strength of prealloyed powder compacts are higher than that of mixed elemental powder compacts on the whole range of density. However, when the sintering temperature increases to 1623 K, the strength of mixed powder compacts are higher than that of prealloyed powder compacts at high density range. Especially from Figure 1 (c), the critical density seems to be around 7.3 mgm^{-3} at which all pores are isolated and close to the spheroidal in the shape.

In order to clarify the different properties of prealloyed and mixed elemental powder compacts, the microstructure and the distribution of alloying elements were investigated by microscopic observation and EPMA. Prealloyed powder compact is consisted of dual phases : ferrite and bainite. On the other hand, mixed elemental powder compact shows the heterogeneous microstructure consisted of ferrite, bainite, pearlite and martensite. Appearance of pearlite and martensite is a distinct difference between the microstructures of both powder compacts. Especially, those martensite have a high nickel concentration as shown in Figure 2. Although nickel content of the martensite decreased with an increase in sintering temperature, it was approximately from 5 to 15 mass % according to the result of spot analysis by EPMA.

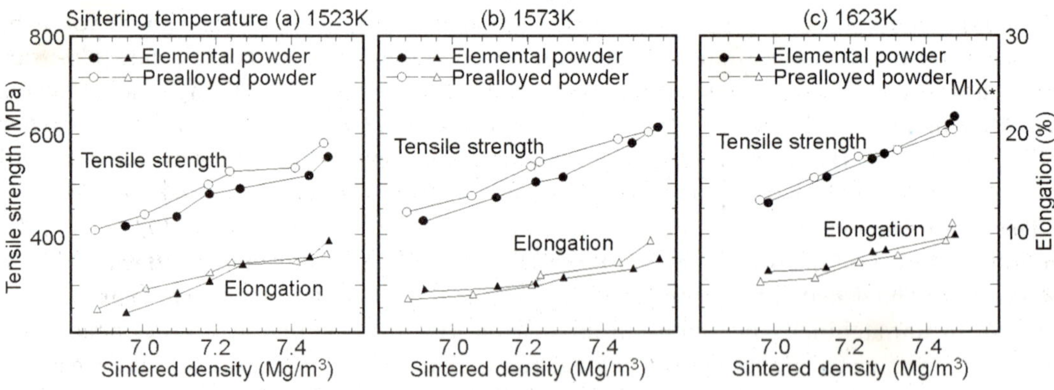

Fig. 1. Tensile properties of the prealloyed and mixed elemental powder compacts sintered at 1523K (a), 1573K (b) and 1623K (c) for 3.6ks.

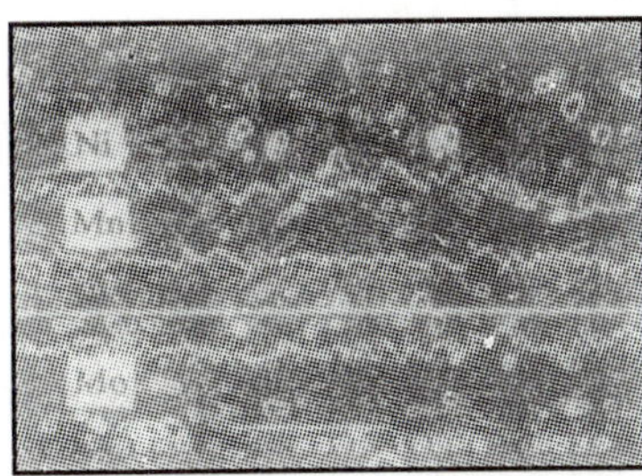

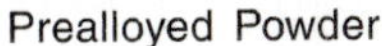

Prealloyed Powder

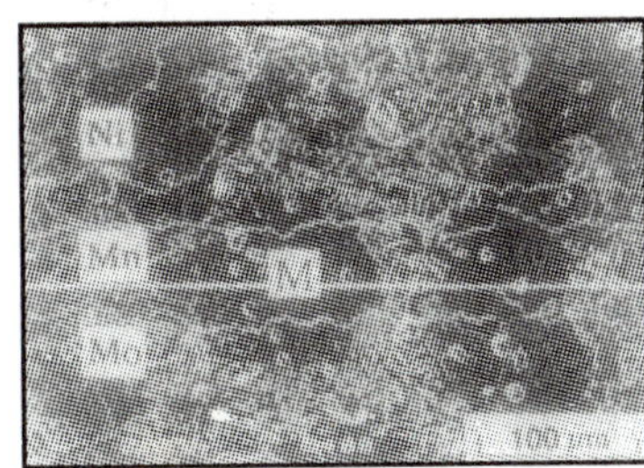

Elemental Powder

Fig. 2. Distributions of alloying elements for the prealloyed and mixed elemental powder compacts sintered at 1623 K for 3.6 ks.

Figure 3 shows the fraction of resultant phases for various sintered compacts. The changes in the fraction of ferrite and bainite for sintering temperatures are not seen in the prealloyed powder compacts. Therefore, the values of tensile properties of prealloyed powder compacts are seemed to be almost unchanged even by varying the sintering temperature as shown in Figure 1 (a), (b) and (c). On the other hand, the fraction of the phases in mixed powder compacts vary considerably with the sintering temperature. The amount of ferrite and pearlite remarkably decrease with an increase in sintering temperature. The amount of martensite also decreases, but the change is not so big. Corresponding to the decrease of those phases, the amount of bainite increases. Appearance of those heterogeneous microstructure is attributed to the diffusivity of alloying elements, especially nickel. In this regard, a carbonyl nickel powder is easy to agglomerate during mixing. In any event, at the sintering temperature of 1523 and 1573 K, substantial amounts of ferrite and pearlite are remained, which diminish the strength. At 1623 K, the presence of martensite makes the mixed elemental powder compact strong at especially high density range as mentioned earlier. The efficacy of this retained martensite is further enhanced by heat treatment.

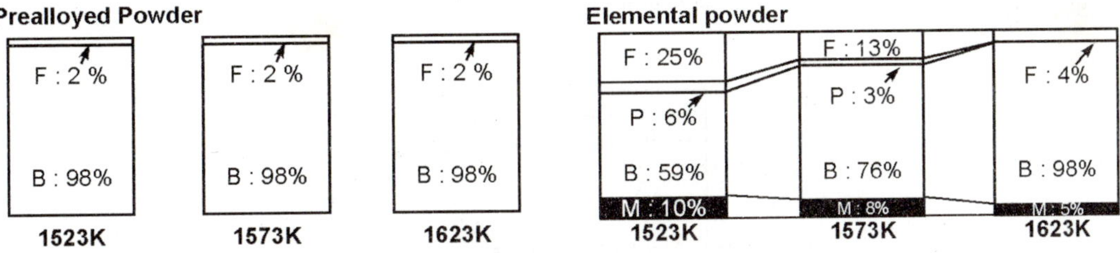

Fig. 3. Fractions of the resultant phases for prealloyed and mixed elemental powder compacts sintered at various temperatures.

Figure 4 shows the effect of tempering treatment (at 473 K) on the tensile properties of prealloyed and mixed powder compacts. The strength of mixed powder compacts are clearly higher than that of prealloyed powder compacts by tempering at low temperature, because the martensites are still remained in the matrix as shown in Figure 5 (a). However, as increasing the tempering temperature, homogenization of the microstructure occur and it leads to the same strength for both powder compacts as shown in Figure 6. Figure 5 (b) shows an example of the homogenization of microstructure for the mixed elemental powder compacts by increasing the tempering temperature.

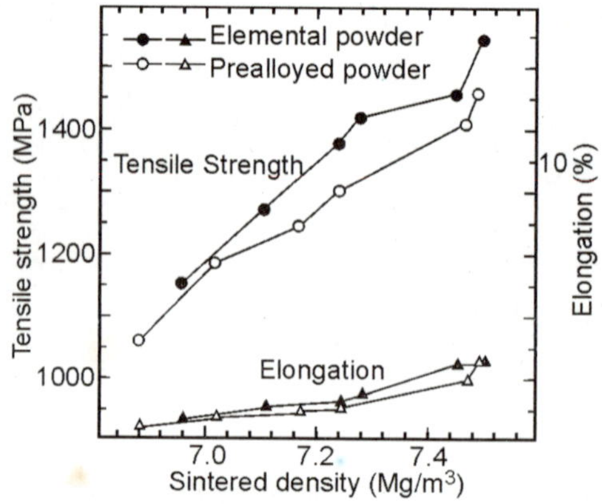

Fig. 4. Effect of tempering treatment (at 473K) on the tensile properties of prealloyed and mixed elemental powder compacts sintered at 1623 K for 3.6 ks.

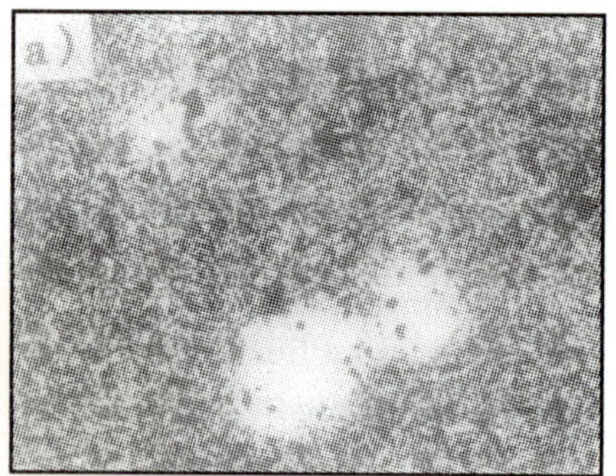

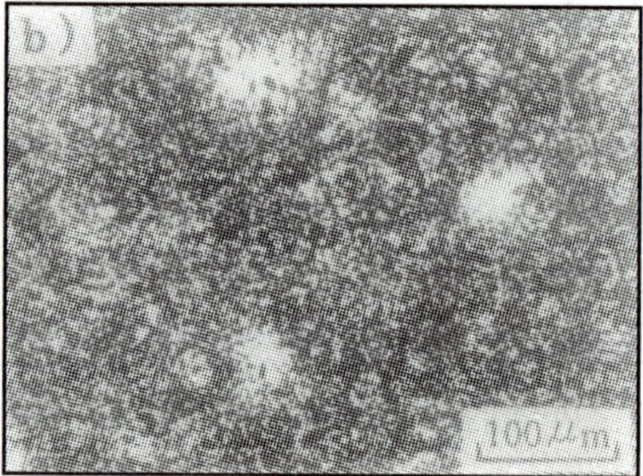

Fig. 5. Effect of tempering treatment (at 473 K (a) and 873 K (b) on the microstructures of mixed elemental powder compacts.

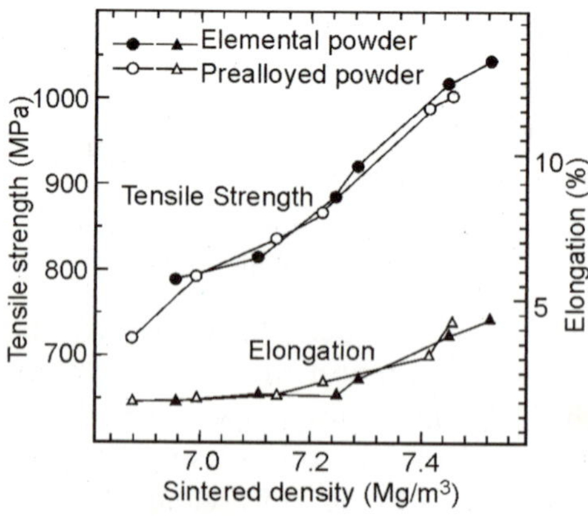

Fig. 6. Effect of tempering treatment (at 873 K) on the tensile properties.

Finally, the heterogeneous microstructure consisted of martensite surrounded by a network of tempered martensite will be effective for strengthening the P/M steels. In such a case, it is more effective that the density of compacts is high and the heterogeneous microstructure is fine as marked the strength of compact produced by M/M process in Figure 1 (c). Moreover, it was confirmed that the fine heterogeneous microstructure was quite useful to increasing the fatigue strength.

Figure 7 shows the fatigue strength of both powder compacts tempered at 473 K. Obviously, mixed elemental powder compact show higher fatigue strength as compared to prealloyed powder compact, even at the same density. Figure 8 shows the route of fatigue crack propagation for prealloyed powder compact. The crack propagates almost straight. On the other hand, for the mixed elemental powder compact, the crack propagates zigzag avoiding the white regions where are hard phase of matensite as shown in Figure 9. In Figure 10, both routes were determined quantitatively by image analysis. Clearly the route of mixed elemental powder compact is longer than that of the prealloyed powder compact, which means large crack propagation energies. Therefore, the mixed elemental powder compact seems to show high fatigue strength.

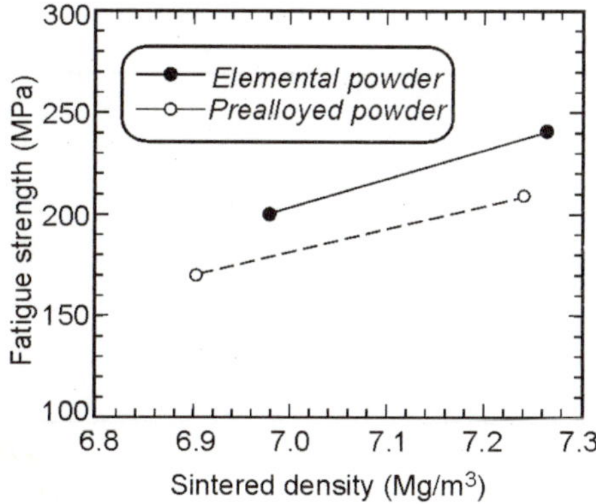

Fig. 7. Fatigue strength of the sintered compacts tempered at 473 K for 3.6 ks.

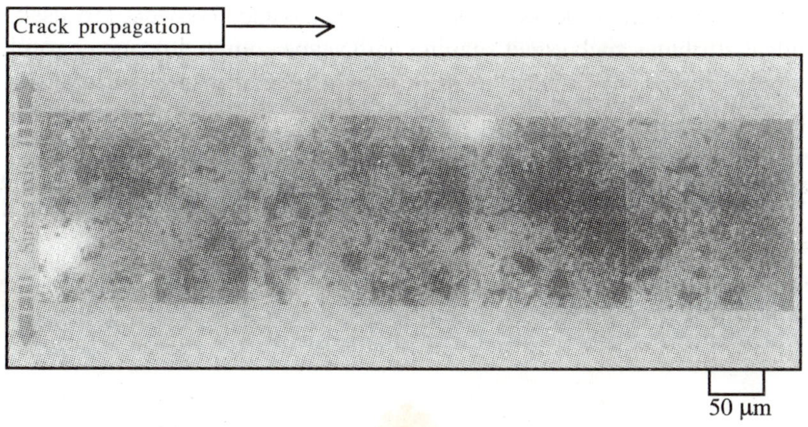

Fig. 8. Fatigue crack propagation of prealloyed powder compact tempered at 473 K.

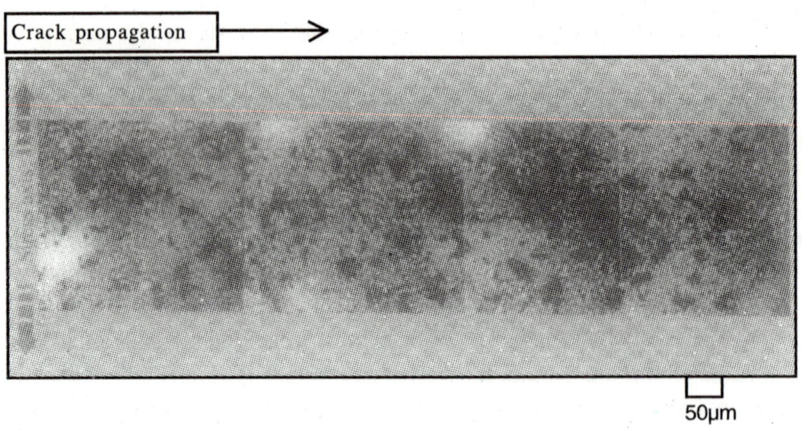

Fig. 9. Fatigue crack propagation of mixed elemental powder compact tempered at 473K.

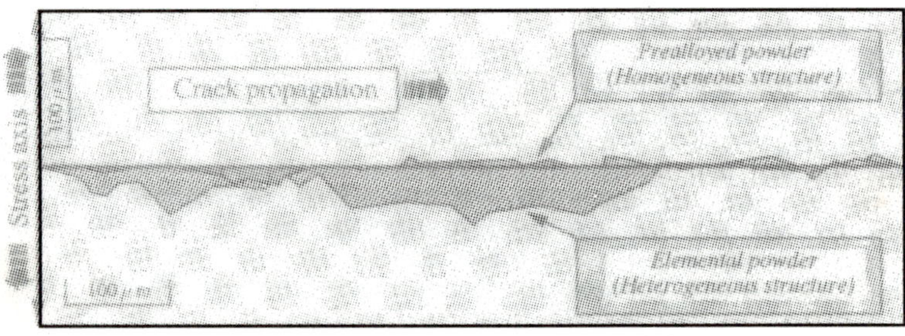

Fig. 10. Different crack propagation route behavior between prealloyed and elemental powder compacts.

2. Metal Injection Molding

Metal injection molding process is hoped to be an advanced powder processing technique because of many excellent attributes such as net shaping with shape complexity and near full density. Therefore, we tried to apply this novel process to 4600 steels to make sure the efficacy of heterogeneous microstructure. Table 2 shows the mechanical properties of injection molded 4600 steels. The tensile strength of mixed elemental powder compact shows approximately 2GPa which is amazing value for sintered low alloy steels. This steel has still 4% elongation. Also the fatigue strength of mixed powder compact is obviously higher than that of prealloyed powder compact as shown in Figure 11, it close to the value of powder forged steels. Of course, this value is extremely higher than that of conventional PIM steels [4]. The higher density and the finer heterogeneous microstructure, the more effective for strengthening the steels.

Table 2 Mechanical properties of MIM 4600 steels tempered at 473

	Prealloyed powder	*Mixed elemental powder*	
Relative density (%)	93~95	93~95*	96~97
Tensile strength (MPa)	1740	1840	1900
Elongation (%)	6	6	4
Fatigue strength (MPa)	472	(500~550)	575
Endurance	0.27	(0.27~0.3	0.3

* Sentering condition 1423 k × 1h

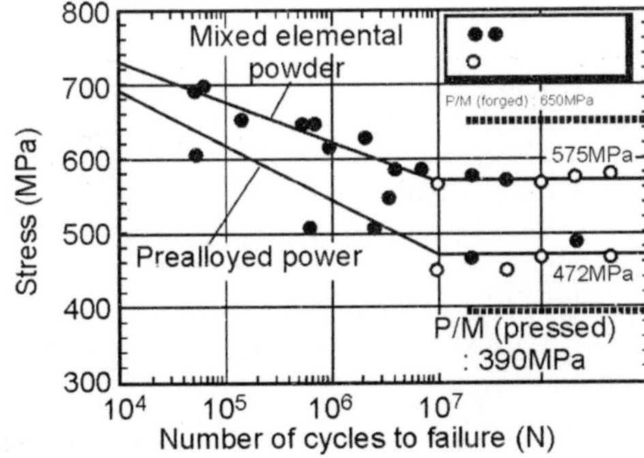

Fig. 11. S-N curves of MIM 4600 steels tempered at 473 K.

3. Optimization of the Heterogeneous Microstructure for Super-High Strengthening

Figure 12 shows the effect of Ni content on the tensile properties of injection molded Fe- Ni- 0.5 Mo-0.2 Mn steels tempered at 473 K. The tensile strength is significantly improved with increasing Ni content, also without loss of the ductility. Especially, it is worthy of remark that the steel added 6mass% Ni indicates a ultrahigh tensile strength of approximately 2000 MPa, and still has a ductility of 5% elongation. However, the tensile strength of the steel added 8 mass% Ni decreases remarkably. Figure 13 shows the microstructures of the injection molded Fe-Ni-0.5 Mo-0.2 Mn steels tempered at 473K. All the matrix are the tempered martensitic structure. There are white regions in the matrix, which consists of a high Ni concentration confirmed by EPMA. These white regions are the martensitic structures which hardness is higher than that of the matrix. Also, white regions having the large pore in the center, are consisted of a high Mo concentration, which were confirmed by EPMA. The formation of large pore is reported to be due to the large difference of the diffusivity between Fe and Mo [5]. Anyway, Ni rich phases increase with increasing Ni content. Accordingly, excellent mechanical properties of the steel added 6 mass% Ni are attributed to the solid solution strengthening of Ni and the fine heterogeneous microstructure which consists of Ni rich martensite surrounded by a network of the tempered martensite. By the way, since some center of the Ni rich phases show the low hardness, the phase was investigated by using X-ray diffraction. As a result it was found to be the retained austenite, because high percentages of Ni depress the transformation (M_s, M_f) range to considerably below room temperature. The amount of retained austenite for various Ni content steels are shown in Figure 14, and which were derived from the equation (1).

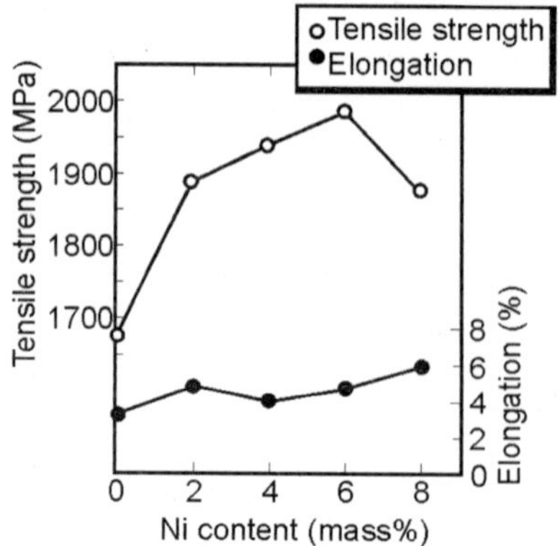

Fig. 12. Effect of Ni content on the tensile properties of MIM Fe-Ni-0.5 Mo-0.2 Mn steels tempered at 473 K.

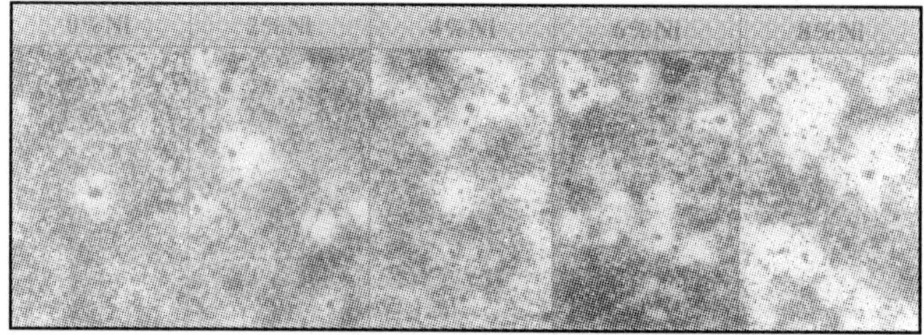

Fig. 13. Microstructures of MIM Fe-Ni-0.5 Mo-0.2 Mn steels tempered at 473 K.

$$V_{\gamma}(\%) = \{1/(1+1/K\,I_{\alpha}/I_{\gamma})\}.100 \qquad ...(1)$$

where V_{γ} is the amounts of retained austenite, I_{α} is α (211) integral intensity, I_{γ} is the average of γ (220) and γ (311) integral intensity, and K is a material constant (1.4). The amount of retained austenite increases with increasing Ni content of 6 or 8 mass%. Therefore, a drop of the tensile and fatigue strength of the steels added 8 mass% Ni may be due to the increase of the retained austenite (6mass% Ni may be a critical value for the strengthening). Table 3 summarizes the mechanical properties of Fe-Ni-0.5 Mo-0.2 Mn steels produced by various processes. All these steels were tempered at 473 K for 3.6 ks in Ar. The properties of injection molded steels with the fine heterogeneous microstructure are significantly higher than those of the conventional P/M steels and similar to the properties of powder forged steels. The microstructure of high temperature and long time sintered steels (at 1623 K for 18 ks) become homogeneous. The tensile and fatigue strength of those steels with homogeneous microstructure are lower than those of the steels with the fine heterogeneous microstructure. Consequently, it is confirmed that the fine heterogeneous microstructure is quite useful for ultrahigh strengthening the sintered low alloy steels.

Table 3 Mechanical properties of the Fe-Ni-0.5Mo-0.2Mn steels produced by various processes.

	P/M (Forged) (2 mass % Ni)	*MIM (6 mass % Ni) heterogeneity*	*MIM (6 mass % Ni) homogeneity*	*P/M(Pressed) (6 mass% Ni)*
Density(%)	99.9	95*	96**	92.4
		1983	1789	1443
Tensile strength (Mpa)	2115	max. 2047	max. 1808	max. 1514
Elongation (%)	11	5	6	2
Fatigue strength (Mpa)	652	488	–	463

Sintering condition : * 1523 K × 3.6 ks

** 1623 K × 18 ks

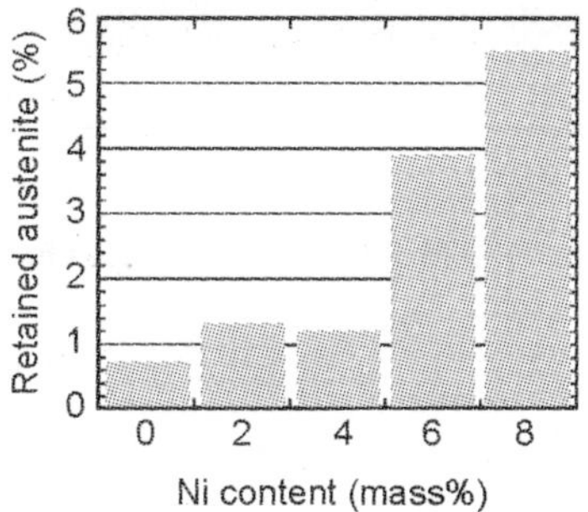

Fig. 14. Fraction of the retained austenite of MIM Fe-Ni-0.5 Mi-0.2 Mn steels tempered at 473 K.

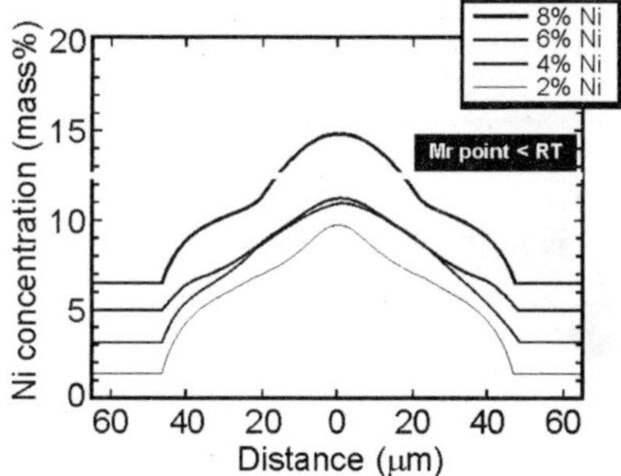

Fig. 15. Profile of Ni concentration in the white region of MIM Fe-Ni-0.5 Mo-0.2 Mn steels tempered at 473 K.

Figure 15 shows the profile of Ni concentration in the white region by EPMA. Ni diffuses in the matrix surrounding of white region, Ni concentration of the matrix is approximately 2 mass% lower than added Ni content. Ni concentration of the white region increases with increasing added Ni content. The size of Ni rich phase (correspond to white region) is about 80 μm in all the steels. When Ni segregates over 12.4 mass%, M_f point become below room temperature [4], which means to induce the increase of retained austenite. A solute concentration C is shown in equation (2) by the Fick's second law on a thin film, when a solute diffuse at nonsteady state.

$$C(x, t) = \{bC_0/2(\pi Dt)^{1/2}\}\exp(-x^2/4Dt) \quad ...(2)$$

where C_0 is the solute concentration on the thin film, D is the diffusion coefficient, b is the thin film thickness, and x is the distance from the center of thin film. Secondary electron and characteristic X-ray images of debound compacts are shown in Figure 16. Ni powders agglomerate and the size is approximately 15 μm. From this, these agglomeration is regarded as the thin film thickness of 15 μm.

When the distance from the center of Ni rich phase is 20 μm the Ni concentration of 8.6% are obtained from the steel added 4mass% Ni as shown in Figure 15 simultaneously, and the diffusion coefficient of Ni is calculated by entering those values to equation (2). When the Ni concentration is 12.4 mass% (M_f point is below room temperature), the value of about 18 μm is obtained as the initial thin film thickness b. Figure 17 illustrates schematically the structure for Ni rich phase. Upper side figure shows that the Ni rich phase consists of only the martensitic structure if the size of agglomerated Ni powders is below 18 μm, whereas in the bottom side figure the retained austenite is formed in the center of Ni rich phase if the size of agglomerated Ni powders is above 18 μm. From the calculation, if the Ni powders of approximately 15 μm particle size is dispersed uniformly at regular intervals of 115 μm in the matrix, superior mechanical properties may be obtained by even same composition of Fe-2Ni-0.5Mo-0.2Mn steels.

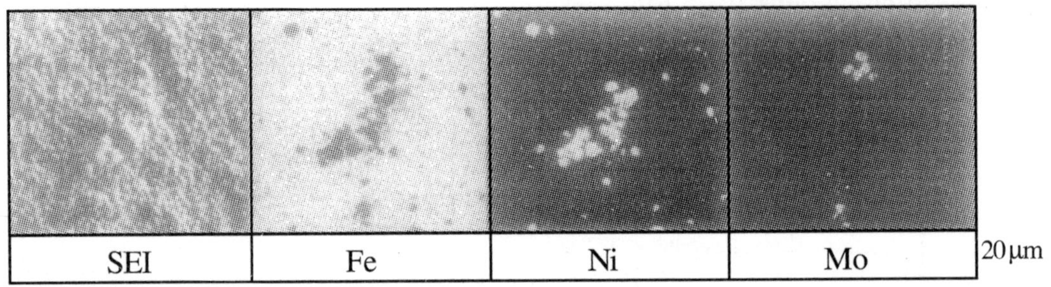

Fig. 16. Secondary electron and characteristic X-ray images of MIM Fe-4 Ni-0.5 Mo-0.2 Mn steels debound at 923 K.

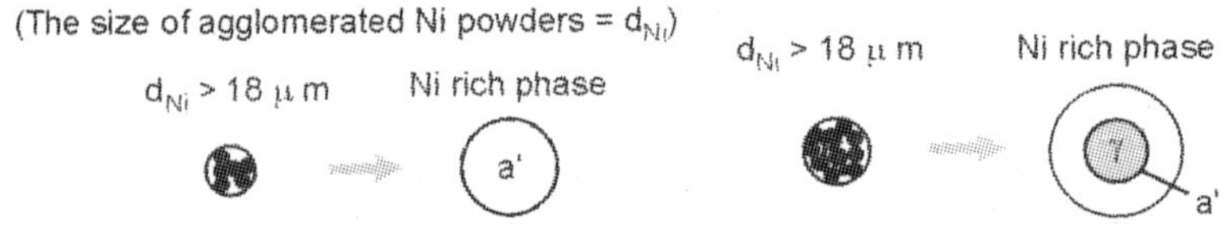

Fig. 17. Schematic illustrations of the structure for Ni rich phase.

CONCLUSION

The mechanical properties of sintered low alloy steel compacts are very dependent on the microstructures as well as the density and chemical compositions. Generally, the homogeneous microstructure has been favorable for strengthening the steels in the conventional P/M processings. The homogeneous structure is normally obtained by using prealloyed powder. On the other hand, mixed elemental powder compacts show the various heterogeneous microstructures because of the different diffusivities between alloying elements, resulting in the poor strength as compared to that of prealloyed powder compacts. However, if the good combination of resultant phases such as bainite and martensite, or transformation products by heat treatment such as retained martensite surrounded by a network of tempered martensite are obtained, the strength of mixed powder compacts will be significantly improved. In such a case, the higher density and the finer heterogeneous structure, the more effective for strengthening the steels. This suggests the coming of new high performance ferrous materials for heavy duty machine parts through the powder metallurgy, especially MIM technique.

For the injection molded Fe-Ni-0.5 Mo-0.2 Mn steels using mixed elemental powders, the tensile and fatigue strength were improved significantly with increasing Ni content, without loss of the ductility. Especially it was worthy of remark that the steels added 6 mass% Ni showed a tensile strength of approximately 2000 MPa and still had a ductility of 5% elongation. This was due to the solid solution strengthening of Ni and the fine heterogeneous microstructure which consisted of Ni (and Mo) rich martensitic structure surrounded by a network of tempered martensitic structure. However, the tensile and fatigue strength of the steels added 8mass% Ni decreased remarkably on account of increasing the amount of retained austenite.

According to the model using the Fick's second law on the thin film, superior mechanical properties might be expected by even same composition of Fe-2 Ni-0.5 Mo-0.2 Mn steels, when the Ni powders of approximately 15 μm particle size is dispersed uniformly at regular intervals of 115 μm in the matrix.

REFERENCES

1. R.M. German, Powder Injection Molding, MPIF, Princeton, NJ, (1990).
2. H. Miura, T. Honda, K.F. Hens and R.M. German, Powder Injection Molding Symposium-1992, Edited by P.H. Booker, J. Gaspervich and R.M. German, MPIF, Princeton, NJ, pp. 203 (1992).
3. H. Miura, T. Honda and R.M. German, J. of the Japan Society of Powder and Powder Tech. (JSPM), Vol. 39, No. 4, pp. 254 (1992).
4. T. Baba, Y. Yamanishi, T. Honda and H. Miura, JSPM, Vol. 43, No. 7, pp. 863 (1996).
5. S. Tanabe and A. Morita, J. Japan Soc. Powder and Powder Metal, Vol. 35, p. 233 (1988).
6. D.K. Bullens, Steel and its Heat Treatment, John Wiley & Sons Inc, New York, vol. 3 (1949).

3

Extruded FE-CR-AL Alloy Honeycombs for Electrically Heated Catalyst to Control Cold Start Emissions from Gasoline Engines

Raja Wusirika, Kevin Brundage and Dave Weiss,
SP-DV-0 1 -9, Corning Incorporated,
Corning, NY 14831, USA

ABSTRACT

The harmful emission from a gasoline vehicle is minimized using a three-way catalyst (TWC) that oxidizes hydrocarbons and carbon monoxide and reduces nitrous oxides. However, it takes about two minutes for the TWC to heat up to 350°C from a cold start before it is effective. During this warm up period, 70% of pollutants are released. An Electrically Heated Catalyst (EHC), typically made of Fe-18Cr-7Al metal alloy, is effective in reducing the emissions during the cold start. Iron-aluminium alloy and chromium metal powders were mixed and extruded into cellular honeycombs using methyl cellulose as a binder. The extruded parts were sliced and fired to 1425°C in hydrogen. The EHC could be heated to 650°C in 20 seconds, and it reduced the harmful emissions such as CO, NO_x and un-burnt hydrocarbons by 75% during cold start. Properties such as microstructure and oxidation resistance of metal honeycomb substrates, used in EHC, are presented.

INTRODUCTION

Typical pollutants in the exhaust of a modern gasoline vehicle are un-burnt hydrocarbons (HC),CO and nitrogen oxides (NO_x,). A three-way catalyst (TWC) installed in the exhaust stream simultaneously reduces NO_x and oxidizes HC and CO. A TWC, however, is effective at temperatures above 350°C. The heat from the exhaust stream is the source for the catalyst to reach 350°C. Consequently, it takes up to 120 seconds for the catalyst to reach the operating temperature from a cold start. During the warm up period, 70% of emissions, known as cold start emissions, are released untreated. An electrically heated catalyst (EHC) is very effective in the remediation of the cold-start emissions [1]. An EHC is made of a high- temperature alloy in the Fe-Cr-Al ternary system. Fe-Cr-Al alloys are known for their oxidation resistance and high electrical resistivity making them suitable for EHC. An EHC assembly is shown in Fig. 1. An EHC is a Fe-Cr-Al alloy metal honeycomb coated with a catalyst. It is placed in front of the regular TWC catalyst as shown in the cut away. The metallic honeycomb has electrical leads on the side. A 20 volt battery is connected to the leads. Slots are cut in the honeycomb to increase the electrical path length and thus increase the electrical resistance. As the gasoline engine is turned on, the battery turns the EHC on and heats to 700°C in 20 seconds and begins to convert the pollutants HC, CO and NO_x to harmless H_2O, CO_2 and N_2 before the main catalyst takes over. The EHC is turned off once the TWC reaches 300°C. Figure 2 shows laboratory testing of EHC powered with a 20 volt battery.

Fig. 1. EHC Assembly showing EHC in front of regular TWC catalyst.

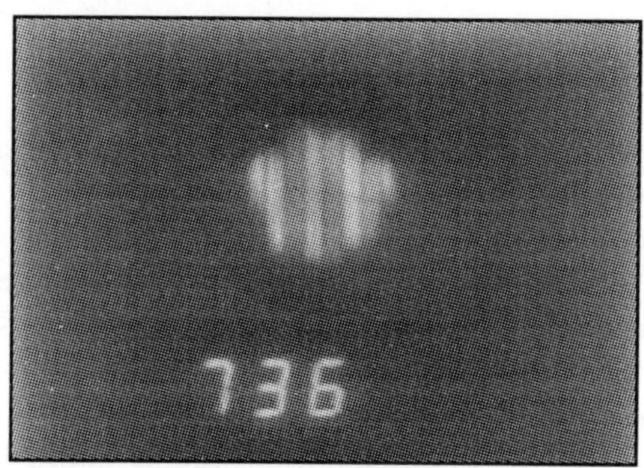

Fig. 2. EHC can be rapidly heated to 736°C in 20 seconds using a 20 volt battery.

As the figure shows, EHC can be heated rapidly, in seconds, to 700°C. Two different strategies can be used during cold start. In the first strategy, known as post-heat, the EHC is powered as soon the ignition to the gasoline engine is turned on. In the second strategy, known as pre-heat, the EHC turns on 20 seconds prior to the start of the engine. The temperature rise as function of time is shown in Figure 3 [2] in both the strategies. The reductions in emissions are shown in both strategies compared to the emissions without any EHC, in Table 1. The following sections describe the composition of the metal alloy and processing of metal powders to make the honeycomb.

Table 1. HC and CO Emissions with and without EHC using Pre-heat and Post-heat Strategies

Almost 75% of the cold-start emissions can be reduced using an EHC.

	HC Emissions in grams	*CO Emissions in grams*
No EHC	2.7	27.2
EHC Post-heat	0.67	7.95
EHC Pre-heat	0.37	4.7

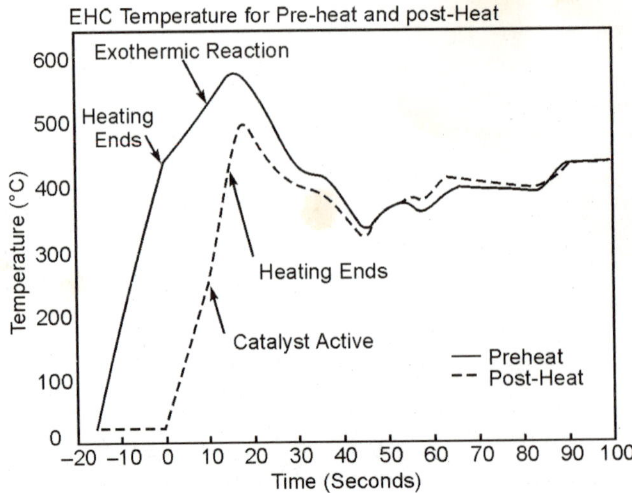

Fig. 3. EHC temperature versus time using pre-heat and post-heat strategies.

COMPOSITION

One of the preferred compositions is Fe-18 Cr-7 Al-0.05 B-0.5 Y_2O_3 [3]. Chromium and aluminium provide oxidation resistance to the alloy. A small amount of boron is added as a sintering aid. Yttrium oxide is added to improve the adherence of aluminium oxide scale to the alloy to prevent spallation of the aluminium oxide layer and thus provide cyclic oxidation resistance. Fine Carbonyl iron powder is used as a source for iron. Chromium can be added as an elemental powder or chromium-aluminium alloy powder. Aluminium being pyrophoric is added as iron-aluminium alloy powder. Boron is added as an iron-boron alloy powder. A typical powder batch composition is shown in Table 2. The metal powders were mixed with 1% oleic acid as lubricant and 5% methyl cellulose as an organic binder. Metal powders along with organic additives were mulled with about 12% water into a paste and extruded into square cell honeycombs with 67 cells per square centimeter and a 0.012 cm wall thickness. The honeycombs were dried and fired in hydrogen to 1425°C. Polished samples fired to different temperature were examined using scanning electron microscopy (SEM). Energy dispersive spectroscopy (EDS) was used for compositional analysis. The sample fired to 1150°C, as shown in Fig. 4, has residual porosity.

Table 2. Metal powder batch for Fe- 18Cr-7A1-0.05B-0.5 Y_2O_3

Component	*Wt%*	*Mean Particle Size in μm*
Carbonyl Fe	74.25	6
CrA1	23.33	20
Cr	1.66	20
FeB	0.25	25
Y_2O_3	0.5	6

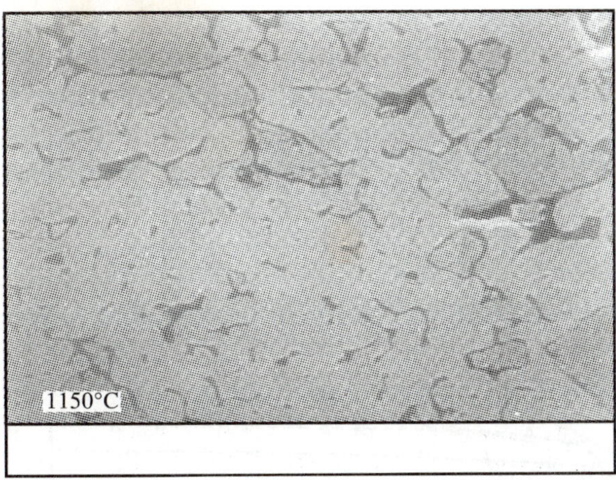

Fig. 4. SEM of polished sample fired to 1150°C. The scale is 10 μpm.

The sample fired to 1425°C (Fig. 5) has very little porosity. EDS analysis shows the matrix to be compositionally homogeneous. Some of the organic binders, such as methyl cellulose, can leave residual carbon during sintering. The residual carbon forms carbides of Cr and Al and, thus, reducing the oxidation resistance as shown in Fig. 6. The lower the residual carbon, the better the oxidation resistance as depicted by lower weight gain loss at 1000°C in air. During sintering of Fe-Cr-Al alloys, it is very critical to control the residual carbon level. The oxidation resistance at 800°C of Fe 18Cr7A1 honeycombs sintered at 1425°C is shown in Fig. 7. The weight gain plateaus after 100 hours in air and very little weight gain is observed thereafter.

The sintered monoliths were pre-oxidized in humid air at 1000°C to form an aluminium oxide coating. The oxide coating provides oxidation resistance. It also anchors the γ-alumina wash-coat. The noble metal catalysts such as Pt, Rh and Pd are supported on the wash-coat.

Fig. 5. SEM of polished sample fired to 1425°C. The scale is 10 μpm.

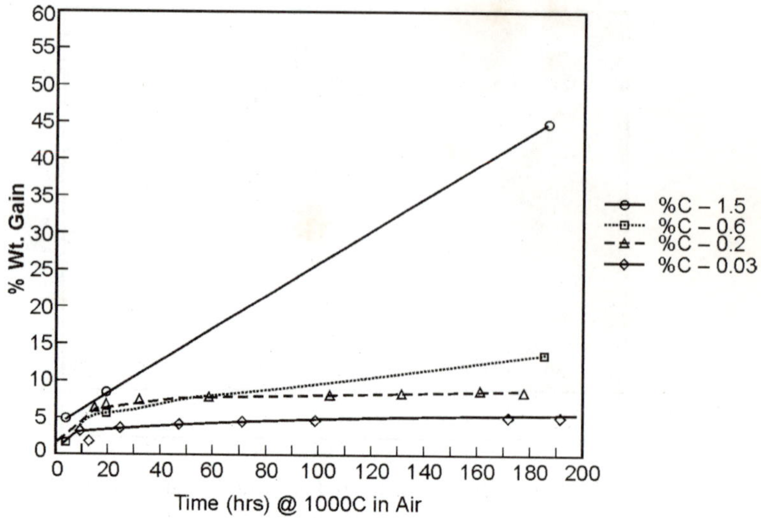

Fig. 6. Residual carbon from binders reduces the oxidation resistance.

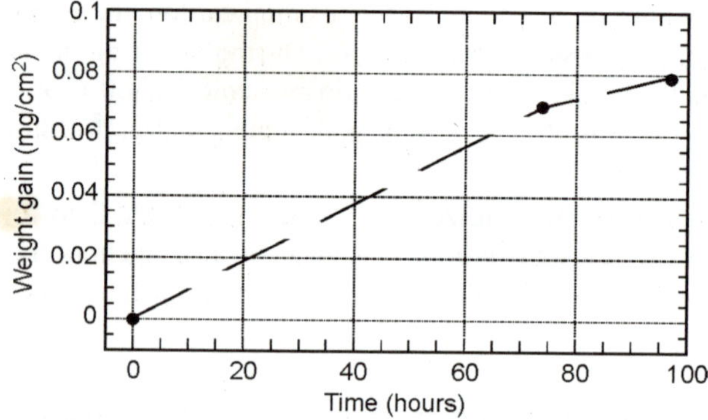

Fig. 7. Oxidation resistance at 800°C in air for Fel8Cr7A1 honeycombs sintered at 1425°C.

DURABILITY TESTING

The EHC was tested in severe electrical cycling test to predict the performance in on-vehicle test [4,5]. In this accelerated test, the EHC was resistively heated to 1000°C in approximately 16 seconds with no air flow. Once the heater reached 1000°C, the power was shut off and cold air was blown through the heater for 20 seconds during which the temperature of EHC dropped to 100°C. The test cycle was repeated until failure occurred. Significant change in electrical resistance, such as short circuiting, is considered a failure. The EHC survived 2400 cycles. EHC was evaluated on vehicle tests using 5.2 liter engines. Even after 20,000 kilometers, no deformation or change in electrical properties were observed.

CONCLUSIONS

(1) Metal powders could be extruded into honeycomb structures using traditional ceramic processing of extrusion with methyl cellulose binders.

(2) Fe-Cr-Al alloy, doped with a small amount of boron as sintering aid and a small amount of Y_2O_3 for enhancing the spallation resistance of the protective aluminium oxide coating, was found to be a durable candidate for Electrically Heated Catalyst to reduce cold-start emissions from a gasoline vehicle.

ACKNOWLEDGMENT

The authors thank Lynne Seeley, Doug Aranda and Greg Dillon for their efforts in fabricating the extruded heaters.

REFERENCES

1. S.R. Nakouza, C.H. Narula, J.R. MeBride, K.E. Nietering, J.H. Visser and A.A. Adamezyk, "Novel Concept Prototype of Low-Power-Consumption Electrically Heatable Catalyst", A1Che Journal, Vol. 44, No. 1, pp 184-187 (1998).
2. L.S. Socha, D.F. Thompson, P.A. Weber, "Reduced Energy and Power Consumption for Electrically Heated Extruded Metal Converters", SAE Paper 930383, February (1993).
3. T. Harada et al., "Sintered Metal Bodies and Manufacturing Method Thereof ", US Patent 5, 427, 601, June 27, (1995).
4. D.S. Weiss, C.L. Buhrmaster, L.E. Hampton, D.F. Thompson, "Durability of Advanced Electrically Heated Catalyst Design", SAE Paper 960345, February (1996).
5. D.S. Weiss, C.L. Buhrmaster, L.E. Hampton, W.R. Powell, L.S. Socha, "Durability of Extruded Electrically Heated Catalysts", SAE Paper 950404, February (1995).

4

Composite Materials for Automotive Brakes

A.N. Tiwari
Department of Metallurgical Engineering & Materials Science
Indian Institute of Technology Bombay
Powai, Mumbai 400 076, India

ABSTRACT

Automotive brakes must provide high and stable friction, noise-and vibration-free performance and low wear for reasons of safety, comfort and durability. A single material has never been successful to meet the numerous performance related requirements of a break material. Only a composite material with a complex composition (10 to 30 ingredients) can satisfy the diverse and conflicting requirements. Although a great deal of research has been going on to develop friction materials, the details on composition and processing are shrouded in secrecy due to proprietary reasons. This paper presents a brief review of the materials technology involved in automotive friction materials. The following aspects have been considered: (i) Friction & wear, (ii) Ingredients of friction materials (classification, functions and description), (iii) Different types of friction composites (organics, semimetallics, ceramic enhanced, sintered and carbon fiber reinforced materials), and (iv) Drum or disc materials. An attempt has been made to simplify the complexities and to develop an understanding of the materials aspects of friction materials while emphasizing current trends.

1. INTRODUCTION

A brake provides a means to decelerate a moving vehicle by converting kinetic energy into friction heat. A brake is a sliding friction couple consisting of a rotor (drum or disc) connected to the wheel and a stator on which is mounted the friction material. The heat generated at the sliding interface is dissipated primarily by conduction through various components of the brake and by convection to the surrounding. A part of the absorbed energy is consumed by chemical, metallurgical and wear processes at the interface. On application of brakes, a stopping force is generated which can be ten times as powerful as the force that puts the automobile in motion. The friction lining is the primary determinant of frictional behaviour and acts as a sacrificial or expendable wearing element of the brake system. If the brake fails, the result can be disastrous. Therefore, the friction material used for brake lining is an essential part of a braking system.

Typically the average operating temperatures at the braking surface are in the range of 40° to 300°C for passenger cars and 300 to 750°C for racing cars. At high temperatures the friction is significantly reduced giving rise to a phenomenon called fade. The ability of the friction material to return to its prefade friction level is known as recovery. A brake must provide safe and reliable stopping power on every application. The main performance-related requirements for a friction lining material are given in Table 1.

Table 1. Requirements of Brake Materials

1.	High Coefficient of friction (COF)
2.	Fade resistance
3.	Good thermal conductivity
4.	Anti-seizure properties
5.	Minimum damage to the counter surface (drum or disc)
6.	Long endurance under heavy loads and high speeds
7.	Freedom from noise. Smoothness of engagement (no snatch or judder)
8.	Acceptable wear rate
9.	Resistance to contaminants

From the beginning of the 20th century until the late 1960s, automobiles used drum brakes on all four which with organic brake linings. These organic linings known as "organics" contained 30-40 wt% organic constituents along with asbestos as a reinforcement. With increasing awareness about environmental pollution, carcinogenic nature of the asbestos was discovered. Beginning in the mid 1970s, due to stringent health regulations a great deal of R & D efforts have been given to produce a new generation of friction materials which are free from asbestos and can sustain high temperatures. Also disc brakes gradually replaced drum brakes at least for the front wheels. Aluminium matrix composites and carbon composites have substituted cast iron or steel discs in demanding applications.

From Table 1 it can be seen that the brake materials must meet conflicting requirements. These are met by incorporating a large number of ingredients in a commercial friction material. As a result the friction material is one of the most complex composite materials and usually contains more than 10 ingredients. It is known that approximately 150 different ingredients are currently being used by the manufactures of friction materials. But the details about ingredients, their concentration and processing of a commercial friction materials are difficult to find in the literature due to proprietary reasons. This paper is an attempt to review automotive brake materials to develop a basic understanding of materials aspects of these complicated multicomponent composites.

2. FRICTION AND WEAR

Unlike mechanical properties, tribological behaviour (i.e. friction & wear) depends on the sliding couple (i.e. brake pad and disc) and the nature of interaction between the friction surfaces. Some important tribological parameters are (*i*) interfacial conditions (roughness and friction film formation), (*ii*) hardness and microstructure, (*iii*) sliding speed, (*iv*) contact pressure, (*v*) temperature etc. Therefore, friction and wear are the properties of a tribo-system. In a frictional braking system, COF is of the order of 0.3 – 0.5. The mechanisms which give rise to friction are

(*i*) Adhesion-tearing

(*ii*) Abrasion (microploughing)

(*iii*) Elastic and plastic deformation and

(*iv*) Asperity interlocking

Mechanisms of wear in friction materials are as follows:

(*i*) Adhesion of the two materials, followed by cohesive failure of one as the two slide past each other.

(*ii*) Ploughing or gouging of one material by a fragment of harder material.

(*iii*) Thermal or mechanical fatigue which causes debris to become detached from the surface.

(*iv*) Oxidation and pyrolysis, leading to gas formation (ablation).

In a typical automotive brake pad, the average wear per brake application is 0.5 μm of thickness loss. It corresponds to a mass loss of about 3 mg. For a fixed amount of braking, the amount of wear (Δ W) of an organic friction material tend to increase slightly or remains practically constant as a function of temperature. But under severe braking conditions, if break temperature riser above 232°C, the wear increases exponentially with increasing temperature. It results in thermal fading i.e. reduction in the coefficient of friction (COF). In the normal operating regime wear rate is primarily controlled by adhesion and abrasion and wear can be described by the following wear equation proposed by S.K. Rhee [1,2] of Bendix Corporation, USA.

$$\Delta W = \alpha P^a.V^b.t^c$$

where,

ΔW = the amount of wear

α = the wear factor

P = the normal pressure

V = the sliding speed

t = the sliding time

Values of exponents a, b, and c are material pair-related. For example in the case of an organic lining against cast iron:

$$\alpha = 9.09 \times 10^{-9}$$

$$a = b = 1.35, \ c = 1$$

Temp. range 90–280°C

In the high temperature regime over heating due to friction can cause the liquescence of organic ingredients, oxidation and pyrolysis leading to ablation. Here the wear rate (WR) can be correlated with Arrhenius plot and the wear loss can be represented as

$$\log \text{WR} = A - B/T$$

Liu and Rhee [3] obtained an activation energy in the range of 8.0 – 8.8 cal/mol for semi-metallic disc brake pads operating in the temperature range of 260-330°C. They concluded that the pyrolysis was the controlling mechanism because the oxidation would require an activate energy in the range 30-50 kcalmol^{-1}.

3. INGREDIENTS OF FRICTION MATERIALS

A large variety of constituents are necessary to fulfill the diverse functional characteristics required. They can be broadly classified into five categories

1. Binder (i.e. matrix)
2. Abrasives
3. Friction Modifiers
4. Fibre Reinforcements
5. Fillers and Miscellaneous

A brief review on the role of various ingredients and their characteristics is presented here. For more details one can refer to an excellent report by P.J. Blau [4].

3.1 Binder

It is the continuous phase (i.e. matrix) in a friction material, which binds the different ingredients together. It determines the fabrication characteristics and the upper operating temperature limit for the friction lining. The binder is not a major component by volume or cost, but it is the most important with respect to the mechanical integrity of the composites, if there is too small a proportion of binder, the friction material may be physically weak, if too great a proportion the COF may fall at high temperatures. Depending upon the matrix, the friction materials can be broadly classified into three types:

(*i*) Organic (usually phenolic resin based)
(*ii*) Metallic or sintered (cermets based on copper or iron)
(*iii*) Carbon-carbon composites

3.2 Abrasives

Abrasives are the most important ingredient after the matrix as they contribute to the following :

(*a*) Increase friction: The contribution to COF is approximately proportional to the Mohs hardness of the abrasive [5].
(*b*) Help in preventing local welding.
(*c*) Control material transfer from the brake lining to the mating surface during braking. Abrasives remove the pyrolysed friction film at the interface. They also remove minor rusting from the brake drum or disk. Thus they help to maintain the cleanliness of mating surfaces and control the build-up of friction film.
(*d*) They tend to stabilize friction level during braking. If the COF of the matrix should fall due to high interface temperature, the abrasive takes a greater proportion of the load and so off-set the fall in COF.

There is a limit to the amount of abrasives that can be used because hard particles tend to wear the countersurface (i.e. drum or disk). An inappropriate selection of abrasives or excess amount often causes brake-induced problems, such as noise, rotor disk thickness variation and judder. Therefore, the choice of the abrasives is very important and often two or three different abrasives are used in a friction material to compromise the weakness of particular abrasives. A brief description of the abrasive ingredients is presented in Table 2.

Table 2. Abrasive Ingredients

Ingredient	*Description/Comments*
Oxides Alumina	(*i*) Hydrated form [AlOOH, $Al(OH)_3$] is added as polishing agent. It provides wear resistance, but can produce fade. (*ii*) Anhydrous form is more abrasive (*iii*) In fused condition it is very hard and very abrasive.
Silicon dioxide	SiO_2: quartz or silica
Silicate Minerals Zircon and Alumino Silicates	$ZrSiO_4$ Mullite ($3Al_2O_3.2SiO_2$), Kyanite or Sillimanite ($Al_2O_3.SiO_2$) (*i*) The minerals are friable and help control frictional behaviour (*ii*) Also control wear on the countersurface due to their own fracturing. (*iii*) Too much mineral filler tends to wear the countersurface.
Ceramic microspheres	(*i*) Basically aluminosilicates containing Fe_2O_3 or TiO_2. (*ii*) Size range 10-350 μm (*iii*) Loading: 5–10 vol% (*iv*) Low density filler (*v*) Reduces rotor wear and control friction (*vi*) It can absorb rotor dust
Metal Oxides	(*i*) Fe_3O_4 improves cold friction (mild abrasive) (*ii*) Cr_2O_3 raises friction (*iii*) ZnO lubricates but can cause drums polishing.

3.3 Friction Modifiers

These ingredients are used to modify and stabilize the COF, reduce wear rate and decrease break noise. Friction modifiers consist of solid lubricants (graphite, metal sulfides PbO etc.), metallic powders (Fe, Cu, brass, Pb, Sb, Bi and Mo) and friction dust.

Solid Lubricants

Solid lubricants are generally used to reduce the wear rate, rather than to reduce COF. They also avoid gross seizure between the friction lining and the mating part by forming a friction film. Obviously a high COF is generally advantageous as it means that the brake actuating force can be reduced for a given torque. The problem therefore, is generally in determining the optimum loading which will give a substantial reduction in wear rate with little reduction in COF. A blend of solid lubricants is normally used in a brake formulation to regulate friction and wear at all sliding speeds and brake temperatures. The friction film plays crucial roles in various aspects of brake performance such as: - Smoothness of brake engagement, rotor wear, noise propensity etc. Therefore, the relative amounts of solid lubricants and abrasives should be carefully chosen for a proper brake performance.

Metallic Friction Modifiers

The powders of Fe, Cu, brass, Zn, Pb, Sb, Bi and Mo may be incorporated in the formulations of heavy duty linings to increase COF and heat dissipation. They reduce the temperature gradient at the friction material surface and reduce high velocity and high load (and thus high temperature) wear. For example, the inclusional brass and zinc contributes significantly to the recovery of friction performance following fade. The inherently poor chip-forming characteristics of copper and brass powders/chips ensure that wear debris can be reincorporated into the mating surfaces, effectively recycling otherwise lost material. Because of high plasticity, especially that of copper, their deposition on the counter surface enlarges the real contact area between the rotor and stator facilitating more uniform temperature generation. It also helps in breaking up undesirable surface films by resisting the subsequent deposition of decomposition products from the organic material.

Depending on the overall formulation of the friction material, metallic constituents may give rise to certain problems with respect to break noise, and slip-stick phenomenon due to their inherent adhesive compatibility with the rotor.

Friction Dust

It is an ambiguous term. The friction dust may contain rubber particles and/or cashew particles or cashew friction dust. The rubber powder is used for noise damping by absorbing the friction excitation at the break interface. Cashew friction dust is a granular, free flowing polymerized resin derived from Cashew Nut Shell Liquid (CNSL). The main component is cardanol, which is a naturally occurring meta-substituted alkenyl phenol. Cardanol is hydrophobic in nature and remains flexible at very low temperatures. The advantage of cashew resin, as compared with plain phenolic resin, is that it produces softer friction material which is more efficient for wear when the brakes are relatively cold. Table 3 summarizes characteristics of friction modifiers.

Table 3. Friction Producers/Modifiers

Material	*Description/Comments*
Carbon (graphite)	(*i*) Cheap and widely used. (*ii*) Burns in air at T > 700°C (*iii*) Friction is affected by moisture and structure.
Metal sulfides [Sb_2S_3, PbS, MoS_2 , ZnS, CU_2S]	(*i*) They act as solid lubricant additives and enhance frictional stability. They reduce pad and rotor wear. (*ii*) Sb_2S_3 is most effective in raising COF but it is toxic. (*iii*) The thermal decomposition products of PbS are also lubricating. It reduces noise. (*iv*) MOS_2 adheres more readily to metal surfaces than graphite and has broad temperature range. (*v*) ZnS is a low cost solid lubricant recommended for high loads and temperatures.
Friction powder	Usually consists Fe sponge
Brass (~62% Cu, 38% Zn)	Machine shop cutting swarf is used. It improves wet friction and recovery
Copper	(*i*) Increases thermal conductivity (*ii*) It can cause excessive wear of cast iron.

(Contd...)

Fluxing metals	Pb, Sb, Bi and Mo These metals serve as oxygen getters to stabilize friction-induced films and help to keep them from getting too thick.
Friction dust	(1) Processed cashew resin [cardanol] (2) Rubber scrap-ground up tires. It must not contain road dirt. (3) It reduces brake noise. (4) It improves resilience in the binder system.

3.4 Fibre-Reinforcements

Fibrous reinforcements exert a much greater influence over the mechanical properties of a friction material then they do over tribometric properties. The following factors can influence the strengthening process

(*i*) Volume fraction of fibre (5-25 vol%).
(*ii*) Tensile modulus and strength of the fibre.
(*iii*) Aspect ratio of the fibre.
(*iv*) Interfacial adhesion
(*v*) Fibre orientation.

There are three types of fibres:

(*i*) Inorganic oxide fibres e.g. crysotile asbestos, glass, rock-wool, Fiberfrax® etc.
(*ii*) Metallic fibres e.g. fibrous products base on steel wool, copper, brass and stainless steel
(*iii*) Organic fibres such as cotton, acrylic, carbon and aramide.

A brief summary of various types of fibres used in friction materials is given in Table 4. Tables 5 and 6 show the dimensions and properties of the fibres.

Table 4. Fibre Reinforcements

Material	*Description/Comments*
Asbestos [$3MgO.2SiO_2.2H_2O$]	Excellent thermal stability and tribological properties combined with low cost. Most common reinforcement in early brake materials.
Ceramic fibres or mineral wool	Composition of fibres: SiO_2 (40-50), $A1_2O_3$ (5-15), CaO (34-42), MgO (3-10) and other oxides (0-7) in wt%. The main function is to control fade and increase braking effectiveness.
Glass fibres (E-glass)	Composition: SiO_2 (53-55), $A1_2O_3$ (14-16), CaO+MgO (20-24), B_2O_3 (6-9), Na_2O <1, TiO_2 (0-2) in wt%.
Potassium titanate [$K_2O.6TiO_2$]	Umt whiskes and structural participant to replace the role of asbestos.
Steel wool	Reinforcement
Copper, Brass	Thermally conductive fibres.
Aramid fibre	Prime candidate for non-asbestos automotive braking applications.

Table 5. Dimensions of Fibre Reinforcements

Fibre	*Average Diameter (μm)*	*Average Length (mm)*
Steel	50	1.5
Brass	60	5.0
Copper	60	5.0
Carbon (PAN)	7	4.5
Ceramic (alumino-silicates)	12	1.0
$K_2O.6TiO_2$ (Whiskers)	0.35	15 (μm)
Aramid (pulp)	12	2.0

Table 6. Properties of Some Reinforcements

Property	*Ferritic steel*	*Asbestos (Chrysotile)*	*E-glass*	*Kevlar-49^R*	*Carbon* (IM)*
Density (gcm^{-3})	7.5-7.7	~2.5	2.62	1.45	1.8
Tensile Modulus (GPa)	200-205	145	72.3	131	270
Tensile Strength (MPa)	500-800	3780	3400	3600	~4000

*From PAN precursor

Since the beginning of auto industry, asbestos has been a popular and versatile reinforcement due to its following characteristics:

(*i*) It is thermally stable to 500°C above which it decomposes to Forsterite (Mg_2SiO_4) and silica (SiO_2).

(*ii*) Silicates produced from asbestos are harder and more abrasive than asbestos.

(*iii*) It helps to regenerate the friction surface during use.

(*iv*) It is thermally insulating.

(*v*) It is strong yet flexible.

(*vi*) It provides wear resistance to friction material.

(*vii*) Its process-ability is good.

(*viii*) It is a cheap and easily available raw material.

The necessity driven research for the replacement of hazardous asbestos has led to many innovations in friction composites as described later.

3.5 Fillers

Low cost materials such as barytes ($BaSO_4$), $CaCO_3$, MgO and clay are used as space fillers whose primary role is to cut the overall cost, and not to deteriorate the required properties of the friction material. Inorganic materials having acicular particles (e.g. $K_20.6\ TiO_2$) or platey materials {e.g. mica, attapulgite ($Mg_5Si_8O_{26}H_{50}.4H_2O$)} can contribute to strength and reduce high temperature wear. Fillers can also comprise of fire proofing substances such as Sb_2O_3 and aluminium hydroxide [AIOOH, $Al(OH)_3$], Lime [$Ca(OH)_2$] is used to avoid corrosion to Fe-additives by ensuring basic pH. Table 7 gives a summary of common fillers used for brake friction materials.

Table 7. Fillers

Material	*Description/Comments*
Barytes ($BaSO_4$)	Basically inert and stable at high temperatures may contribute to wear resistance
$CaCO_3$	A low cost alternative to barites, but not quite as stable at high temperatures.
Lime [$(CaOH)_2$]	(*i*) It is used to avoid corrosion in Fe-additives. (*ii*) Helps raise fade temperatures.
Anti-Oxidants	(*i*) Help to maintain the proper oxide film thickness (*ii*) A thick oxide film can lead to unstable COF (high at low speeds) (*iii*) Graphite is a common additive in cermet brakes.
Rubber (diene or nitrile)	(*i*) It promotes cross-linking in polymeric brake materials. (*ii*) It modifies compressibility.

4. COMPOSITES FOR FRICTION MATERIALS

The composites for brake linings can be classified into three categories based on the matrix phase. These are as given below:

I. Resin Bonded:
   (*i*) Asbestos based (Organics)
   (*ii*) Non asbestos based (NAO)
   (*iii*) Semi-metallic materials (or 'Semimets')

II. Sintered Materials (or Cermets):
   (*i*) Cu based
   (*ii*) Fe based
   (*iii*) AI based

III. Carbon-Carbon Composites

4.1 Resin Bonded Composites

These friction materials have organic matrix usually comprising of phenolic thermosetting resins along with some modifying resins. They are the most commonly composites for brake linings/pads in auto industries. Kane & Mowrer have taken a patent on combining phenolic resins with silicone to form phenolic siloxane resins having enhanced impact resistance without compromising thermal and chemical resistance [6]. The modification by silicon also prevents the friction layer from absorbing water due to its improved water repellency [7]. The main processing steps in the manufacture of resin matrix friction materials are as follows:

(*i*) Mixing of moulding compound with additives.

(*ii*) Compression moulding at a temperature ~ 150°C under a pressure of 2 kg/mm^2 for 20 minutes.

(*iii*) Post curing of briquets at 220-300°C for 4 to 8 hours.

(*iv*) Machining and finishing.

In the past asbestos was used extensively in organic formulations before it was discovered that breathing dust containing asbestos fibres can cause cancer. Today, the use of asbestos is illegal and brake manufactures have shifted towards non-asbestos organic (NAO) formulations and semimetallics. As no single fibre can entirely substitute for asbestos, a combination of fibres (hybrid reinforcement) been used in NAO. Gopal et al [8] have shown that short aramid fibres (Kevlar® pulp) can impart excellent friction stability to hybrid phenolic composites containing steel fibres. The addition of Kevlar® pulp was also found to reduce high frequency (>5kHz) squealing noise of glass/phenolic composites at higher speeds (~40 km/s). Kim et al [9] have shown that the combination of aramid pulp and $K_2O.6TiO_2$ whiskers have a strong synergistic effect providing friction stability and wear resistance. A review on NAO friction materials has been published by J. Bijwe [10]. Table 8 gives typical formulations for 'organics' and NAO materials [11,12].

Table 8. Organic Friction Materials

Ingredients	*Organics (wt%)*		*NAO (wt%)*	
Matrix	Acrylonitrile Butadiene Copolymer Cardolite No. 753 Resin	13.30 8.85 1.75	Phenolic resin Cashew particles	20 10
Reinforcement	Asbestos	55.17	Ceramic fiber Aramid pulp	3 6
Metal	Zine dust	17.20	Cu fiber	3
Abrasives	ZnO	2.42	$ZrSiO_4$ $K_2O.6TiO_2$	4 10
Fillers	Sulfur	1.31	$BaSO_4$ $Ca(OH)_2$ Vermiculite Mica Rubber	20 3 3 3 3
Solid Lubricants	– –		Graphite MoS_2 Sb_2S_3	5 3 4

Semi-metallic materials were introduced in the late 1960s and has gained widespread usage in the mid-1970s. The semimetalic generally refers to the presence of a large amount (>40 wt%) of iron/steel and Cu powder in the formulation. Table 9 shows a typical composition of a semi-metallic lining [13].

Table 9. Formulation of a Semi-metalic Friction Lining

Ingredient	*wt%*
Steel wool	10-25
Porous Fe powder (sponge Fe)	40-55
Total Fe	65
Graphite	15
Additives (Zircon/Mullite/Al_2O_3)	10
Phenolic binder	10

The presence of porous iron powder in the composition imparts the following characteristics:

(*i*) Fe powder with low apparent density reduces both brake weight and total material usage.

(*ii*) High internal porosity and large surface area delivers the most effective friction characteristics, providing improved brake surfaces and excellent pad and rotor wear properties.

(*iii*) High surface area and high internal porosity minimize noise.

(*iv*) Little or no segregation.

The practice of adding brass chips and copper powder in friction material helps in breaking up surface film and thus improves fade performance [4]. Recently Fibretech Ltd. has developed thermally conductive fibres of brass and stainless steel for semi-metallic formulations. These short fibres are produced using the rapid solidification process with patented melt overflow technology [14].

Semi-metallic brake linings are visibly different from organics. Semi-metallic pads are darker in colour with visible metallic fibres and have a rough texture. Organic pads possess a lighter shade of gray and have a smoother texture. Semi-metallic pads generally have high coefficient of friction at higher temperatures ('hot' friction) making them performs well under extreme braking conditions. Larger cars and cars that need to stop quickly, generally, requires semi-metallic pads while organic pads are more suitable for light and medium weight vehicles. On the other hand, semi-metallies tends to be noisier (squeal or groan sound) than organics. Organic pads are often less expensive but may also be less durable then semi-metallics.

Some limitations of organics, NAO and semi-metallics are shown in Table 10. To improve upon these short comings "ceramic" friction materials are being developed since 1985. Here ceramic is not a generic term but a marketing buzz word. Actually these are ceramic- enhanced formulations where steel wool or fibres are substituted by ceramic materials and copper fibres. Semi-metallics contain greater than 40 wt% of metallic constituents whereas it is reduced to about 15 wt% in ceramic friction materials.

Table 10. Limitations of Conventional Friction Materials

Type of Friction Material	*Drawbacks*
Organics	Health issues due to Asbestos
NAO	(*i*) Better suited for low temperature applications and Drum brakes. (*ii*) Cannot meet a wide range of braking requirements
Semi-metallics	(*i*) Generate noise (*ii*) Increase rotor wear

The ceramic formulations can be very complex consisting of 18 to 20 ingredients such as ceramic fibres, potassium titanate, China clay, SiC, B_4C, BN etc. Table 11 gives an idea about the ceramic enhanced formulation used in the brakes of Honda's 'Accord' model.

Table 11. Ingredients of a Cermic-Enhanced Friction Lining

	Function	*Ingredients*
I.	Binder	Acrylic Modified Phenolic Resin Butadine-Styrene Rubber
II.	Abrasives	$A1_2O_3$, MgO
III.	Reinforcement (fibres)	Carbon Aramid (pulp) Ceramic fibre (chopped) Copper
IV.	Friction Modifiers (Powders)	Copper Brass Stainless Steel Friction Dust
V.	Fillers	$BaSO_4$ $Ca(OH)_2$

This composite does not contain asbestos, low carbon steel and glass fibres but gives a pad life greater than 50,000 km [15]. Some of the positive characteristics of ceramic enhanced brakes are:

(*i*) Stable and predictable friction characteristics

(*ii*) Quieter braking because NVH (noise, vibration and harshness) is less

(*iii*) Absence of noticeable dust

(*iv*) Extended brake life compared to conventional lining materials.

4.2 Sintered Brake Pads

These friction materials are used in heavy duty vehicles like earth movers and transport aircrafts. The powder metallurgical process for manufacturing involves following steps:

(*i*) Ball milling of constituent powders

(*ii*) Powder compaction

(*iii*) Assembly of green compact on a nickel plated alloy steel (AISI-4340) back plate

(*iv*) Sintering cum diffusion bonding in reducing atmosphere. Sintering temperatures in °C are: AI matrix 550-620°, Cu matrix: 800-950° and Fe matrix 980-1100°

(*v*) Cooling in forced air

(*vi*) Finishing of brake pad.

Table 12 shows iron, copper and aluminium based formulations for sintered friction materials [11,16,17]. At Sumitomo Japan, mechanical alloying has been used to produce FeMo intermetallic reinforced Cu-9% Sn matrix composites. This mechanically alloyed sintered friction material exhibited superior friction stability and five times less wear then the conventional one [18].

Table 12. Sintered Friction Materials

Ingredients	*Fe based (wt%)*	*Cu based (wt%)*	*Al based* (wt%)*
Fe	59-78	8.8	Al = 77.0
Cu	5	46.2	4.5
Graphite	20-25	8.8	Pb = 4.5
Abrasives	Mullite ≤ 2	SiO_2 = 4.4, Calcined Kyanite = 22.0	–
Others	MoS_2 = 2.6 Pb ≤ 3	Sn = 3.3, Zn = 5.5	Si = 11.0, Ni = 2.0, Cr = 0.5, Mg = 0.5

* Sintered aluminium alloy for railway brakes.

4.3 Carbon-Carbon Composites

A carbon-carbon (C-C) composite is an advanced composite with superior specific strength and thermal properties. It is currently the ultimate material for disc brakes used in racing cars and aeroplanes. A summary of processing steps required for 'C-C' composites are shown in Fig. 1 [19].

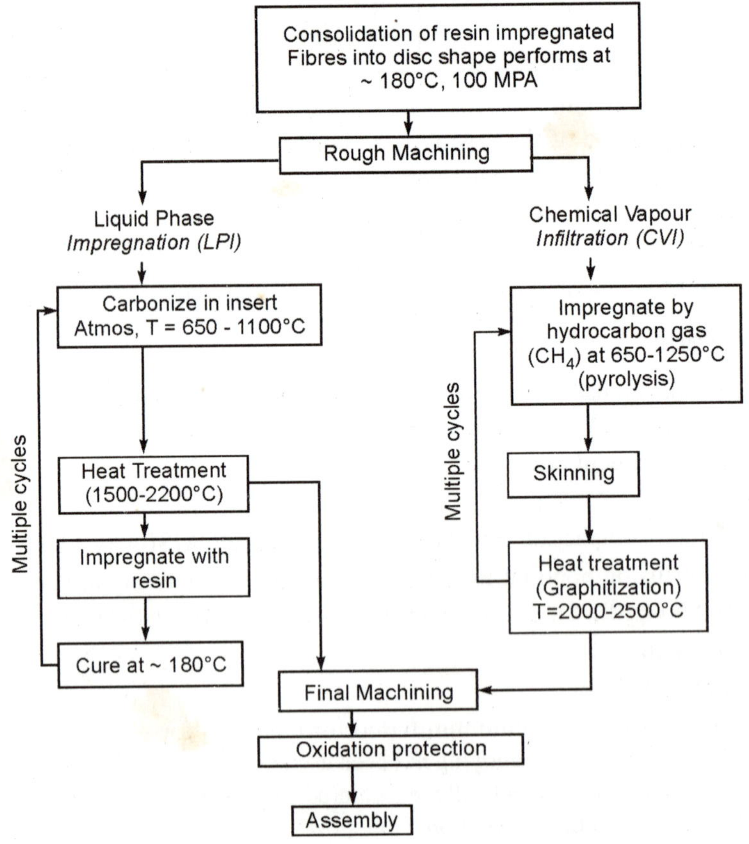

Fig. 1. Flow chart for the Manufacture of C-C Composite Brake Discs.

The advantages of 'C-C' composites as a brake material are as follows:

(*i*) Refractory (m.p. > 3000°C)

(*ii*) Coefficient of friction changes little with surface temperature

(*iii*) High thermal conductivity (to get frictional heat away from the brake surface)

(*iv*) High heat capacity (2.5 times greater than that of steel)

(*v*) High strength at elevated temperatures (twice that of steel)

(*vi*) Very low coefficient of thermal expansion

(*vii*) Good thermal shock resistance

(*viii*) Low density (~1.9 g cm^{-3}) (replacement of metal brakes by C-C composites result in ~40% savings in weight)

(*ix*) They provide both frictional and structural material within the disc brake

(*x*) Service life is about 1.6 times greater than that of metal brakes.

There are two main disadvantages of 'C-C' composites viz. very high cost and oxidation problem.

5. MATERIALS FOR DRUM/DISC

The materials used for drum or disc should possess a combination of properties. It should have good strength to resist all types of non-homogeneous stresses. In an automobile during normal use, the brake temperature can reach about 250°C. The rotor should work as a heat sink and a good radiator. It is required to absorb as much as 80% of the frictional heat. Therefore, the rotor material should have high thermal conductivity and fatigue strength. The material should possess enough hardness to resist wear (Brinell hardness: 170-280 kg/mm^2). Compositions of some of the commonly used materials for brake drum or rotor disc are given in Table. 13 [20,21].

Inspite of higher initial cost, aluminium matrix composites are replacing cast iron or steel parts in a brake system. They are already being used in the following vehicles: Lotus-Elise, Plymouth-Prowler, Audi-A2, and Volkswagen-Lupo3L. The advantages are:

(*i*) Weight saving ~50-60%

(*ii*) Higher thermal conductivity ~ 2.9 times

(*iii*) Uniform friction characteristics

(*iv*) Reduced brake noise

(*v*) Reduced rotor wear

SGL Carbon AG of Germany [22] has developed carbon-ceramic brake discs. This material is a carbon fibre reinforced silicon carbide (C/SiC) composite. It is manufactured by infiltrating carbon fibre reinforced carbon (C-C) with molten silicon. A major portion of carbon matrix reacts with the infiltrated Si to form SiC. However, the reinforcing carbon fibre remain unreacted due to a special treatment. C/SiC composites exhibit excellent braking performance. They have high thermal stability up to 1350°C and lower density ~2.7 g/cm^3. C/SiC brake discs are designed to last the entire lifetime of a car (e.g. Porsche car). Their use in high speed trains [e.g. Inter City Express (ICE-1 & 2)] has resulted in (*i*) Half the number of brake discs per train, (*ii*) Weight reduction by more than 5 tons and (*iii*) Extended life time of the brakes with reduced wear on wheels [22].

Table 13. Materials for Brake Drum of Rotor Disc.

Material	*Composition (wt%)*	*Remarks*
Cast iron	C = 3.6-4.0 Si = 2.5-4.0% Pearlite stabilizing element upto 2% Fe = balance Carbon equiv. ≥ 4.5 wt%	Hypereutectic Pearlitic gray Cast iron BHN~250
Al-alloy	Si = 19.0 Cu = 0.6 Mg = 1.0 Mn = 0.4	Hypereutectic Al-Si alloy
Steel Timken 17-22 (A) S	C = 0.30 Cr = 1.25 Mo = 0.5 V = 0.25 Si = 0.65 Mn = 0.55 Fe = bal.	Steel Rotor disc
Cast A1 Matrix	DRISC™ Brake system Matrix: A1-8 wt% Si Dispersions: SiC_p = 10 vol% (mean size 14 μm) Ni coated Graphite Powder = 5 vol% [Thickness of Ni coating = 5-10 μm Mean size of graphite powder = 85 μm]	Racing cars Max. service temp. ~ 450°C
Ceramic Composites	Carbon fibre reinforced SiC matrix composite	High wear resistant even at high temperatures

SUMMARY

1. A brief review of composite materials and functions of constituents used in automotive friction materials has been presented.
2. No single reinforcement can match the performance of hazardous asbestos. A combination of reinforcements incorporating aramid/steel/ceramic fibres is being used for non-asbestos friction materials.
3. Ceramic-enhanced brake formulations are replacing semi-metallics for brake pad applications.
4. Mechanical alloying has been shown to enhance the performance of sintered metallic friction materials.
5. Carbon-carbon composites are the ultimate materials for disc brakes.

6. Aluminium matrix composites and ceramic matrix composite (SiC/C_f) are being developed to replace conventional cast iron and steel rotors in high performance and fuel efficient vehicles. The new rotor materials can reduce the mass of brakes by as much as 40%.

REFERENCES

1. S.K. Rhee, Wear Equation for Polymers Sliding against Metal Surfaces, Wear, Vol. 16, pp 431-445, (1970).
2. S.K. Rhee, Wear of Metal Reinforced Phenolic Resins Wear, Vol. 18, pp 471-477, (1971).
3. T. Liu and S.K. Rhee, High Temperature Wear of Semi-metallic Disc Brake Pads, Wear, Vol. 46, pp 213-218, (1978).
4. P.J. Blau, Compositions, Functions, and Testing of Friction Brake Materials and their Additives, Technical Report ORNL/TM-200 (Sept. 2001). http://www.ornl.gov/-webworks/
5. R.T. Spurr, Fillers in Friction Materials, Wear, Vol. 22, pp 367-372, (1972).
6. J.F. Kane and N.R. Mowrer, Phenolic Resin Compositions with Improved Impact Resistance, US Pat. 5736619 (1998).
7. Y. Hara, T. Oyama, M. Inoue, T. Mibe and H. Nakanishi, Non-asbestos Disc Brake Pad for Automobiles, US Pat. 6220404 (2001).
8. P. Gopal, L.R. Dharani, F.D. Blum, Hybrid Phenolic Friction Composites containing Kevlai®' Pulp, Part 1: Enhancement of Friction and Wear Performance", Wear, Vol. 193, pp 199-206, (1996).
9. S.J. Kim, M.H. Cho, D.S. Lim and H. Jang, Synergistic Effects of Aramid Pulp and Potassium Titanate Whiskers in the Automotive Friction Material, Wear, Vol. 251, pp 1484-1491, (2001).
10. J. Bijwe, Composites as Friction Materials: Recent Developments in Non-Asbestos Fiber Reinforced Friction Materials — A Review, Polymer Composites, Vol. 18, pp 378-396, (1997).
11. M.G. Jacko and S.K. Rhee, Brake Linings and Clutch Facings in Encyclopedia of Composite Materials and Components (ed. M. Grayson), John Wiley, N.Y., pp 144-154, (1983).
12. H. Jang and S.J. Kim, The Effect of Antimony Trisulfide (Sb_2S_3) and Zirconium Silicate ($ZrSiO_4$) in the Automotive Brake Friction Material on Friction Characteristics", Wear, Vol. 239, pp 229-236, (2000).
13. A.E. Anderson, Friction and Wear of Automotive Brakes, in ASM Handbook, Friction, Lubrication and Wear Technology, Vol. 18, ASM Int., Materials Park, Ohio, pp 569-577, (1992).
14. Filial Brass, Brass Fibres from Fibretech. http://www.azom.com/details.asp? Article ID=2411.
15. J. Yamaguchi, Honda's Non-asbestos, Non-metallic, Non-glass Brake Pad Composite, Automotive Engineering, Vol. 94, No. 12, pp 45 -46, (1991).
16. I.M. Fedorchenko, Materials for Friction Assemblies in Powder Metallurgy: Recent Advances, (eds. V. S. Arunachalam and O.V. Roman), Oxford & IBH Pub. Co., New Delhi, pp 121-170, (1989).
17. L.B. Newman, Friction Materials: Recent Advances, Noyes Data Corp. N.J. USA, pp 196-197, (1978).
18. K. Kondoh, Y. Takano, and Y. Takeda, Friction and Wear Properties of Integrated Composite Copper-Based Friction Materials in Transmission and Drive Line Systems Symposium: Transmission Systems, Components, Gears and Friction and Fluid Materials, SAE, Warrendale, PA, USA, pp 185-194, (1997).
19. G. Savage, Carbon-Carbon Composites, Chapman and Hall, London, pp 3 23 -346, (1993).

20. G. Cueva, A. Sinatora, W.L. Guesser and A.P. Tschiptschin, Wear Resistance of Cast Irons used in Brake Disc Rotors, Wear, Vol. 255, pp 1256-1260, (2003).

21. J.A.E. Bell, A.E.M. Warner, T.F. Stephenson and E. Sierist, Some Automotive Applications for Cast Graphitic Aluminium Nickel Silicon Carbide MMC, in Processing Properties and Applications of Cast Metal Matrix Composites, (ed. P.K. Rohatgi) TMS, Pennsylvania, USA, pp 247-258, (1996).

22. Carbon-Ceramic Brake Discs. http://www.sglcarbon.com/sgl_t/brakedisc/index.html

5

Sinter Hardening of Some Ferrous PM Materials

Yang Yu
Höganäs AB, 263 83 Höganäs, Sweden

ABSTRACT

Compared to the conventional sintering and heat treatment techniques, sinter hardening has the following advantages: lower total cost, equivalent/higher mechanical properties, improved dimensional control and more environmental friendliness.

The present paper discusses the mechanisms of sinter hardening in terms of material and process. Factors influencing the sinter hardening process and their influences on the mechanical properties are discussed. Typical sinter hardening materials, Distaloy DH, Astaloy CrM, Astaloy 85Mo with Cu addition and Astaloy CrL with Cu addition are presented with respect to both process parameters and mechanical properties. Resultant microstructures are discussed in relation to different cooling rate and carbon content.

INTRODUCTION

Powder metallurgy (PM) is one of the most competitive methods to produce components with complicated shapes and large in quantity. During the last twenty years the PM industry, esp. the PM steel industry, has been developed greatly thanks to the highly and increasing requests of the automobile industry. In practical PM production, many sintered components need to be heat treated in order to improve their mechanical performances. Due to the fact that PM parts contain many pores that can accelerate the corrosion effect inside the treated parts in case of water quenching, oil quenching has to be applied during heat treatment. The assimilated oil during quenching process has to be removed from the treated parts of effective tempering, which costs much not only for the equipment and energy but also for the environment.

Sinter hardening combines both sintering and hardening (quenching) in one process to produce martensite as the major phase in microstructure. It is achieved by a gas quenching setup in the cooling zone of the sintering furnace, which can generate cooling rates of 2C/S-5C/S for the sintered components in the temperature range of 950C-200C. The main advantages of sinter hardening lie in cost saving and environmental protection.

In general, two conditions should be fulfilled to reach the sinter hardening effect: materials suitable for sinter hardening and furnaces equipped with gas quenching in the cooling zone. As shown in

Fig. 1(a), if less alloying element is involved in the material the bainitic nose of the TTT diagram locates more to the left. In such case, a conventional cooling rate will make the system generate large amount of bainite or even some pearlite. If alloying elements such as Mo, Cr, Mn, Ni, Si, ect., which are effective to the hardenability of steels, are added in the system as prealloy, the TTT curve will move to the right. Therefore, more martensite can be produced under the same cooling rate. The same effect can be achieved by increasing the cooling rate in the cooling zone of the sintering furnace, as shown in Fig. 1 (b). In order to obtain the best sinter hardening effect, we need to balance the two factors discussed above. Without or with too little amount of the above mentioned alloying elements, very little martensite can be obtained even the cooling rate is over 5 C/S. On the other hand, if too much above mentioned alloying elements are added in the system as prealloyed, there comes two problems although fast cooling can be avoid to a certain extent: Firstly, the system become harder due to the solution hardening effect, which decreases the compressibility of the base powder. Secondary, the cost of the base powder will increase due to the more expensive alloying elements such as Mo and Ni. Therefore in practical application, both factors (base powder and cooling rate) related to sinter hardening effect have to be carefully selected and balanced. in addition to these, factors such as density and carbon content also affect the sinter hardening.

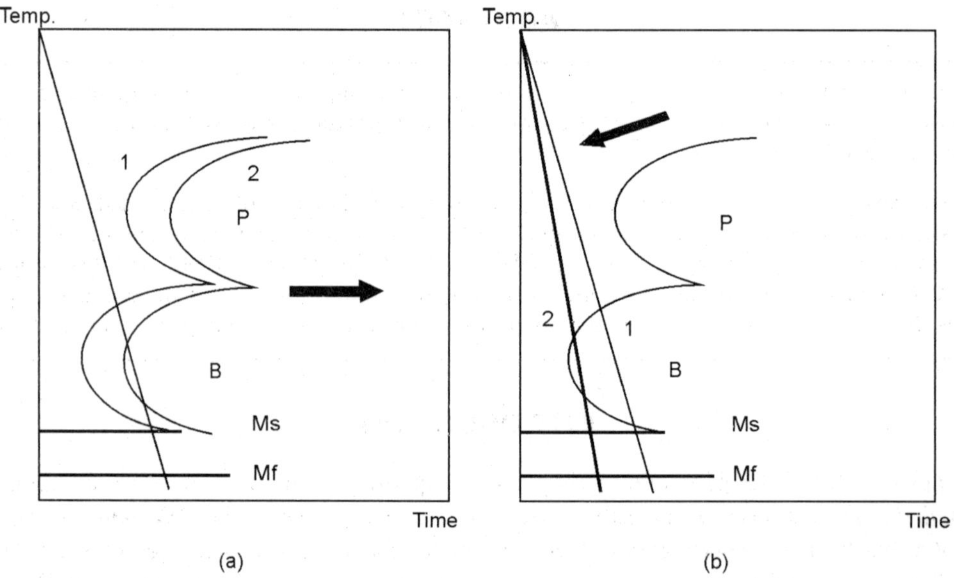

Fig. 1. Schematic TTT diagram of PM steel with varying alloying elements (a) and cooling rate (b), where 1 represents low alloying amount in (a) and low cooling rate in (b); and 2 represents high alloying amount in (a) and high cooling rate in (b).

EXPERIMENTAL

Materials

Four different base powders are used. Distaloy DH is based on a fully pre-alloyed powder with 1.5% Mo to which 2% Ni has been diffusion bonded. Astaloy 85Mo is a fully pre-alloyed powder with

0.85% Mo. Astaloy CrL is a fully pre-alloyed powder with 1.5% Cr, 0.2% Mn and 0.3% Mo whereas Astaloy CrM is pre-alloyed with 3% Cr and 0.5% Mo. In order to reach a better sinter hardening effect, 2% Cu (-325 mesh) is add mixed to Astaloy 85 Mo and 1% and 2% Cu to Astaloy CrL. No Cu was added to Astaloy CrM. The graphite used in all cases was UF-4 and lubricant was Amide wax. The compositions of the mixes prepared are shown in Table 1.

Table 1. Compositions of Powder Mixes

Base powder	*Cu (%)*	*Mo (%)*	*Cr (%)*	*Gr (%)*	*Lub. (%)*
Astaloy 85 Mo	2	0.85		0.65/0.8	0.8
Distaloy Dh	2	1.5		0.65/0.8	0.8
Astaloy CrL	2	0.2	1.5	0.6/0.8	0.8
Astaloy CrM	–	0.5	3	0.4/0.5	0.8

Processing conditions

Both tensile testing (TS) and impact energy (IE) specimens according to the ISO standards were compacted under pressure between 400MP and 800MPa to reach densities in the range 6.8-7.2 g/cm using a laboratory press. Sintering was carried out in a laboratory belt furnace from Abbot. The furnace was equipped with a VARICOOL cooling system enable the cooling to desired cooling rates. Sintering temperature was 1120°C for 30 minutes and the sintering atmosphere was of 90% nitrogen and 10% hydrogen. After sintering, the samples were tempered at 200°C for 60 minutes in air.

Testing

Rockwell hardness (HRC) was measured on the surfaces of each of the tempered samples. Tensile testing was performed in a MTS Sintech 20D tensile testing machine. Impact energy testing was carried out in a Charpy tester. Dilatometer studies were performed in the cooling range between 0.2C/S to 10C/S. Both the TS and the dilatometer samples were studied metallographically.

RESULTS AND DISCUSSION

Effect of cooling rate

The effect of cooling rate on hardness (HRC), tensile strength (UTS) and impact strength (IE) was summarised in Table 2. It is clearly seen that both HRC and UTS increase with increasing cooling. This is because higher cooling rate produces more martensite which has positive effect on both HRC and UTS. The influence of cooling rate on the material toughness (IE) is not as pronounced as on the hardness and tensile strength. In case of Distaloy DH and Astaloy CrM as base powder, higher cooling rate even decreases the impact strength.

Among the four proposed materials, Distaloy DH and Astaloy CrM general enough amount of martensite to make the materials a direct hardening effect even in conventional cooling rates (furnaces without fast cooling).

Table 2. Mechanical Properties of the Tested Materials (Sintered Density 6.8-6.9g/cc) Under Different Cooling Rates.

Material	*CR (C/S)*	*HRC*	*UTS (MPa)*	*IE (J)*
Astaloy 85Mo + 2%Cu + 0.7%C	2.5	27	834	13,2
	5	37	896	14,3
Distaloy DH + 0.7%C	2.5	31	916	17,1
	5	39	1019	15,8
Astaloy CrL + 2%Cu + 0.6%C	2.5	30	821	13,4
	5	32	837	13,4
Astaloy CrM + 0.4%C	2.5	33	1026	15,5
	5	35	1067	11,4

Effect of density

As shown in Fig. 2, under all cooling rates, the hardness (HV10) increases with increasing density. Here, there are two effects with increasing density: one is that the porosity decreases with increasing density, resulting in a higher hardness which is a straight forward effect; the other is that the increase in density improves the thermal conductivity of the material, which gives a better cooling effect within the material under the same cooling rate of the furnace. According to Fig. 3, the first effect was much higher than the second one. The same principle applies also on the other base powders. In practical production, the combination of warm compaction and sinter hardening gives the best mechanical performance suitable for many components for automobile industry.

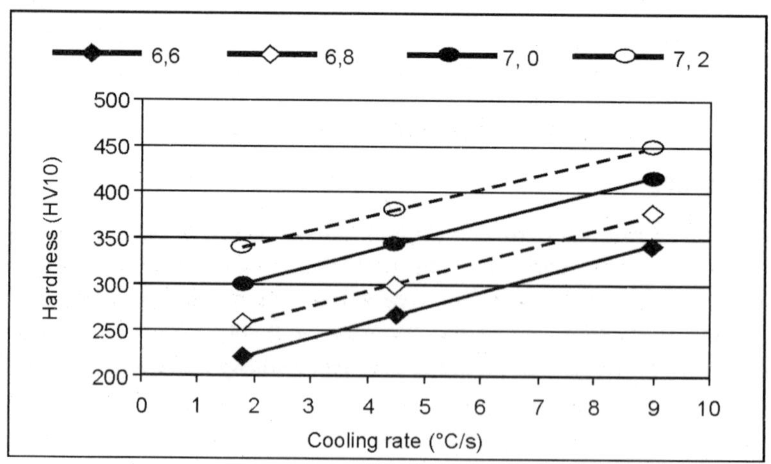

Fig. 2. Hardness vs. cooling rate of Distaloy DH + 0.65% C with different sintered densities.

Effect of carbon content

Carbon plays a very important role in the process of sinter hardening. In many systems, esp. the prealloyed system, a small variation of the carbon content can make a significant change of the sinter hardening effect. Take Astaloy CrM, for instance. As seen in Fig. 4a, when the carbon content in the sintered material is 0.4%, the martensite/bainite ration is around 40/60 for the cooling rate of 1 C/S. If teh combined carbon is increased to 0.5%, however, the martensite/bainite ration is changes to 95/5 under the same cooling rate, as shown in Fig. 4b. This indicates that the carbon content is very sensitive to the sinter hardening effect of this material.

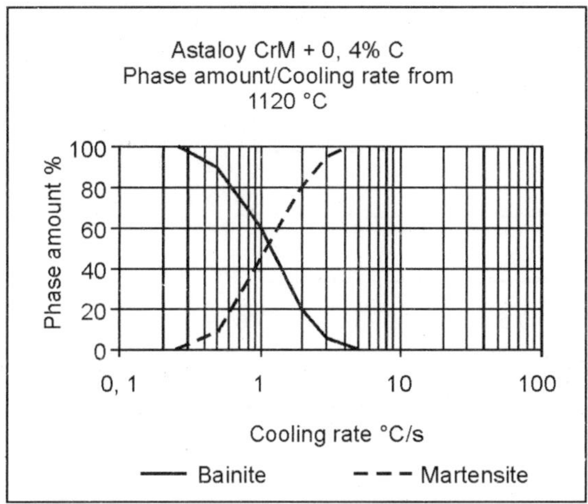

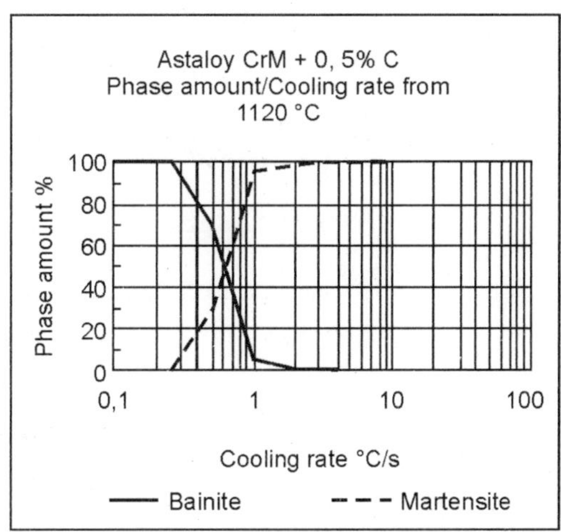

Fig. 4. Phase variation vs. cooling rate for Astaloy CrM + 0.4%C (a) and Astaloy CrM + 0.5%C (b)

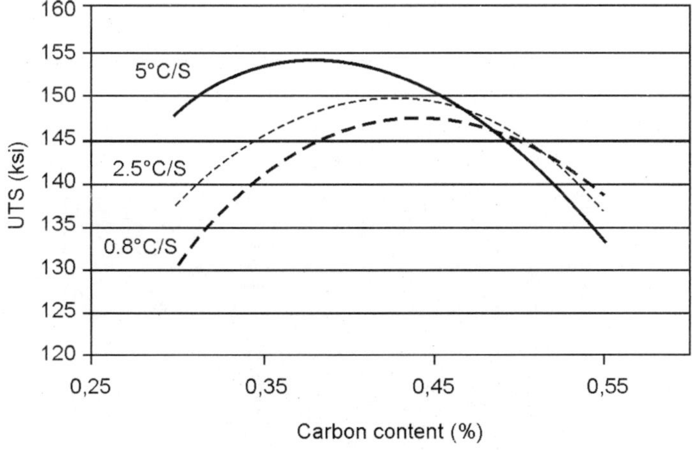

Fig. 5. UTS uv carbon content of Astaloy CrM (7.0 g/cc) at different cooling rates

There is a certain interaction between the cooling rate and the carbon content in most sinter hardening materials. Fig. 5 shows the relation between the UTS and the carbon content of Astaloy CrM base material under different cooling rates. It can be seen that the optimum carbon content for the highest UTS is around 0.45% for cooling rates of 0.8-2.5C/S. If the cooling rate increases to 5C/S, the optimum carbon content decreases to 0.38%. The same is true also for Distaloy DH. The optimum carbon content locates at 0.7% for cooling rates of 0.8-2.5C/S. If the cooling rate increases to 5C/S, the optimum carbon content decreases to 0.6%, as seen in Fig. 6 In practical work, the carbon content has to be decided according to the alloying system and the cooling rate in order to reach the best mechanical properties of the material.

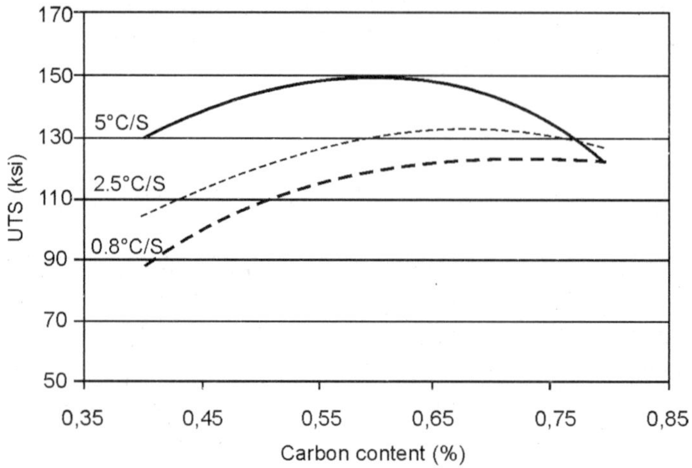

Fig. 6. UTS vs carbon content of Distaloy DH (7.0g/cc) at different cooling rates.

Microstructures

Since Astaloy CrM is one of the most typical materials sensitive to both carbon content and cooling rate, its as-sintered microstructures are presented in this paper. Fig. 7 shows Astaloy CrM + 0,56% C with cooling rate of 0,5 C/S (a), Astaloy CrM + 0.45%C with cooling rate of 0.8°C/S (b) and 5°C/S (c). It is interesting to see even the cooling rate in (a) is lower than in (b), the amount of martensite in (a) is much more than in (b). This is because the carbon content in higher in (a) than in (b). In the case of (c), even the carbon content is not very high, due to the high cooling rate (5°C/S) the final microstructure is composed of fully martensite. This agrees with the results of the mechanical properties in terms of the influence of cooling rate and carbon content on the sinter hardening effect.

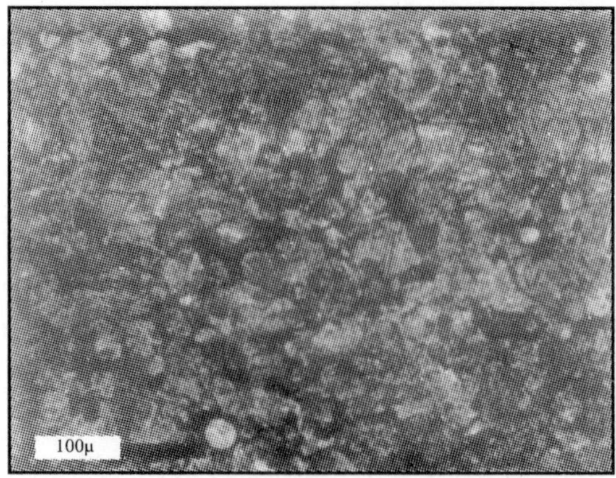

Fig. 7 (a) Astaloy CrM + 0,56% C, cooling rate ~ 0,5 c/s: 25% Bainite + 75% Martensite

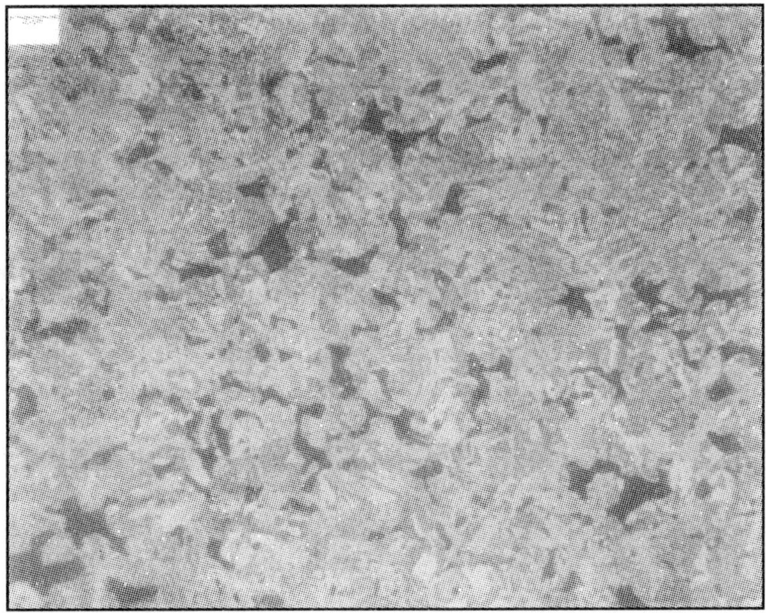

Fig. 7 (b) Astaloy CrM + 0.45%C, cooling rate ~0.8°C/S: Bainite + 10% Martensite

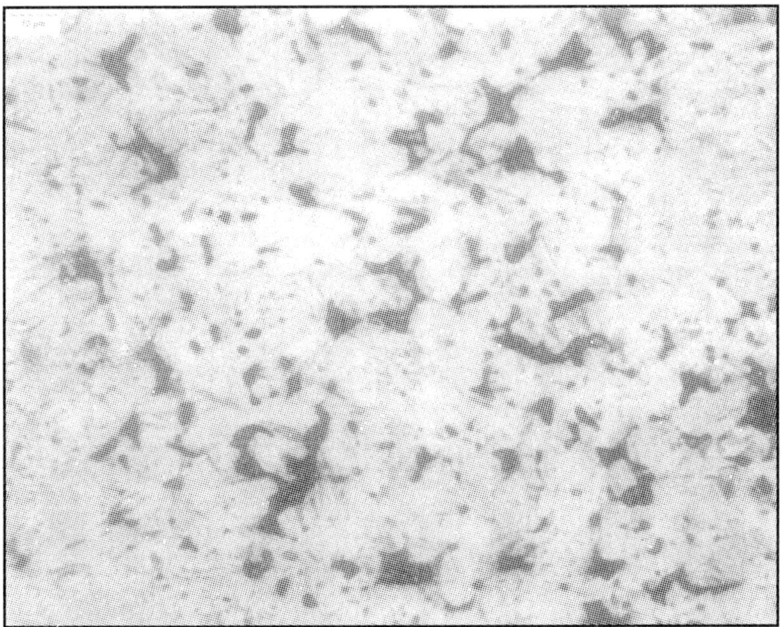

Fig. 7 (c) Astaloy CrM + 0.45%C, cooling rate ~5°C/S: 100% Martensite

CONCLUSIONS

1. All the base materials presented above effective to sinter hardening
2. Cu has a positive effect to sinter hardening
3. Astaloy CrM and Distaly DH are even suitable for direct hardening (conventional cooling rate)
4. Higher density improve the sinter hardening effect
5. There is an interaction between the cooling and the carbon content regarding sinter hardening effect. In practical work, the optimum caron content should be adjusted according to the alloying system and the cooling in order to reach the highest performance.

6

Titanium and Nanocrystalline Titanium for the Automotive Industry

D.K. Pattanayak
Dept. of Metallurgical Engineering and Materials Science, Indian Institute of Technology Bombay, India

B.B. Panigrahi
Dept. of Metallurgical and Materials Engineering, Indian Institute of Technology Kharagpur, India

T.R. Rama Mohan
Dept. of Metallurgical Engineering and Materials Science, Indian Institute of Technology Bombay, India

M.M. Godkhindi
Dept. of Metallurgical and Materials Engineering, Indian Institute of Technology Kharagpur, India

V.V. Dabhade
Dept. of Metallurgical Engineering and Materials Science, Indian Institute of Technology Bombay, India

P. Ramkrishnana
Dept. of Metallurgical Engineering and Materials Science, Indian Institute of Technology, Bombay, India

ABSTRACT

Amongst the light metals, titanium possesses outstanding properties such as high strength (even at moderate temperatures), low density and excellent corrosion resistance leading to expectations of a wide range of applications in the automotive industry. In order to increase the use of titanium in standard production vehicles, however, this material will have to become significantly less expensive *i.e.* both the cost of producing raw titanium and the cost of processing. Possible solutions to the technical challenges have been considered including the futuristic nanotechnology.

In the automotive industry, nanotechnology applications are manifold. They reach from power train, light-weight construction, energy conversion, pollution sensing and reduction, interior cooling, wear reduction, driving dynamics, surveillance control, up to recycle potential and much more. Additional, visions of "nano in cars" reach from contributions for CO_2-free engines, safe driving, quiet cars, self-heating body and windscreens, up to a mood-depending choice of colour and a self-forming car body. All this will meet the present society trends and customer demands for improved ecology, safety and comfort, often summarized by the term sustainability. Powder metallurgy is the major client of automotive industry.

Sintering is a critical stage in the powder metallurgy process and hence the sintering behavior of nano sized Ti powders investigated by dilatometry has been reported. The nano sized Ti powders (40nm) were produced by the attrition milling of micron sized Ti powders (12μm) in Argon atmosphere. For comparison purpose the micron Ti powder was also sintered in similar conditions as that of nano Ti powder. Sintering was carried out in Ar atmosphere in the temperature range of 450-1250°C for nano Ti and 650- 1250°C for

(*Contd...*)

micron Ti and. Sintering was carried out by constant rate of heating (10°C/min) to a fixed temperature followed by isothermal holding for 1 h. The nano Ti powders exhibited a lower sintering onset temperature, larger shrinkage, larger shrinkage rate and lower activation energy for sintering as compared to the micron Ti powders.

Keywords: Automotive, Titanium, Nanotechnology.

1. TITANIUM FOR AUTOMOTIVE APPLICATIONS

To attain weight and functional advantages in automotive production through the use of alternative materials in the future, only one element truly remains: titanium. Since the beginning of industrial production of titanium and titanium alloys in the 1950s, this material has been very attractive to car manufacturers because of its high specific strength, its highly elastic energy absorption capacity, and its corrosion resistance. Despite these high expectations and the numerous advantages of titanium materials, they have yet to establish themselves outside the motor racing industry *i.e.* in the conventional automotive industry. Their use has always failed for a single reason: price. Yet, because greater demands in safety and efficiency are being placed on coming generations of automobiles and as the aerodynamic and, with many components, the design options have been exhausted, the willingness to consider the use of more expensive materials is growing (1).

1.1 Properties and Cost

Titanium reveals an attractive applications spectrum. High strength, low density and excellent corrosion behavior makes this metal an ideal structural material. It still displays good strength values at temperatures in excess of 300°C and good oxidation resistance up to 600°C. Combined with its high strength, its modulus of elasticity, which is approximately 40% lower than that of steel, also means that titanium is an ideal material for springs. Figure 1 (a) compares the specific characteristics of the most frequently used titanium alloy, Ti-6A1-4V, with those of the high-strength AIZnMg alloy X-7075, the high -strength MgAIZn alloy AZ80 and high-strength quenched and tempered 34CrMo4 steel. The yield strength, fatigue strength and modulus of elasticity have been standardized to the relevant density. In the case of lightweight design, titanium will be the material which is selected if a component has to be dimensioned as regards strength. If, however, high stiffness is primarily required, light aluminium and magnesium metals, in particular, become more attractive. One of titanium's great advantages should again be emphasized at this point. No corrosion protection measures whatsoever are required in the case of titanium. Another very important aspect, which is vital as regards the use of a material in automobiles today, is environmental compatibility. Titanium is recyclable and non-toxic (2).

If the costs of titanium are compared with the other structural materials conventionally used in automotive engineering, steel, aluminium and more recently also magnesium, titanium always reveals the highest material price. In this case, it is irrelevant whether the price refers to mass, volume or specific strength (Fig. 1(b)). Titanium will not therefore achieve success in automotive engineering as a pure substitution material. Titanium is only employed when the advantage which it offers has a major impact in every regard. For example, an important technical advantage may be a reduction in weight via the exploitation of so-called secondary effects. This means that other components or entire assemblies may be lighter thanks to the use of a titanium component. Another positive effect would be if production of a component using titanium were to lead to the removal of several process steps such as, e.g. corrosion protection measures (2).

In recent years research has been directed towards a low cost extractive process for titanium. Traditionally the Kroll and Hunter processes have been employed for the extraction of titanium. Newer technologies such as vapor-phase processes, molten salt process, electrolytic reduction, plasma processes and simultaneous non-consumable arc- melting/plasma-refining process (SNAPP) are been tried out to reduce the extraction cost of titanium as well as to produce cheaper titanium powders (3-5).

Net-shape processing methods are well known as a means for reducing scrap; minimizing machining, grinding, and finishing; and providing a basis for high-speed and efficient production. Among net-shape processes, precision casting and P/M methods could be used for automotive applications such as valves, valve-spring retainers, and connecting rods. The impediment to the growth of P/M processing of titanium powders is the high cost of the powder (3,6). Titanium powder injection moulding is an attractive emerging approach for the production of titanium automotive parts of titanium alloy and titanium aluminides. The only limitation of this process is the high cost of fine titanium powders of high purity (less oxygen content) which are used as the raw material (7).

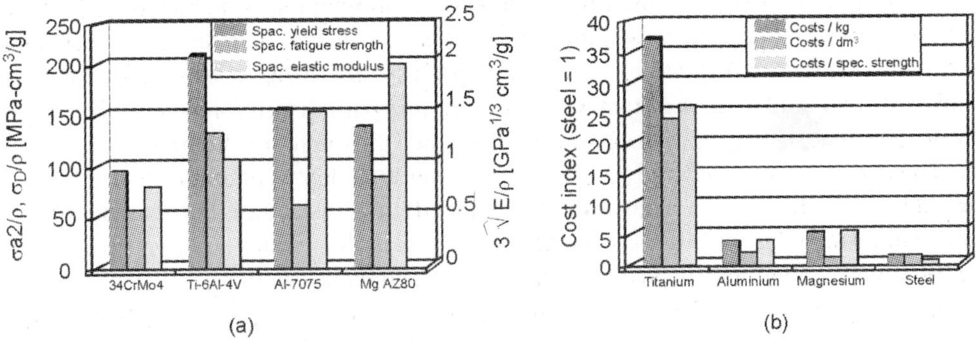

Fig. 1. (a) Specific characteristics of typical structural materials, (b) Comparison of costs: Ti *vs.* Al, Mg and steel (steel = 1).

1.2 Potential usages of Titanium

1.2.1 Applications in Engine

Connecting Rods

Manufacturers are ready to utilize more expensive materials in the engine if this contributes to reductions in rotating and, more importantly, oscillating masses (1). A titanium connecting rod therefore appears to be a very attractive application for reducing moved masses. No problem arises when designing for strength, as the strength level of the Ti-6A1-4V alloy is higher than that of conventional steel connecting rod. However, a connecting rod has to be designed for stiffness, particularly as regards its big end. In this case, design measures have to be implemented to compensate titanium's lower modulus of elasticity. Considerations must also be given to the fact that titanium has a lower coefficient of thermal expansion than that of steel used in the crankshaft or the piston pin. Greater bearing tolerances are necessary in this case. This may lead to increased wear. Titanium's wear resistance, which is in any case lower, has to be increased by coating the connecting rod at the appropriate points, for example with a PVD-CrN layer (1,2).

Pistons

As pistons manufactured from lighter aluminium have already been in use for many years, manufacturing pistons from titanium does not appear necessary. The recent trend towards increased combustion chamber temperature and pressures is revealing the limits of aluminium-based pistons. In comparison with aluminium, titanium's considerably higher stiffness and strength make it a more interesting material for pistons in order to increase efficient engine operation.

Piston Pins

In addition to high resistance to wear and corresponding strength, the piston pin must also reveal high stiffness. The V-titanium aluminides are suitable for this purpose, even under consideration of the fact that this involves a relatively simple geometrical component.

Valve Gear

The continuing trend towards high -speed engines necessitates the increased use of light metal alloys in the valve gear. In particular, the valves purely oscillating mass has to be reduced. The level and diversity of the strains to which the valves are subjected are extensive. In addition to mechanical strain, the exhaust valves are also subjected to very high thermal strain of up to 900°C, *i.e.* they have to be durable, and creep and oxidation- resistant. Only near- α and particle reinforced ($\alpha + \beta$) alloys such as Ti-6A1-4V are adequate. However, the opportunities afforded to γ-titanium aluminides as a valve material are greater. The density of this material group is even lower than that of titanium. γ -TiAl has very high thermal resistance, high creep resistance and high oxidation resistance but low ductility at room temperature. Unfortunately, the high costs and even greater difficulties involved in producing and processing γ-titanium aluminides negate the materials attractiveness again.

Further mass reduction can be achieved by using titanium valve springs. However, only the expensive β-titanium alloys can be used for this purpose. Titanium valve spring retainers are possible. However, a special coating has to be applied due to the high level of wear to which they are subjected, which leads to additional costs. The use of titanium or γ-TiAl in the valves, springs and retainers leads to a 50 to 70% reduction in mass. These measures enable the maximum engine speed to be increased by approximately 10%.

Turbo Charger Wheels

Titanium or γ-TiAl turbo charger wheels are used occasionally. Significant problems exist as regards fatigue strength in the case of cast and thin-walled wheels. The method to forge these components is extremely uneconomical.

1.2.2 Applications in the Brake System

Brake fluid is a very aggressive medium. In order to ensure corrosion protection sealing rings in the brake line connection flange and brake pad guide pins in certain car models have been made of titanium. Titanium's low thermal conductivity, which protects the brake fluid, would recommend its use in further brake system applications.

1.2.3 Applications in the Chassis

The reduction of unsprung masses in the chassis is of fundamental interest. In addition to the desired lightweight effect, low masses particularly lead to comfort and vehicle dynamics advantages in this regard. Typical forged components such as pivot bearings, wheel carriers, the steering linkage and axle stubs would be ideal components for Ti-6Al- 4V or other high-strength ($\alpha + \beta$) alloy. Once again, one great advantage of titanium is the fact that no corrosion protection measures are required.

1.2.4 Applications in the Exhaust Systems

Exhaust systems are probably the most promising application area for titanium. The use of titanium in exhaust systems is particularly attractive because of the high weight saving both in percentage (45 to 50%) and absolute terms, and the necessary semi-finished titanium sheet is relatively inexpensive. Titanium exhaust systems should last throughout the vehicle's service life. As a rule, high-quality exhaust systems are manufactured from stainless steel, which is also supposed to last throughout the vehicles service life. However, investigations have shown that corrosion problems occurred at the weld seams. The composition of stainless steel changed at these points and was susceptible to corrosion. Weld seams in titanium exhaust systems were not affected (2).

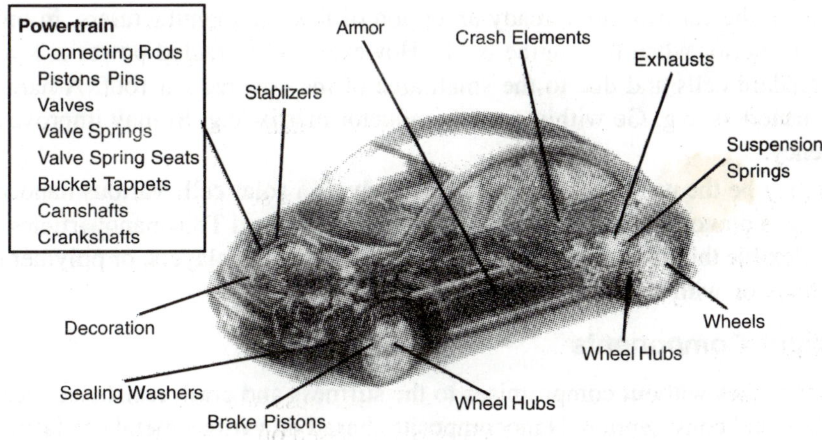

Fig. 2. Examples of possible automotive applications titanium.

2. NANOTECHNOLOGY FOR AUTOMOTIVE APPLICATIONS

Nanotechnology is already in mass production for automobile components. Antireflection coating based on multiple nanolayers on glass is used for instruments in trucks. Sun protecting glazing with infrared reflecting nanolayers embedded into sheets of glass is presently already used in buses. Thermoplastic nanocomposite with nanoflakes is used for stiff and light exterior parts like the step-assist by GM.

2.1 Components for Power Train

Fuel cells are regarded as the engines of the future with the strongest impact on the emission. They will consume either conventional fossil fuel, e.g. methanol that will be converted to hydrogen, or directly hydrogen. Though cars powered by fuel cells will soon be available on the market in a limited number, many components still require further improvements. Nanotechnology can contribute to hydrocarbon (gasoline, methanol) injection and reforming, to hydrogen storage, to the cell electrodes and to the proton exchange membrane (PEM). An efficient aersol formation in the mechanical reformer, but also of fuel in the traditional combustion engine, can be achieved by means of nanojet arrays.

Porous compounds like foams with irregular nanopores may improve sub- components of the fuel cell. A further approach to improve the fuel cell can be the substitution of the traditional Nafion PEM by a polymer - inorganic nanocomposite with improved conductivity, permeability, water management and interfacial resistance to the electrode.

Improved power train also concerns friction and wear reduction of transatory or rotating parts. The largest mechanical losses at high revolution rates come from piston, piston rings and crankshaft gear, at low revolving rates, instead, mainly from the walve train. Nanocomposites with SiC, SiO_2, TiO_2, BN_3, C, diamond, even Teflon within matrix material such as Ni, AI, Fe, and alloys suppress dislocation gliding and crack formation by pinning at the diluted particles. The improved mechanical stability means less wear, better gliding, thinner coating, less lubricants, longer service intervals and pronounced fuel saving.

2.2 Components for Energy Conversion

Solar cells on the car roof are already an option of few car manufacturers, in order to provide a ventilation in the interior when the engine is off. However, the extracted power is low due to the low efficiency of standard cells and due to the small area of the covered car roof. A nanocomposite with semiconductor nanodots, e.g. Ge within a semiconductor matrix, e.g. Si may improve absorption, and increase efficiency.

One idea may be the use of the complete car body as a solar cell. Various nanocomposites came into the game, e.g. a photovoltaic paint composed of dye-sensitized TiO_2-nanoparticles embedded in an electrolyte or a flexible thin film semiconductor cell with multi-nanolayers, or polymer cells either with carbon bucky balls or with semiconductor nano-rods.

2.3 Light-weight Components

Lighter car bodies without compromises to the stiffness and crash resistance means less material and indirectly less fuel consumption. Nanocomposites based on various metal or plastic matrix material strengthened by metal or ceramic nanoparticles or nanoplatelets can improve the strength by 100%. However, a real breakthrough will occur when carbon Bucky fibers will become available in huge quantities. These giant one-dimensional molecule promises a tensile strength of ~ 150 MPa, about 50 times that of steel with 1/5th of the weight. Even a polymer nanofiber nanocomposite would yield considerably thinner, stiffer and lighter parts for cars. That opens a huge research field for the future (8).

3. NANOCRYSTALLINE TITANIUM

Titanium is an important metal and the design choice for the automotive applications. It offers low density, good corrosion resistance and good mechanical properties at room and moderately elevated temperatures (9,10). Titanium powder metallurgy offers the potential for true net shape capability, and, efficient material utilization combined with mechanical properties that are equal to or exceed cast and wrought products (10, 11). The properties of titanium can be further improved by processing it in a nanocrystailine form. Nanocrystailine titanium is expected to exhibit superior properties as compared to conventional grain sized titanium (12).

In recent years there have been considerable efforts to produce titanium nanopowders/ nanocomposite powders by mechanical milling/alloying and in subsequent consolidation (13). This process has emerged as a popular method because of the possibility of producing large quantities of titanium nanopowders/ nanocomposite powders by repeated welding, fracturing and rewelding of powder particles subjected to high energy ball milling (14). Furthermore, mechanical alloying (MA) has gained considerable attention as a non-equilibrium process resulting in solid-state alloying beyond the equilibrium solubility limit and the formation of amorphous or nanostructured titanium. The unique properties of nanocrystailline titanium may be fully exploited if the titanium nanopowders are densified into bulk parts that retain the initial nanostructure. It is therefore necessary to achieve densification with minimum microstructural coarsening. Restrictions on the consolidation process, particularly, the minimum high temperature exposure is therefore necessary. Although lower sintering temperatures are desirable, full interparticle bonding and pore free structures must be achieved. Since, at least some departure from metastability will occur during sintering, it is important to define this metastability and the conditions upon which it is lost (15).

4. EXPERIMENTAL

4.1 Raw Material

4.1.1 Micron Sized Ti Powder

In the present investigation micron sized titanium powder (99.68% purity) obtained from M/S Alfa Aesar (Germany) was used. The average particle size was 12 μm , determined by scanning electron microscopy (SEM). The shapes of the particles were irregular as seen from the SEM photograph Fig. 3. SEM studies were carried out using a FEIE make instrument (model 200 HV Quanta). The X-ray diffraction (XRD) pattern of the Ti powder is shown in Fig.4. All major peaks of Ti are seen along with two minor peaks of TiO_2. The TiO_2 peaks are most probably from the thin oxide film present on the Ti particles. A Philips make X-pert pro XRD with X'celerator instrument was used in the present investigation with Cu target. The chemical analysis of the powder (obtained from the supplier) is shown in Table 1.

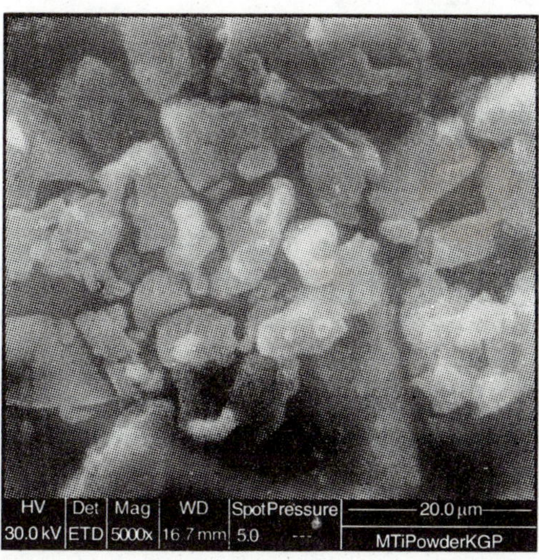

Fig. 3. SEM photograph of micron size titanium powder.

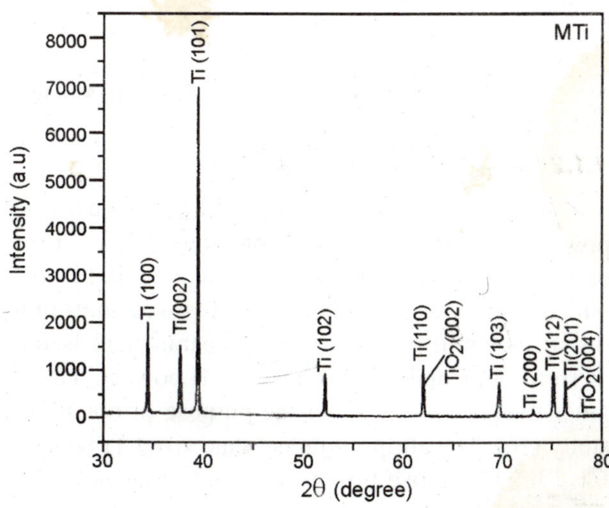

Fig. 4. XRD pattern of micron size titanium Powder.

Table 1. Chemical Analysis of the Titanium Powder (M/S Alfa Aesar, Germany)

Element	*Ti*	*O*	*H*	*N*	*Fe*	*C*	*Mn*
Weight (%)	99.68	0.22	0.04	0.01	0.03	0.01	0.01
Element	Al	S	Mg	Na	Si	Cl	P
Weight (%)	0.01	<0.01	<0.01	<0.01	<0.01	<0.01	<0.01

Figure 5 shows the differential thermal analysis (DTA) curve of Ti powder. The DTA was carried out in a Du pont instrument, (9900 model) using α-alumina as the reference. DTA was carried out in high purity Ar atmosphere which prior to introduction in the DTA furnace was passed through a getter furnace with Ti turnings to remove the traces of oxygen. The DTA curve showed two exothermic peaks one at 790°C and the other at around 880°C. The peak at 790°C is most likely due to oxidation of the Ti powder while that at 880 due to the $\alpha-\beta$ transformation.

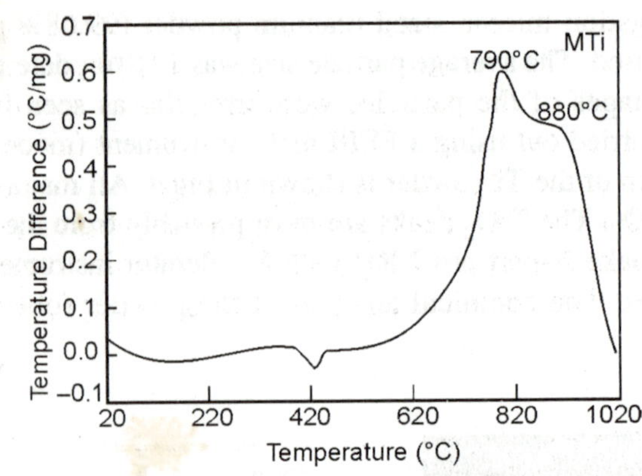

Fig. 5. DTA of micron Ti powder.

4.1.2 Nano Sized Ti Powder

Nanosized Ti powder was synthesized on a laboratory scale by attrition milling of micron sized Ti powder under high purity argon flow. The micron sized Ti powder (M/S Alfa Aesar, Germany, as described above) of average particle size of 12 μm was charged into a stainless steel attrition mill which comprises of a stainless steel vessel, shaft impeller assembly and milling media. Table 2. shows the attrition mill details and milling parameters used in the present investigation to synthesize nano sized Ti from elemental micrometer size Ti powder. The attrition mill was run for a total time of 40 hrs, the latter time was decided for the purpose of obtaining nano sized Ti powder on the basis of a separate study (16). The attrition milled nano sized Ti powder was heated under vacuum (< 10^{-4} torr) at 400°C for 4 hrs to remove the process control agent (PCA).

Table 2. Attrition Mill Details and Milling Parameters used in the Present Investigationfor the Synthesis of Nano Sized Ti Powder.

Attrition mill details	*Milling parameters*
Vessel capacity: 1 L	Milling atmosphere : argon at 1 atm
Milling media: hardened steel balls	Charge to media ratio: 1:10
Vial materials : stainless steel (AISI 304)	Milling speed (rpm): 450
Vial wall cooling: water cooled.	Amount of PCA (wt%) : 2.5
PCA: Stearic acid.	Total time of milling (h) : 40
	Weight of initial charge (g): 40

The average particle size of the nano sized Ti powder produced was 40 nm as determined from TEM. The particle sizes of over 100 particles were determined individually from the TEM, and their average was taken as the average particle size. The shape of the particles was irregular as seen from the TEM photograph Fig. 6. The XRD pattern of the nano Ti powder is shown in Fig. 7. It shows all major peaks of Ti along with some minor peaks of TiO_2. The TiO_2 peaks are of more intensity than that observed in the micron Ti powder Fig.2. This is most likely due to the smaller particles size (large surface area) and presence of minute quantity of O_2 in the argon gas used during milling. The chemical analysis of the powder as obtained by using the inductively coupled plasma- atomic emission spectroscopy (ICP-AES) is shown in Table 3. The Major impurities were that of Fe, Cr and Ni.

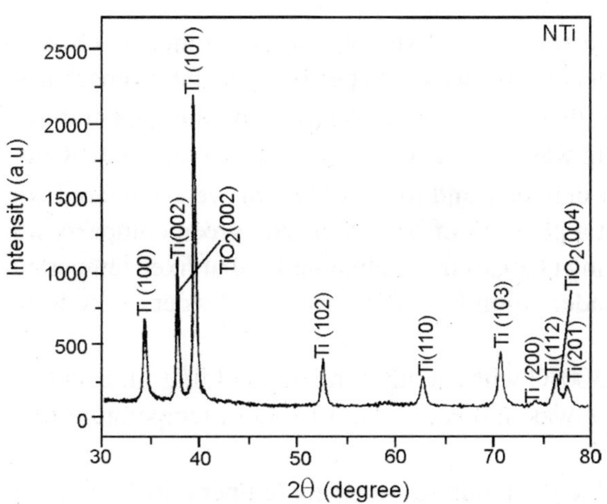

Fig. 6. TEM photograph of nano sized titanium powder.

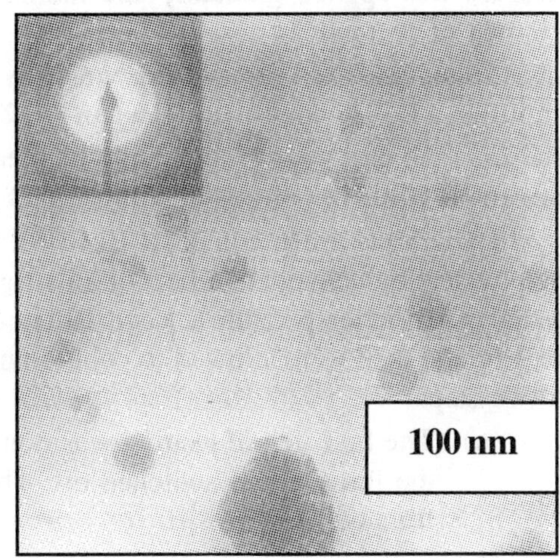

Fig. 7. XRD pattern of nano sized titanium powder.

Table 3. Chemical Analysis of the Nano Sized Titanium Powder

Element	Fe	Cr	Ni	Mn	Mg
Weight (%)	0.62	0.16	0.10	0.04	0.02

Figure 8 shows the DTA curve of the nano Ti powder. Two exothermic peaks are seen, one at 264°C and one at 574°C. The peak at 264°C is most likely due to the relieving of the stresses in the powders while that at 574°C is most likely due to the $\alpha - \beta$ transformation.

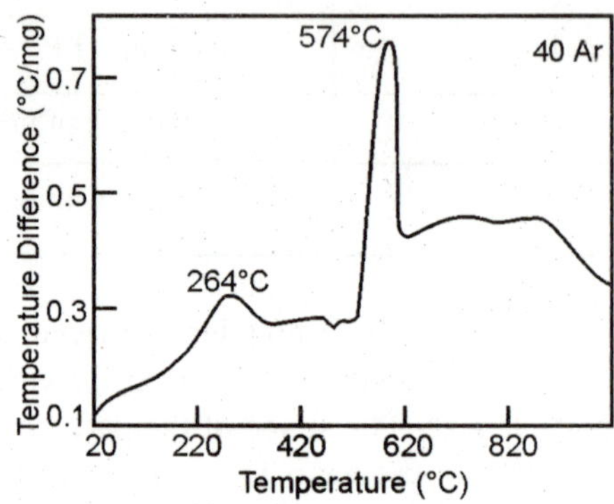

Fig. 8. DTA of nano sized Ti powder.

4.2 Compaction and Sintering Procedure

Compaction of the powders was carried out in a uniaxial press using a cylindrical die after proper die wall lubrication with stearic acid. Compaction was carried out in a motor driven hydraulic press. Pressure was applied gradually and maintained for 1 min for all the compacts followed by gradual removal of pressure.

Shimadzu TMA-30 system, which is a combination of a TMC-30 sample holder, a TMA- 30 amplifier, a DT-30 control unit, and an R-22T recorder, has been used to perform the thermomechanical analysis. The change in length could be measured with an accuracy of $\pm 1\mu m$ and the temperature of the sample could be controlled within ±1°C. Sintering was carried out using a heating rate of 10°C/min.

Cylindrical green compacts (0.4g) of 6mm in diameter and 4mm in height were prepared from the powders by uniaxial compaction in a rigid die at a pressure of 600 MPa. The green compacts were heated to a fixed temperature at a constant rate of heating (10°C/min), maintained at that fixed temperature for 1 hr and then were allowed to cool to ambient under argon flow. The nano Ti powders were heated in two steps:

(*a*) The first step of heating was done up to 300°C without allowing any holding time and then the furnace was switched off. The sample was allowed to cool to room temperature in the furnace itself.

(*b*) In the second step, the sample was heated to the required, sintering temperature and held for 1 hr before furnace cooling.

The first step of beating was carried out to remove the adsorbed gases in Ti nano powders (which lead to huge expansion during the first beat) and the stresses induced in the specimens during compaction. The 300°C temperature was chosen as to avoid sintering that can take place (this temperature is below the sintering onset temperature for nano Ti). Sintering was carried out at various temperatures in the range 450-1250°C for nano titanium and in the range 650 – 1250°C for micron titanium.

5. RESULTS AND DISCUSSION

5.1 Sintering Onset Temperature

As explained earlier the sintering onset temperature for nano Ti was found to be 415°C (0.35T_m) while that for micron Ti to be approximately 606°C (0.45T_m). The reported onset temperatures are the average of all the samples heated in the temperature range of 650-1250°C for micron Ti and 450-1250°C for nano Ti. The lower sintering onset temperature for nano Ti compared to micron Ti could be due to the numerous interfaces which provide short circuit paths for diffusion.

5.2 Linear Shrinkage and Shrinkage Rate

During the constant rate of beating the linear shrinkage was determined as a function of temperature while during isothermal holding it was determined as a function of time. Figures 9 and 10 show the dilatometric curves obtained for beating at a constant rate of 10°C/min up to various sintering temperatures followed by holding at that temperature for 1 hr for specimens of micron Ti and nano Ti respectively. The left portion of Figs. 9 and 10 represent the dilatometric data during constant rate of heating while the right side portion represents dilatometric data during holding at sintering temperature.

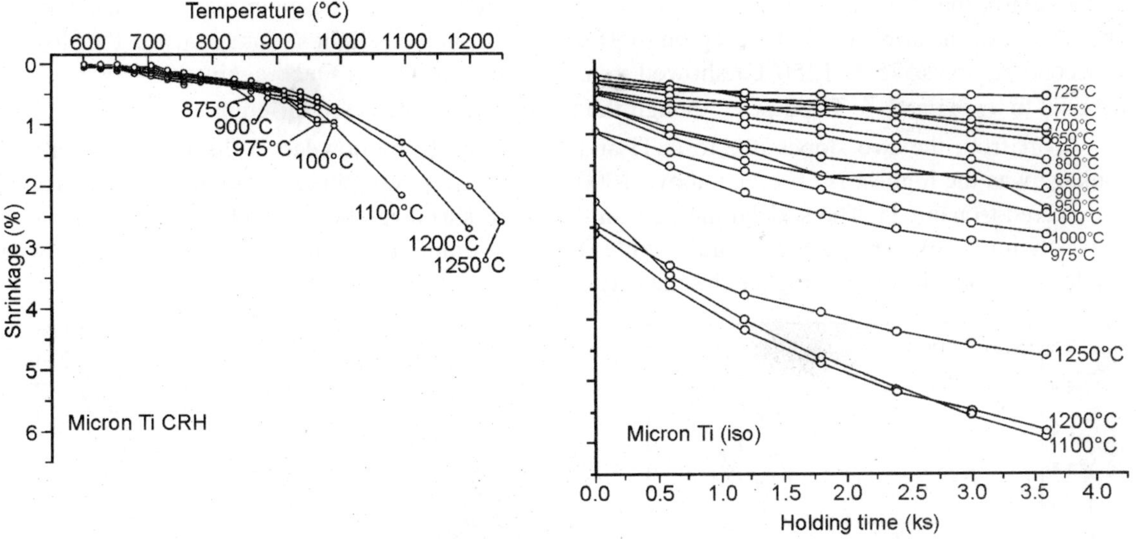

Fig. 9. Variation of linear shrinkage with temperature during constant rate of heating at 10°C / min (left side) and linear shrinkage with time during isothermal holding (right side)for specimens made from micron Ti.

The data obtained during initial constant rate of heating is separately shown in Fig. 11 for the sake of clarity. It can be noticed that the observed shrinkage for specimens made from micron titanium Fig. 11 (a) showed two distinct slopes; the slope in the first region (606°C-850°C) corresponds to α-Ti where the shrinkage was lower, and the slope **in the** second region (900°C-1250°C) corresponds to β-Ti where the shrinkage was higher. This is most likely due to open structure of the β-Ti phase (bcc) compared to the α-Ti phase (hcp) which is close packed.

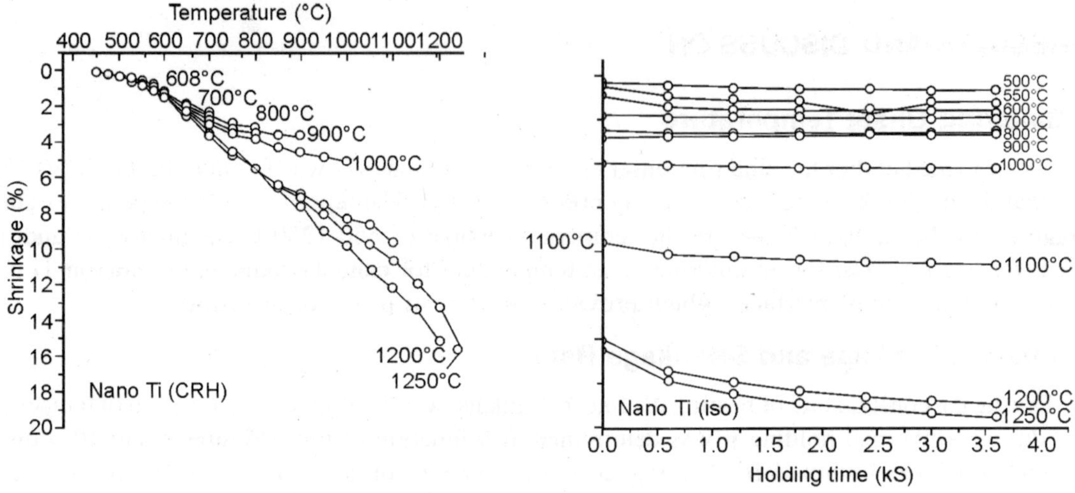

Fig. 10. Variation of linear shrinkage with temperature during constant rate of heating at 10°C/min (left side) and linear shrinkage with time during isothermal holding (right side),for specimens made., from nano Ti.

Similarly the shrinkage dependence for samples made from nano titanium Fig.11(b) showed two distinct slopes. The slope in the first region (450°C-550°C) showed slight shrinkage while the slope in the second region (575°C-1250°C) showed considerable shrinkage. Once again this slope can be attributed to α-Ti and β-Ti phases although the temperature ranges are different.

Apart from the two slopes due to α-Ti and β-Ti in both the powders a third slope where the shrinkage was the highest is observed above 1000°C. This slope was more predominant in micron Ti as compared to nano Ti. The sudden increase in the linear shrinkage beyond 1000°C could be due to the dissolution of oxygen (of the titanium oxide film) into titanium which brings the titanium powder particles in contact leading to higher sinterability.

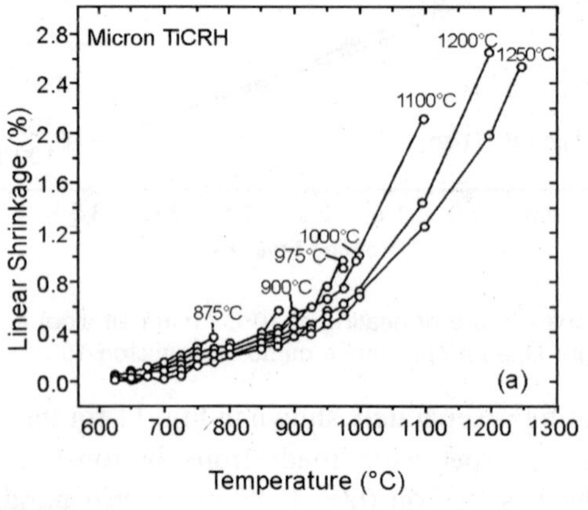

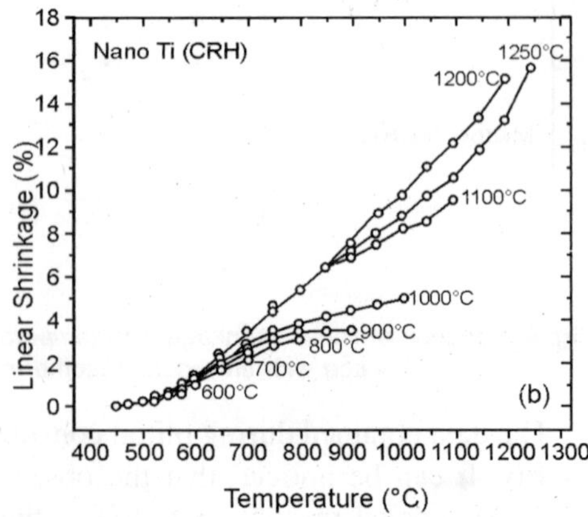

Fig. 11. Variation of linear shrinkage (%) with temperature during constant rate of heating at 10°C/min for specimens made,from (a) micron Ti and (b) nano Ti.

Figures 11(a) and 11(b) indicate that for both micron Ti and nano Ti powders the shrinkage rate during constant rate of heating varied with respect to temperature and therefore this parameter was plotted as a function of temperature. Since the individual curves shown in Figs. 11(a) and 11(b) represent the shrinkage obtained at a single constant rate of heating (10°C/min) ideally all these should represent a single curve. Therefore average of all the shrinkage values at a given temperature were obtained and were used to determine the shrinkage rate. The latter is shown as a function of temperature in Figs. 12(a) and 12(b) for micron Ti and nano Ti respectively.

The shrinkage rates were found to be an order higher in nano Ti as compared to those observed in micron Ti at any given temperature.

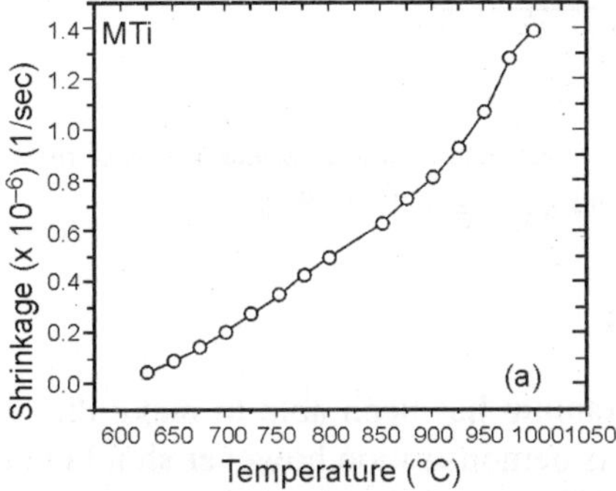

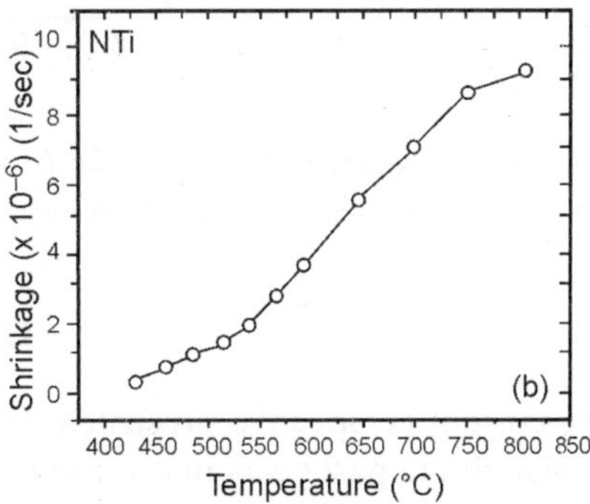

Fig. 12. Variation of linear shrinkage rate with temperature during constant rate Of heating at 10°C/min for specimens made.from (a) micron Ti and (b) nano Ti.

3.2 Activation Energy

Logarithm of shrinkage rates corresponding to α-Ti region and β-Ti regions are plotted as a function of inverse absolute temperature in Fig. 13. The curves obey the equation:

$$(y/t)_\alpha = r_\alpha = r_{\alpha o} e^{-(Q\alpha/RT)}$$

and $$(y/t)_\beta = r_\beta = r_{\beta o} e^{-(Q\beta/RT)}$$ respectively.

Using these equations the activation energy values for sintering were obtained and are shown in Table 4. It is evident that the sintering required lesser energy in β-Ti region with low activation energies of 70 ± 4 kJ / mol for micron Ti and 30 ± 4 kJ 1 mol for nano Ti respectively.

Table 4. Activation energy values obtained from the plots of ln r vs 1/T for micron Ti and nano To

Micron Ti (α) (kJ/mol)	*Micron Ti (β) (kJ/mol)*	*Nano Ti (α) (kJ/mol)*	*Nano Ti (β) (kJ/mol)*
92±6	70±4	72±3	30±4

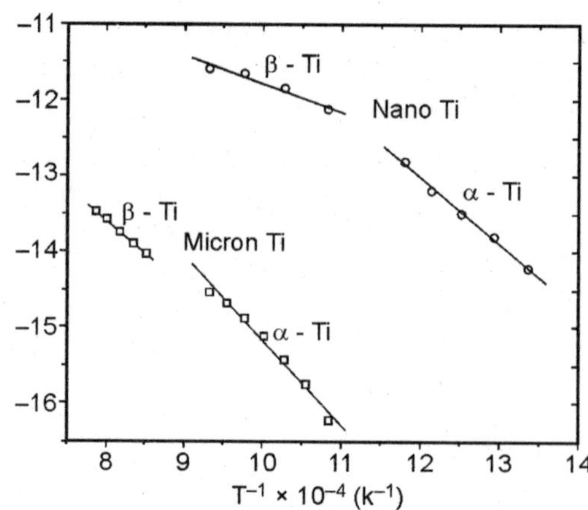

Fig. 13. Variation of logarithm shrinkage rate (r) with inverse absolute temperature during constant rate of heating for specimens made from micron Ti and nano Ti in the α-Ti and β-Ti range.

CONCLUSIONS

The current perspective presented demonstrates that titanium has been able to make the leap to serial production in past years in a variety of applications. This demonstration however should not hide the fact that the overall scope of application is very small and that given the potential components and automobiles, use ultimately pertains to quite small components and/or niche applications. Further, in only very few applications did the use of titanium result from weighing the technical and economic aspects. In order to be established even in small series production, significant cost-cutting is necessary for semi-finished titanium products. New processes for cheaper titanium and titanium powder production are therefore urgently needed. Extensive work is required before these visions can be implemented.

Dilatometry studies revealed a considerably lower sintering onset temperature for nano Ti (415°C) as compared to micron Ti (606°C), and considerably larger densification in nano Ti as compared to micron Ti (for nano Ti it was 19% while for micron Ti it was 6% at the end of isothermal holding at 1250°C). In nano Ti densification mostly took place during CRH while in micron Ti it mostly took place during isothermal holding.

The shrinkage rates (r) were also higher by an order in nano Ti as compared to micron Ti. Nano Ti exhibited lower levels of activation energies (obtained from the plots of In r vs 1/T) of 71 ± 12 kJ/mol and 29 ± 2 kJ/mol (in the α-Ti and β-Ti range) while micron Ti exhibited activation energies levels of 106 ± 26 kJ/mol and 89 ± 13 kJ/mol (in the α-Ti and β-Ti range).

In automobile industry, nanotechnology applications are manifold. Nanotechnology, especially nanocomposites have a striking impact on car components. Visions of "nano in cars" reach from contributions for CO_2 - free engines, safe driving, quiet cars, self- cleaning body and windscreens, up to a mood-depending choice of colour and a self- forming car body. All this will meet society trends and customer demands for improved ecology, safety and comfort.

REFERENCES

1. O. Schauerte, Titanium in Automotive Production, Titanium and Titanium Alloys, Ed. Christoph Leyens and Manfred Peters, WILEY-VCH Veriag GmbH & Co. KGaA, Weinheim, (2003), p. 467.
2. H. Friedrich, J. Kiese, H.G. Haidenwanger, A. Stich, Titanium in Automotive Applications-Nightmare, Vision or Reality, Ti-2003 Science and Technology, Vol. V, Proceeding of the 1 0'h world conference on Titanium, Germany, Ed. G. Lutjering and J. Albrecht, WILEY-VCH Veriag GmbH & Co. KGaA, Weinbeim, (2004), p. 3393.
3. A.D. Hartman, S.J. Gerdemann, and J.S. Hansen, Production of Lower-Cost Titanium for Automotive Applications, JOM, September (1996), p. 16.
4. S. Jackson, How to Process Cheap Titanium, Advanced Materials and Processes, Vol. 161, No. 11, (2003), p. 51.
5. S. Amarchand, T.R. Rama Mohan, P. Ramakrishnan, A novel chemical solution technique for the preparation of nano sized titanium powders from titanium dioxide. Advanced Powder Technology, Vol. 11, No. 4, (2000), p. 415.
6. M.M. Keller, P.E. Jones, W.J. Porter III, and D.Eylon, The Development of Low- Cost TiAl Automotive Valves, JOM, (1997), p. 42.
7. F.H. Froes and R.M. German, Cost Reductions Prime Ti PIM for growth, Metal Powder Report, (2000), p. 12.
8. H. Presting and Ulf Konig, Future nanotechnology developments for automotive applications, Materials Science and Engineering C, 23, (2003), p. 737.
9. K.H. Kramer, Titanium applications-a critical review, Proceedings of The Sixth World Conference on titanium 1988, Part 1, Edited by P. Lacombe, B. Trieot, G. Beronger, (1988), p. 5231.
10. P.C. Eloff, Sintering of titanium, Metals Handbook (Ninth Edition), Vol. 7, Powder Metallurgy, ASM Handbook Committee, Coordinator Erhard Klar, (1984), p. 393.
11. V.S. Arunachalam, Powder Metallurgy of Titanium, Titanium and Alloys,Vol. 3, The Metallurgical Society of AIME Proceedings, Edited by J.C.Williams and A.F. Below, (1976), p. 2305.
12. F.H. Froes,C. Suryanarayana,G.H.Chen,A.Frefer and G.R. Hyde, Nanostructure processing for titanium-based materials, Journal of Metals, May (1992), p. 26.
13. B.S. Murty and S. Ranganathan, Novel materials synthesis by mechanical alloying/ Milling, International Materials Reviews, Vol. 43, No. 3, (1998), p. 101.
14. C. Suryanarayana and C.C. Koeh, Nanostructured Materials, Non-equilibrium processing of materials, EdO C. Suryanarayana, Pergamon Materials Series, Amsterdam, (1999), p. 313.
15. J.R. Groza, Powder Consolidation, Non-equilibrium processing of materials, Ed. C. Suryanarayana, Pergamon Materials Series (1999), p. 347.
16 V. V. Dabhade, T. R. Rama Mohan and P. Ramakrishnan, Synthesis of nanosized titanium powder by high energy milling, Applied Surface Science, Vol. 182, Issues 3-4, (2001), p. 390.

7

Vacuum Hot Pressing and Hot Rolling of Aluminium based MMCs through Powder Metallurgy Technique

V.V. Bhanu Prasad
Defence Metallurgical Research Laboratory, Kanchanbagh, Hyderabad- 58

S. Sri Devi
Osmania College of Engineering, Hyderabad

B. Naga Prasada Rao
Osmania College of Engineering, Hyderabad

S. M. Hussain
Defence Metallurgical Research Laboratory, Kanchanbagh, Hyderabad- 58

K. Soma Raju
Defence Metallurgical Research Laboratory, Kanchanbagh, Hyderabad- 58

ABSTRACT

Aluminium base metal matrix composites (MMCs) a class of light weight high performance aluminium system reinforced with stiffer and stronger ceramics in the form of fibers, whisker or particulates. Properties of MMCs can be tailored to meet the structural and functional applications in aerospace, defence, and automotive industry by suitable combinations of matrix, reinforcement and processing route. In the present study MMCs were made through PM route by consolidating the pure Al powder blended with 0-20%SiC_P using a vacuum hot press (VHP). The hot pressing temperature was found to have a definite effect on the density of the MMC billet and the optimum temperature is found to be the function of reinforcement content. The pressed billets are hot rolled at three different temperatures using optimized rolling schedule. Density of stock material, rolling temperature and reduction per pass are the key parameters controlling the integrity of the rolled composite sheets. The mechanical properties such as elastic modulus, tensile strength and hardness are experimentally determined for the hot rolled sheets. These values are compared for the composites having different volume fractions of reinforcements rolled at different temperatures. Microstructures have been studied for the composites and correlated with the processing parameters and mechanical properties.

1. INTRODUCTION

Discontinuously Reinforced Aluminium (DRA) matrix composites is the family of aluminium matrix composites reinforced with particulates or whiskers or short fibres of different ceramic materials, so as to realise the attractive properties such as high specific strength and stiffness [1]. Unlike continuous fibre reinforced composites they have isotropic properties, amenability to conventional metal working processes, along with the availability of comparatively low cost and high volume production methods [2]. DRA composite billets processed by primary processing routes such as liquid metallurgy route or powder metallurgy route, invariably consist of defects and non uniform microstructure, which leads to poor ductility and other mechanical properties [3]. Secondary processes such as extrusion, forging and rolling not only results in improvement of the mechanical properties due to healing of the defects, improved distribution of the reinforcement particles and refinement in the matrix microstructure, but also provides the final required shape of the component. Extrusion is suitable for shaping the axi-symetric cross sections while the forging is used to achieve non-uniforrn reductions in different sections of a component [4]. Rolling is also equally important as it is capable of producing products like sheets with large surface area and low thickness. BRITE/EURAM. programme studied the forging of Al base MMCs for automotive components such as brake drum/disc (A359 Al/20 SiC_p), suspension arm (6061 Al/10 Al_2O_{3p}) and bearing cups (6061A1/20 SiCp). Evangelista [51 suggested use of lower forging speeds to improve the quality of the MMC forging. Extrusion limit diagram by Davis [6] proposes shifting of safe operating window to lower extrusion speeds due to stick-slip tearing. Harrigan et al. [7] studied the hot rolling of 6061 Al matrix composites and reported improvement in properties with increasing reductions during rolling and attributed the same to the improved metal to metal bonds. Similar observation was made in 2124 Al matrix composites [8]. Lee and Subramanian [4] studied the effect of rolling the 6061 Al/10SiC extruded composites along a direction perpendicular to the extrusion direction of the as-received sample. Such rolling resulted in spreading of the fibrils along the rolling direction, depending upon the amount of reduction and after about 70% reduction the fibrils disappeared totally. Stone et al. [9] reported that hot rolling of HIP'ed and forged Al- Cu/20SiC composites, having different RPS ratios, at 475°C, resulted in edge cracking. The intensity of edge cracking was found increased with increasing RPS ratio. Such cracking could be avoided by rolling the MMC disc after providing a soft Al ring around the MMC disc [10]. Generally, secondary processing of MMCs demands isothermal conditions and the same can be provided with much ease in extrusion and forging. While the difficulty involved in creating isothermal conditions during rolling makes the rolling process challenging.

Most of the studies reported in published literature are on characterisation of rolled products. Where as none of these papers talk about the suitable rolling conditions for one to adopt. While optimal forming temperatures and deformation rates for composites differ from those of conventional alloys, they need to be optimised for each composite system. The present investigation is aimed at developing a rolling process for Al/SiC_p (0-20%) composite VHP stock material by understanding the effect of hot pressing and rolling parameters on their integrity, microstructure and mechanical properties.

2. EXPERIMENTAL

In the present study commercially pure Al powder (–100 mesh) procured from M/s MEPCO, Madurai, India and abrasive grade Silicon Carbide particulates (–600 mesh) procured from M/s Carborundum Universal, Ernakulam, India were used as matrix and reinforcement materials, respectively. Shape and morphology of matrix and reinforcement powder were studied using scanning electron microscope (Jeol, Japan). Particle size of both the powders was analysed by particle size analyzer. These raw materials were used for synthesizing Al/SiC_p composites having different fractions of SiC_p (0-5-10–15-20%). The processing steps involve de-agglomeration of Sicp powders, blending them with matrix Aluminium powder, vacuum hot pressing of the blended powder into a dense billet and followed by hot rolling using conventional equipment.

SiC_p powder can be de-agglomerated by either treating with an organic polar solvent like n-butyl alcohol (n-BuOH) or heating the powders to 120°C. In the present study SiC_p powder was heated to 120°C for 2 hours prior to mixing with matrix powder [11]. The powders were blended in a cylindrical plastic container on a roller mill. These powders were intermittently sieved during blending to ensure homogenous distribution and also to avoid agglomeration of SiC_p particulate powders in the final composite mix. The blended composite powders were loaded in the steel dies of vacuum hot press having a compact size of 125 mm diameter and about 20-23 mm thickness by placing alloy steel separators with graphite foil on both sides. The blended powders were placed in a die and slowly heated to and was held at 450°C for 1-2 hours to ensure complete de-gassing of the blend powder. Then on reaching 500°C temperature, the pressure to the extent of 100 M Pa was applied and kept constant for 1 hour to obtain a fully dense compacts in the vacuum hot press. Using a split-die would make easy stripping of the compacts.

VHP compacts (125-mm diameter and 23-mm thick) were hot rolled in a 12′ two high laboratory rolling mill "DEMAG AKTIENESELLSCHAFT" with a roll speed of 20-30 m/min. The rollers are flat and capable of rolling the hot feed stock. The VHP billets were heated to different temperatures (400, 500 and 600°C) in a resistance-heating furnace for 2 hrs. The bite (reduction per pass) is deliberately put smaller at 4-8%. After 2-3 passes, the partly rolled sheets were re-soaked for 10 minutes at rolling temperatures and rolled further down to 4-5 mm thickness.

Densities of the as VHP composite samples were measured by Serotorious electronic balance with density measurement kit. The method works on Archimedes' principle based on water displacement method. The hardness of the samples was measured using a Vickers Hardness Testing machine (FIE make model: VM50) at a load of 10 kg. Elastic modulus was measured using Elasto-Sonic instrument (ESP-1000, NEPL, Bangalore) working on pulse-echo technique. Tensile testing of the rolled sheet of the composite and base material was carried out on a standard flat tensile specimen as per ASTM E8M having 25 mm gauge length, 6 mm width and 4 mm thickness. Tensile tests were carried out on INSTRON Testing machine (Model 5 500) at a strain rate of 6.67×10^{-4} S^{-1} to determine 0.2% Offset yield strength (YS), ultimate tensile strength (UTS) and % elongation.

3. RESULTS AND DISCUSSION

Microstructural Features

Fig. 1.a and b shows the SEM photomicrographs of Al and SiC powders, respectively. Fig. 1e and d. shows the particle size analysis of both the Al and SiC_p powders. It can be observed that the Al powders are nodular and irregular in shape. This irregularity is a major contributing factor for integrity and strength of the compacts due to mechanical interlocking in between the particles. Surface topography of Al powders was found to be dendritic. The shape of the SiC_p particles appears regular and crystalline and the surface morphology of these materials consists of flat surfaces and sharp edges. The overall geometry of the SiC_p particles can be described as angular. SiC_p particles have Gaussian size distribution with an average particle size of 14.5 μm with some fine particles. Similarly Al powder also has Gaussian distribution, but the average particle size is 50 μm. The relative particle size (RPS) ratio between the matrix particles and reinforcement particles is approximately 3:1. As per the published literature such RPS ratios resulted in uniform distribution of reinforcements [12].

(*a*)

(*b*)

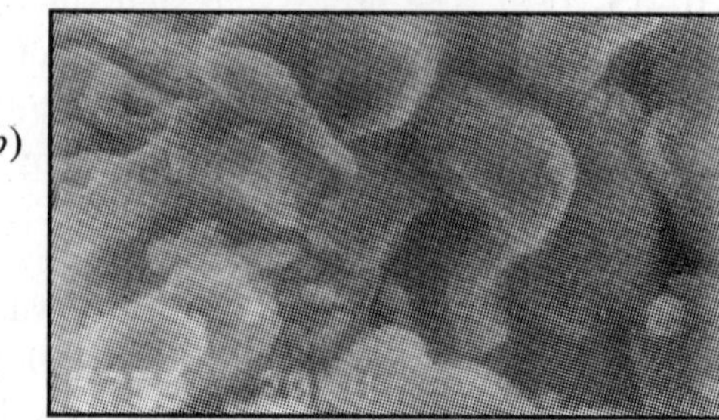

Fig. 1. (a) and (b) SEM micrographs.

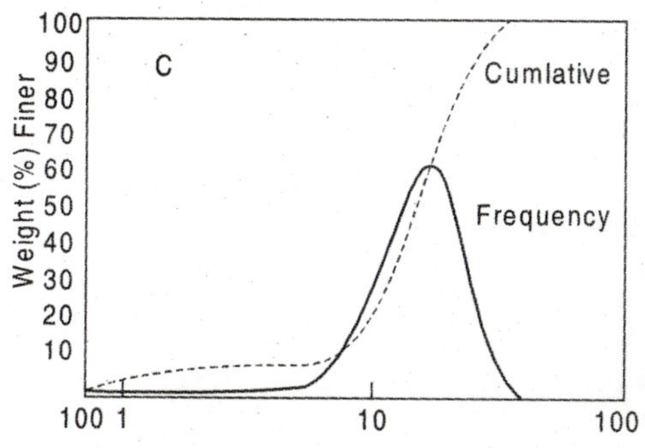

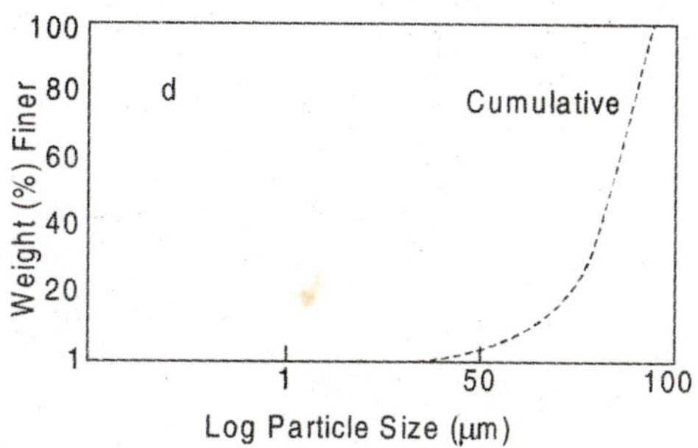

Fig. 1. (c) and (d) particle size distribution curves of Al and SiC_p.

Table 1. Density and Normalised Densities of Al/SiC_p Compacts VHP at 500°C

SiC_p Content in volume per cent	*Density g/cc*	*Theoretical density g/cc*	*% Theoretical density*
0	2.69	2.700	99.78
5	2.72	2.725	99.95
10	2.75	2.750	100.00
15	2.64	2.775	95.11
15*	2.77	2.775	99.87
20	2.57	2.800	91.78
20*	2.79	2.800	99.77

* VHP at 600°C

Primary Processing by Vacuum Hot Pressing

All the powders were vacuum hot pressed at 500°C and the densities of the compacts are given in Table 1. The actual densities are normalized with the respective theoretical densities and reported as % TD for easy comparison. Fig. 2a shows the typical shape of the VHP composite compacts which are used as feed stock for rolling into sheets. Pure Al, Al/5SiC and Al/10 SiC_P samples could be compacted to full density (>99.78%). Where as Al/15 SiC_P and Al/20 SiC_P could be compacted to 95% TD and 92% TD, respectively. These two samples resulted in opening up at the centre of the compact as shown in Fig. 2b during hot rolling. This type of cracking looks like the crocodile mouth opening and is called alligatoring cracking. It is attributed that those VHP compacts having lower density (~92-95% of TD) exhibit this type of cracking due to internal defects or density variation or inhomogeneous blending. Alligatoring is avoided by rolling the fully dense Al/15 SIC_P and Al/20 SiC_P compacts, which were obtained by hot pressing a new batch of powders at higher temperature *i.e.* 600°C.

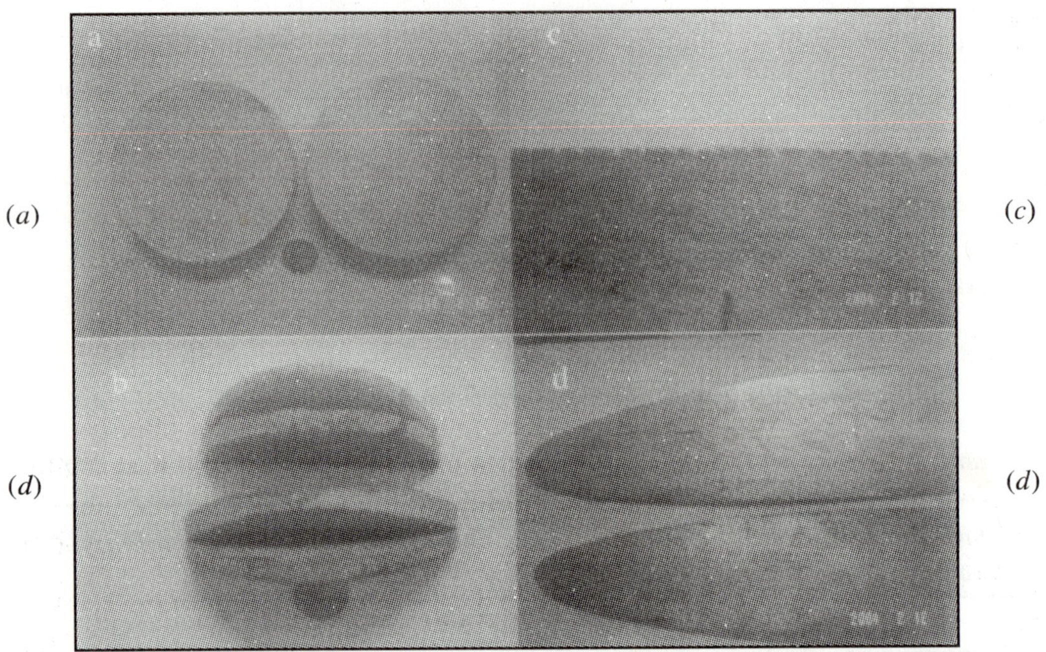

Fig. 2. Photographs of (a) VHP compacts (b) Alligatoring (c) Edge cracking and (d) Integral rolled sheets.

Secondary Processing by Hot Rolling

The VHP compacts were hot rolled at different temperatures (400, 500 and 600°C). It was reported that necklace type microstructure is present in as VHP compacts (12). The necklace type microstructure present in the as VHP compacts is due to the geometric size difference between the matrix particle and reinforcement particle. The small SiC_p particles packed around the larger AI particles during primary processing make this type of microstructure. This results in poor mechanical property [13] and they could be improved by the secondary processing. Secondary processing of VHP compacts not only refine the microstructure but also imparts the necessary shape to the composite material.

The hot rolled sheets of Al/SiC_P composites shown in Fig. 2c were found to be integral and sound. However, there are some edge cracks in the samples rolled at 400°C and more prominently in the AI/20SiCP composite as shown in Fig. 2d. The edge cracks are due to the lower temperatures prevailing and the secondary tensile stresses acting at the edges. The chilling effect by rolls and atmosphere could be minimised by higher rolling temperatures, where relatively softer matrix aluminium flows around the hard and non-deformable SiC particles. Due to this reason, the deformable material could overcome the secondary tensile stresses at the edges and thereby avoiding the edge cracking in the Al/SiC_p sample rolled at 600°C. It is also known that due to differential deformation characteristics of matrix and reinforcement in a composite, these materials are deformed at much lower strain rates than those used for the corresponding base matrix materials [2,4,8]. Because of this reason, 1 mm bite (reduction per pass 4%) was used for the composites, even though wrought AI could be rolled at 30-40% reduction per pass. It is to conclude that Al/SiC_P composites are to be subjected to deformation at higher temperatures and lower strain rates than those used for monolithic wrought matrix materials. Figs.3 a-d show the SEM microstructure of the Al/5-20 volume % of SiC_P composites in rolled condition.

The region with bright contrast corresponds to the matrix Al phase and the dark phase to SiC_P reinforcement. A uniform distribution of the second phase particles's revealed in the matrix phase. It can be attributed that the optimized blending technique adopted for making the composite powder mix, VHP temperature and pressure and refinement of necklace microstructure using hot rolling.

Fig. 3. Microstructure of Al/SiC rolled sheets (a) 5%SiC (b) 10% SiC (c)15% SiC and (d) 20% SiC

Table 2. Hardness Values (VHN) for Al and Al/SiC_P Composites in as VHP and Rolled Condition

S. No.	*Composition*	*As VHP*	*Condition* — *As VHP + Rolled* *Rolling Temperature (°C)*		
			400	*500*	*600*
1	Al	31.6	42.1	37.4	36.8
2	Al/5%SiC_P	37.9	45.6	42.7	39.0
3	Al/10%SiC_P	39.8	50.4	46.4	45.0
4	Al/15%SiC_P	43.5	53.9	48.5	47.3
5	Al/20%SiC_P	46.4	56.2	52.4	50.5

Mechanical Properties

Table 2 shows the hardness values obtained for Al, Al/5SiC$_P$ Al/10SiC$_P$, Al/15SiC$_P$ and Al/20 SiC$_p$ composites in as VHP condition and in rolled (at 400°C, 500°C, 600°C) condition. The hardness values of composites increase with the increasing volume per cent of SiC$_p$ content is evident in the VHP composites. It is due to higher hardness of the SiC$_P$ (2500 VHN) dispersed in the Al matrix material. The decreased inter particle spacing also offers more resistance to the dislocation movement and thereby strengthening the material. However, the improved hardness is much lower than that expected by the rule of mixtures (150-200 VUN). Similarly, the secondary deformation by rolling also improved the hardness of the VHP composites due to improved distribution of the reinforcements. As the deformation temperature is increased from 400°C to 600°C the hardness has come dfown probably due to domination of recrystallisation and grain coarsening mechanism of Al matrix over the work hardening of the material.

Table 3. Tensile Properties of the Hot Rolled AlSiC$_P$ Sheet Material Tested in Longitudinal Direction

Composition	*Rolling temperature (°C)*								
	Yield strength (M Pa)			*Ultimate Tensile Strength (M Pa)*			*% Elongation*		
	400 C	*500 C*	*600 C*	*400 C*	*500 C*	*600 C*	*400 C*	*500 C*	*600 C*
Pure Al	118.5	110.5	104.5	138	131	125	23	23.5	27.5
Al/5% SiC$_P$	125	123.5	106.6	145	142.5	130	19.5	23	24
Al/10%SiC$_P$	131	129.5	125	158.5	150.5	147.5	18.5	19	19.2
Al/15%SiC$_P$	142	133	127	160.5	155	151.5	14	15.7	17.2
Al/20%SiC$_P$	143.5	138	128.5	170	166.5	165	9.5	12.5	14.7

All the VHP and rolled samples are tensile tested to determine the tensile strength, yield strength and % elongation. Table 3 presents the tensile properties of the samples tested in longitudinal direction (rolling direction). Fig. 4 shows the typical stress strain curves obtained for Al/10 SiC$_p$ rolled at 500°C. Each stress strain curve is characterised by initial rapid increase in the stress till the material reaches its elastic limit and then gets strained due to the plastic deformation. Little later the material reaches its peak stress followed by necking in the sample. Once the neck forms, the stress comes down due to the reduced cross sectional area of the sample. At the end the sample breaks and stress comes to zero. Except for change of UTS, YS and % Elongation values the basic nature of the stress strain curve is similar for all the composites. Both the Yield Strength and UTS showed an increase with addition of SiC$_P$ particles (Table 3). At the same time, ductility of the material decreases with the increasing volume fraction of reinforcement. This is attributed mainly due to following reasons: the load is transferred from the matrix to the reinforcement through the interface and as such the composites becomes stronger by the simple rule of mixtures. SiC$_P$ being a very hard material has a better load carrying capacity than the pure Al and this result in increased strength in composites. It was found that with increase in the volume fraction of the reinforcement, the strength value showed an increase. The YS,

UTS of the composite samples rolled at higher temperatures (Δ UTS = 40, 32% improvement) showed higher degree of dependence on SiC volume fraction compared to those of the composites rolled at

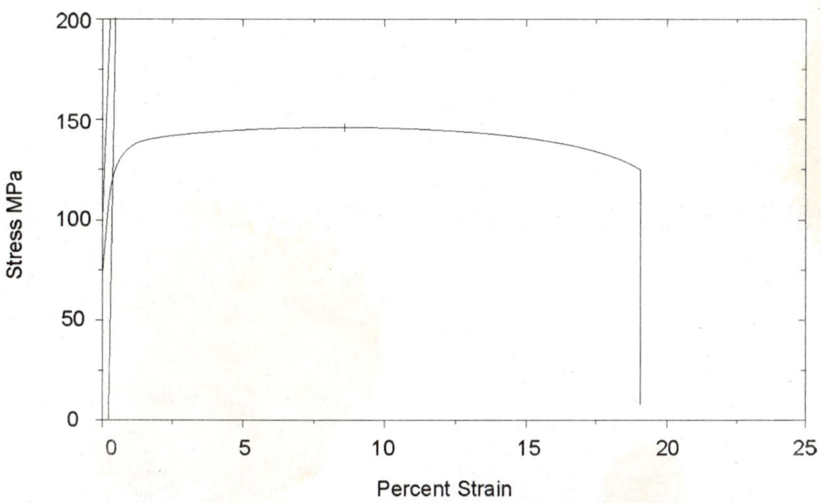

Fig. 4. Typical stress strain curve of Al/10 SiC composite rolled at 500 C.

lower temperatures (Δ UTS = 32, 23% improvement). This indicates better rolling efficiency at higher temperatures. Generally, the fracture during tensile testing occurs due to nucleation of micro-voids at the hard inclusions and void growth and finally by micro void coalescence. In case of Al/SiC_P composites the SiC_P particles act as nucleation sites for the voids and with the increased volume fraction of SiC_P the void concentration goes up. The ductility of the material depends upon the mean free path between the voids and the same depends on inter particle spacing in case of composites. The decreased % elongation in case of composites with higher SiC content can be attributed to the decreasing inter particle spacing in the composites with the increasing SiC content. Rolling temperature has a definite role to play in successful rolling of MMC sheets. Where as there is a 5-10% strength drop by increasing the rolling temperature from 400 to 600°C.There is a general improvement in the % elongation values as a result of increased temperature from 400°C to 600°C however the variations are inconsistent. Where as in case of composites containing 20% SiC_P the improvement in % Elongation is significant. Fig. 5 shows the fractographs of the fractured Al, Al/5SiC_P, Al/15SiC_P and Al/20SiC_P tensile tested samples. Pure Al sample showed well-developed and deep dimples indicating the higher elongation. Where as the fractograph of Al/SiC_P shows the shallow dimples and mostly around SiC_P particles. Fractures of SiC_P particles indicate the highly integral interface between the matrix and the reinforcement. Fig. 6 shows the effect of rolling on fracture surface of Al and Al/15 SiC_P composites. These fractographs observed at 1500X revealed that, in VHP condition the Al fracture surface has shallow dimples and in rolled condition the same as well developed deep dimples. In case of Al/15SiC_P composite the fracture mode changed from interface cracking in case of as VHP composite (Fig. 6c) to Particle cracking in case of rolled composites (Fig. 6d), indicating the development of strong interfacial bonding between Al and SiC_P during rolling.

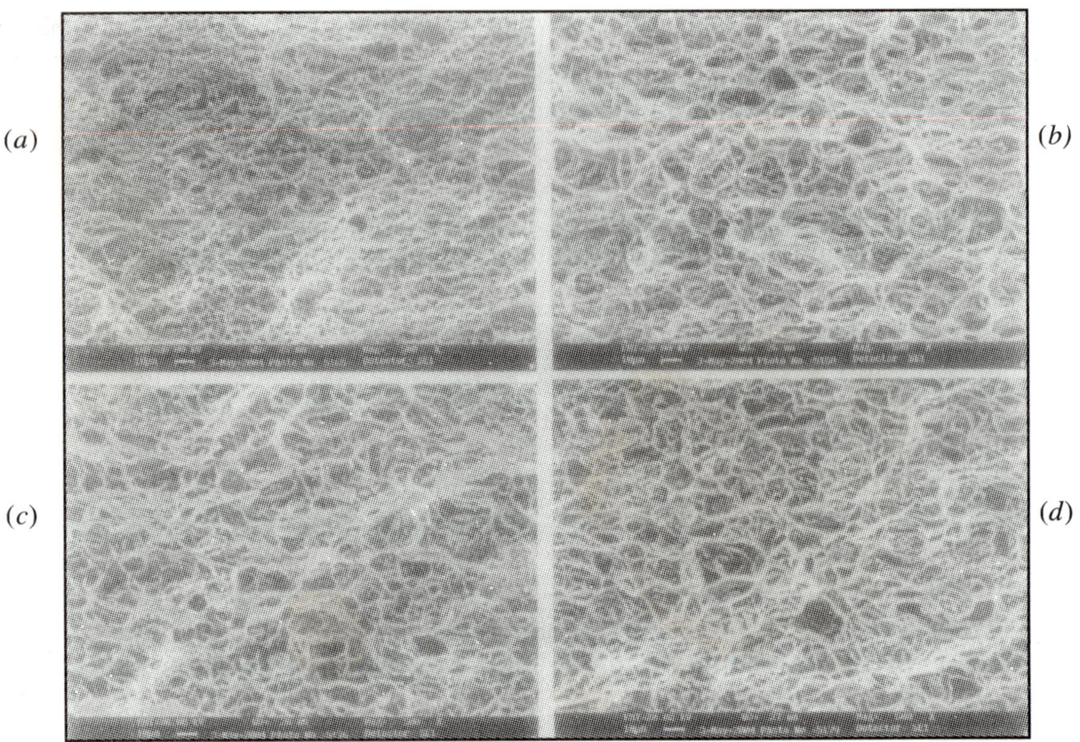

Fig. 5. Fractographs of Al/SiC rolled composites (a) 0% SiC (b) 5% SiC and (c) 15% SiC and (d) 20% SiC.

The elastic modulus values obtained by elastosonic testing were reported in Table 4. It is observed that with increasing SiC_p content E value also increases and the same is due to higher modulus of the SiC_P (480GPa) and effective load transfer from the matrix to reinforcement due to perfect interface. The E values so obtained are plotted in Fig.7 along with those values calculated using various models listed in Table 4. It can be observed from the plot that the iso-strain model gives higher bound modulus values, where as iso-stress model gives lower bound model. The experimental values are matching with those values obtained by Tsai-Halpin model [14] where the particle aspect ratio was taken into consideration.

Table 4. Comparison of Measured Elastic Modulus Values with those Calculated Using Various Models

SiC Content	*Measured E values (GPa)*	$E_{isostrain}$ *(GPa)*	$E_{isostress}$ *(GPa)*	$E_{Tsai-Halpin}$ *(GPa)*
0%	72.168	70.0	70.00	70.00
5%	79.800	90.5	73.12	78.48
10%	86.098	111.0	76.53	87.49
15%	94.140	131.5	80.28	97.10
20%	105.190	152.0	84.42	106.91

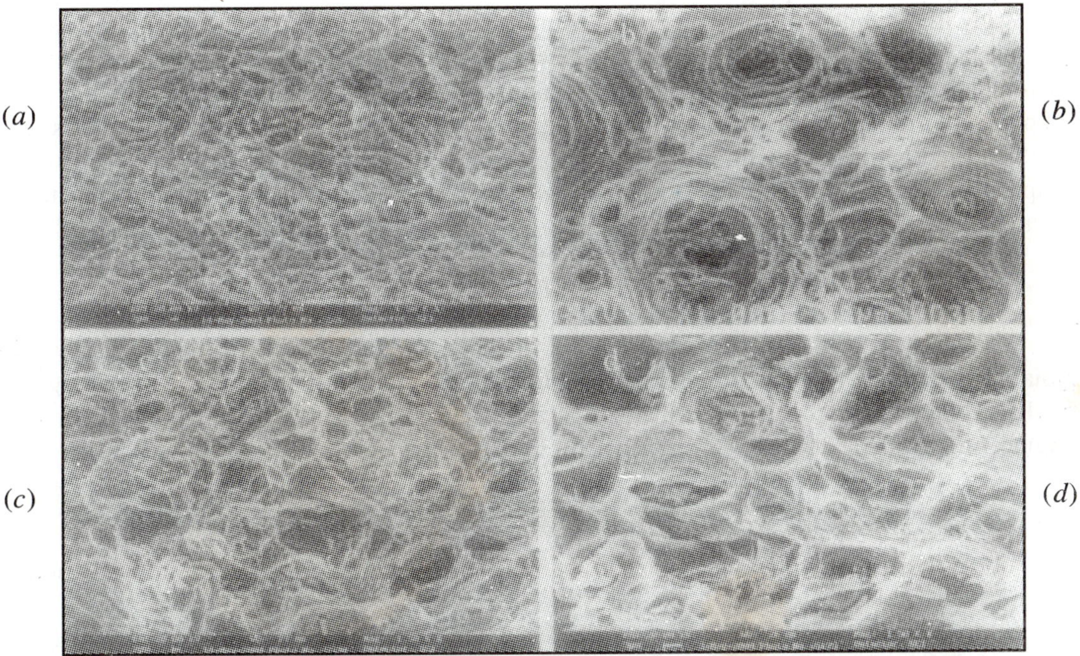

Fig. 6. Fractographs of (a) as VHP Al (b) Rolled Al (c) as VHP Al/15 SiC and (d) rolled Al/ SiC.

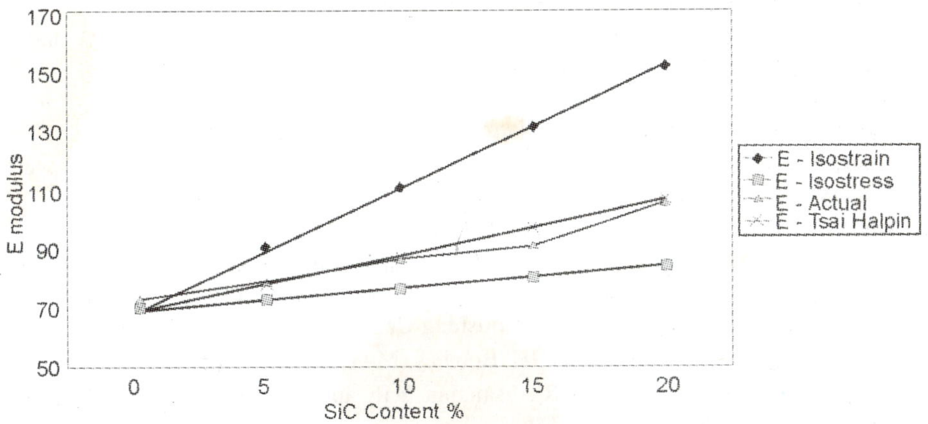

Fig. 7. Effect of SiC content on Elastic modulus of Al/SiC Composites.

CONCLUSIONS

PM route has been successfully adopted to Vacuum hot press Al/SiC$_p$ (0-20%) composites. The VHP compacts have been hot rolled successfully at 400, 500 and 600°C. Dry blending of Aluminium metal and SiC$_p$ resulted in homogeneous composite blend to give uniform microstructure in the final composites compacts or rolled sheets. Optimised VHP temperature and pressure resulted in attaining full density for the composite compacts, which are found to be more suitable for the secondary processing like rolling. Alligatoring crack during rolling in Al/ 15 – 20 SiC composites is attributed to lower density (92 – 95% TD) and the same was eliminated by fully densifying the VHP billets by pressing at high temperature. The Elastic modulus increased with increasing SiC_P content as per Tsai-Halpin model. Whereas the effect of rolling temperature is marginal. The hardness, yield strength and U.T.S showed an increase with increasing SiC_P volume fraction and decrease with increasing rolling temperature. Whereas increased SiC_P content has significant effect on ductility while rolling temperature has marginal effect.

ACKNOWLEDGEMENTS

The authors thankfully acknowledge the financial support received from the DRDO, Govt. of India in order to carry out the present research study. The authors gratefully acknowledge Director, DMRL, Hyderabad and Head, Mechanical Engineering Department, Osmania University College of Engineering, Hyderabad for giving permission to present this paper.

REFERENCES

1. Schoutens J.E., Introduction to Metal Matrix Composite Materials, MMCIAC Tutorial Series, No.272 (1982).
2. Rack H.J., Metal Matrix Composites: Processing and Interfaces, Ed. Everett R.K., and Arsenault, R.J Academic Press, San Diego, p. 83 (1991).
3. Lloyd D.J., Intl. Mater. Rev., 39, p. 1 (1994).
4. Lee J.C. and Subramanian K.N., J. Mater. Sci., 29, p. 4901 (1994).
5. Evangelista E., Forcellese A., De Sanctis A. and Wang, Y. ICCM-10, Canada, Poursati P.A., and Street K., Woodhead pub. Ltd., England, p. 19 (1995).
6. Davis, C.H.J.: Key Engg. Materials, Ed. Newaz, G.M., Neber-Aeschbacher, H. and Wohlbier. F.H., Trans. Tech. Publications, Switzerland, Vol. 104-107, p. 447-458 (1995).
7. Harrigan, W.C., Gaebler, Davis, E., and Levin, E.J.: Mechanical Behaviour of MMCs Ed. Hack, J.E. and Amateau, M.E., TMS/AIME, Warrendale, PA., USA, 169 (1982).
8. Bhanu Prasad V.V., Bhat B.V.R., Mahajan Y.R and Ramakrishnan P., Materials and Manufacturing Processes, 16(6), p. 841 (2001).
9. Stone I.C. and Tsakirroponlos P., Mat. Sci. & Tech., 11, 1995, 222.
10. Mahajan, Y.R., Bhanuprasad, V.V., Bhat, B.V.R. and Pandey, A.B.: Proceedings of Indo-US Workshop on Principles of Solidification and Materials Processing (SOLPROS) Ed. Trivedi, R., Shekar, J.A. and Majumdar, J., Oxford and IBH Publishing Co. Pvt. Ltd., New Delhi, INDIA, 613 (1988).
11. Bhanu Prasad V.V. : Ph.D. Thesis, IIT Bombay, Mumbai, India (1998).
12. Bhanu Prasad V. V., Bhat B.V.R., Mahajan Y.R. and Ramakrishnan P.: Materials Science and Engineering 'A', 337 (2002) 179-186.
13. Soma Raju, K.: M Tech Thesis, IIT, and BHU, INDIA (1994).
14. Halpin, J.C. and Tsai, S.W.: Environmental Factor Estimation in Composites Materials Design, AFML TR 67-423, 1967.

8

An Accurate Compaction Model for Industrial P/M Processes

S. Shamasundar and K. Narendra Babu
Prosim
326, 3rd Stage 4th Block, Basaveshwarnagar,
Bangalore- 560 079, India

ABSTRACT

The FEM modeling and analysis of the P/M processes has a wide range of advantages in the industries. But for the FEM modeling and analysis to be useful to the design and process development engineers in P/M industries, the models have to be accurate and account for the density prediction, damage/fracture, ejection. The capability to handle the tool damage and wear analysis will be helpful for the engineers. Final objective of the modeling should he a reduction in process / product development life cycle time for complex shaped p/m parts. In this paper, some details of the compaction modeling, damage modeling and tool stress modeling are given. Comparisons are made with the benchmarking experiments conducted (and reported) under European Union DIENET project. The results of modeling by the authors match closely with the experimental work reported by the DIENET team.

INTRODUCTION

From last two decades there have been significant R & D efforts towards developing the models for the compaction and sintering processes that will help the design engineers in the P/M industry to reduce the time in product development cycle. As observed by by Ernst and Maassen [1], the compaction model to be applied for industrial process should account for:

- Interaction of tool and powder
- Effect of ejection
- Predict the damage during the process.

In this paper authors have discussed the modeling methodologies that are available for the modeling the p/m processes and the compaction model that is suitable for the industrial applications.

MODELING METHODOLOGIES

The classification of the model methodologies based on the length scales: [2]

- Continuum models
- Micromechanical models
- Multi-particle models
- Atomistic/molecular dynamic models

Continuum Models

Continuum models consider the entire product dimensions geometrically for the simulation. Hence these models are been most attractive for the industrialists as the entire product can be viewed. In the continuum models the external variables such as stresses, temperature are expressed in terms of the specific material properties. The evolution of these external variables is based on the constitutive laws and the fundamental equations. Material properties can also be considered as functions of external variables and functions of a few internal variables that describe the structure. The continuum model developed will be implemented in an FEM programme to apply it for the required product.

Micromechanical Models

These models provide a means to derive macroscopic model parameters from information on a smaller length scale, particle level. A unit problem is defined and solved. The micro-to-macro transition is performed usually by homogenization techniques. Here a unit problem was used to describe the evolution of the center-to-center distance between two particles; it was then related to the macroscopic density. In many models micro-macro transition is based on geometric relations between density and interparticle distances. Micromechanical methods offer a rigorous approach to describing the behavior of porous materials, but their predictive capability can be limited. The accuracy of the solution of the unit problem and the homogenization technique affect the performance of models derived micro mechanically.

Multi-particle Models

This approach involves the numerical simulation of individual particles that interact, based on a set of prescribed contact conditions are taken into account to formulate the motion equations. This technique is frequently used in soil mechanics because it can provide significant information on particle motion, contact forces, and several macroscopic quantities in an assembly of particles subjected to external loads or displacements. The limitations of this method are, 10,000 particles in a 3D problem can distribute only 20-30 particles in each direction and computation difficulty to keep the interaction simple.

Molecular Dynamics

Molecular dynamic models have been used extensively to simulate micro structural evolution in materials, addressing phenomena such as grain growth, recrystalization, and phase transformations. The advantage of this approach is its capability of simulating the evolution of local microstructure without making the mean field assumptions required in analytical models in order to obtain the solutions. The response is stochastic locally but if the volume addressed is large, the macroscopic response is quite stable. A disadvantage of this approach is the lack of between true time and simulation time steps.

POWDER COMPACTION MODELS

The yielding and flow of powder will be treated based on the yield function developed by Doraivelu. Et al. [3]. This yield function is similar in the form to those derived earlier by Kuhn, Shima and their co-workers [4,5], according to which yielding is a function of both deviatoric and hydrostatic stress.

Doraivelu approach is based on the assumption that yielding and densification begin when the deformation energy $\sigma_{ij}\,\varepsilon_{ij}$ imparted to the porous body reaches a critical tensor and the usual convention of summation over repeated indices is followed. The yield function thus derived is

$$[2(1+\nu)]J_2' + [(1-2\nu)/3]J_1^2 = Y_\rho^2 = \delta Y_0^2 \quad \text{...(1)}$$

in this expression, ν represents the Poisson's ratio of the porous compact determined from the uniaxial upset tests; the dependence of ν on relative density ρ is assumed to be the same for all the stress paths. The term J_2' is the second invariant of the deviatoric stress tensor, $\sigma_{ij}'\sigma_{ij}'/2$, where $\sigma_{ij}' = (\sigma_{ij} - \sigma_{kk}/3)\delta_{ij}$ (δ_{ij} = kronecker delta); J_1 is the first invariant of stress tensor, σ_{kk}. Term Y_ρ is the apparent uniaxial yield stress of the porous body. The δ term represents a function whose value is always less than or equal to unity, which modifies the yield stress of the fully dense material, Y_0, to reflect the lower yield stress of the porous body. At full density $\delta - 1$ and ν = 1/2 (incompressible flow conditions), and the von Mises yield criteria, $3J_2' = Y_0^2$ is obtained from eq. (1), two function of ν are defused as follows:

$$A = 2\,(1 + n) \text{ and } B = (1 - 2\nu)/3 \quad \text{...(2)}$$

to give the final form of the yield function

$$AJ_2' + BJ_1^2 = Y_\rho^2 = \delta Y_0^2 \quad \text{...(3)}$$

It should be noted that the Doraivelu formulation and similar continuum approaches postulate ν (and hence A and B) and δ are unique functions of relative density ρ. But this assumption is the major drawback of such formulations, one which can only be overcome by taking microstructural/material effects into account.

An associated flow rule may be derived from the yield function according to Eq. (3), assuming that $Y_\rho = Y_\rho(J_2', J_1)$ represents a plastic potential whose gradient is proportional to the macroscopic strain increments $d\varepsilon_{ij}$. It will be assumed that the gradient of the plastic potential is proportional to the strain rate $d\varepsilon_{ij}/dt = \dot{\varepsilon}_{ij}$, thus

$$\dot{\varepsilon}_{ij} = \lambda(\partial Y_\rho/\partial\sigma_{ij}) = (\lambda/Y_\rho)\{(A/2)\sigma_{ij}' + B\sigma_{kk}\delta_{ij})\} \quad \text{...(4)}$$

The proportionality constant λ, is found by assuming a function for the constitutive behavior of the fully dense material

$$Y_0 = D_0\dot{\bar{\varepsilon}}_0^m \quad \text{...(5)}$$

where D is a constant (at fixed temperature), $\dot{\bar{\varepsilon}}_0$ denotes the effective strain rate of the matrix (in a porous body, this is a microscopic quantity that cannot be measured), and m is the strain rate sensitivity (= $1/n$, where n is the stress exponent that describes the pertinent micro-mechanism of flow).

The power dissipated by the macroscopic stress through the macroscopic strain rates, $\sigma_{ij}\,\dot{\varepsilon}_{ij}\,V = Y_\rho\dot{\bar{\varepsilon}}V$, where $\dot{\bar{\varepsilon}}$ is the macroscopic effective strain rate of the porous body, and V is the volume of the porous body, is equal to that dissipated by the matrix, $Y_0\,\dot{\bar{\varepsilon}}_0(\rho V)$, viz.

$$\sigma_{ij}\,\dot{\varepsilon}_{ij}\,V = Y_\rho\dot{\bar{\varepsilon}}\,V = Y_0\,\dot{\bar{\varepsilon}}_0(\rho V) \quad \text{...(6a)}$$

or,

$$\sigma_{ij}\,\dot{\varepsilon}_{ij} = Y_\rho\,\dot{\bar{\varepsilon}} = \delta^{1/2} Y_0\,\dot{\bar{\varepsilon}} = \rho Y_0\,\dot{\bar{\varepsilon}}_0 \qquad ..(6b)$$

where the relation between Y_ρ and Y_0 has been used in Eq. (6b). Multiplying both sides of Eq. [4] by σ_{ij} and setting the result equal to $Y_\rho\,\dot{\bar{\varepsilon}}$ leads to the result (after some algebra) $\lambda = \dot{\bar{\varepsilon}}$. and thus the strain rates:

$$\dot{\varepsilon}_{ij} = (\dot{\bar{\varepsilon}}/Y_\rho)\{(A/2)\sigma'_{ij} + B\sigma_{kk}\delta_{ij}\} \qquad ...(7)$$

Using the relation between $\dot{\bar{\varepsilon}}$ and $\dot{\bar{\varepsilon}}_0$ as given in eq. (6b) and the constitutive relation as given in eq. (5), the final expression for the macroscopic strain rates is obtained:

$$\dot{\varepsilon}_{ij} = \{\rho Y_\rho^{\,n-1}/\delta^{(n+1)/2} D^n\}\{(A/2)\sigma'_{ij} + B\sigma_{kk}\delta_{ij}\} \qquad ...(8)$$

FEM SIMULATION OF POWDER COMPACTION AND DISCUSSION

The powder compaction model is implemented in a commercial non-linear FEM analysis package DEFORM™ [6]. The model has been used to simulate the compaction process for different components and the results are discussed.

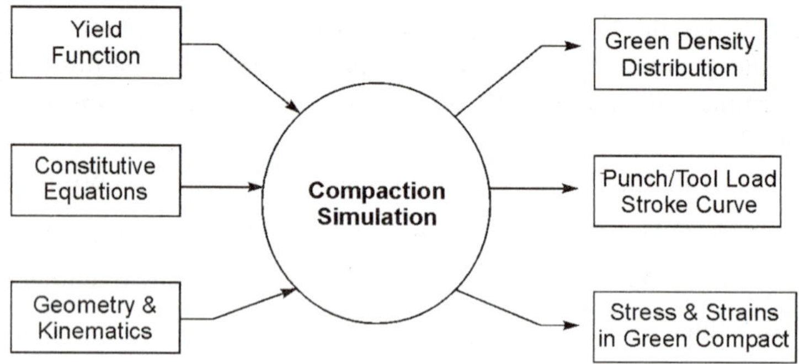

Fig. 1. Systems approach to Compaction Simulation.

Compaction of a Flanged Bush

Compaction of a flanged bush is modeled using the model thus developed. Experimental details of the same are given by Tweed et al [7], in which damage of the component was observed at the corner radius. Figure 2 shows the tool set up and the component geometry used for the experimental work. Figure 3 shows the comparison of FEM predicted load - time curve with the experimental measurements for top punch and lower inner punch. It is observed that the load predictions for both punches are matching quite well with the experimental measurements. Figure 4 shows the comparison of FEM predicted density distributions with the experimental measurements. It has to be noted that the experimental density predictions are following the trends of experimental predictions, although there is an over estimation of about 4-8% density. This mismatch could be the result of getting the proper material constitutive property model, which has not been discussed in the experimental data [7].

Accumulation of high strain in the corner radius region can create potential damage. Figure 4 shows the FEM predicted damage showing the high level of damage accumulation in the corner radius. Figure 5 also shows the damage reported experimentally in [7].

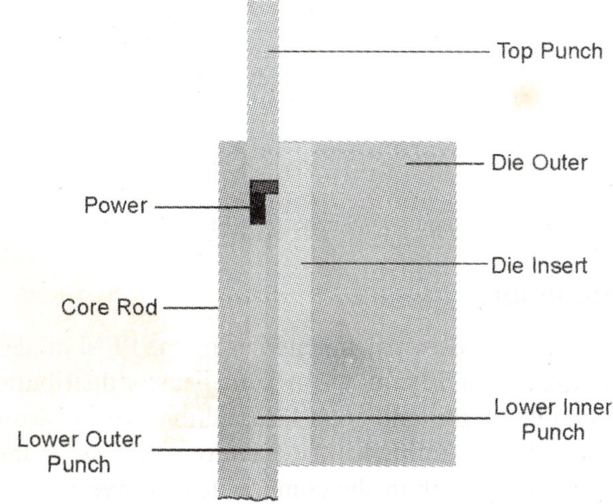

Fig. 2. Tool Setup.

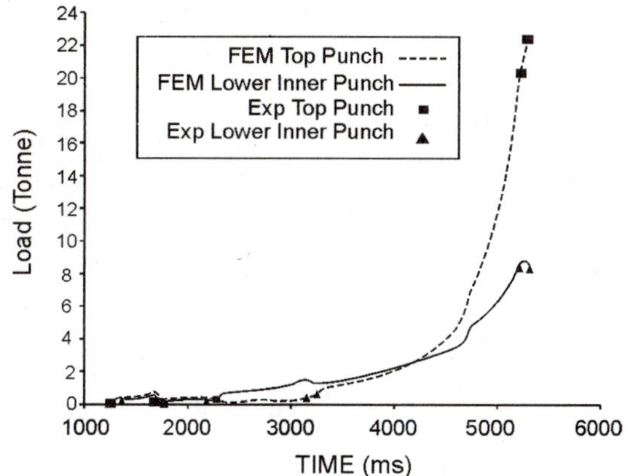

Fig. 3. Comparison of Experimental and FEM Load – Stroke Curves.

Experimental		FEM Simulation	
7.73	7.67	8.264	8.46
7.77	7.57	7.98	7.94
	7.41	8.12	
	7.57	8.15	

Fig. 4. Comparison of experimental and FEM simulated density distribution.

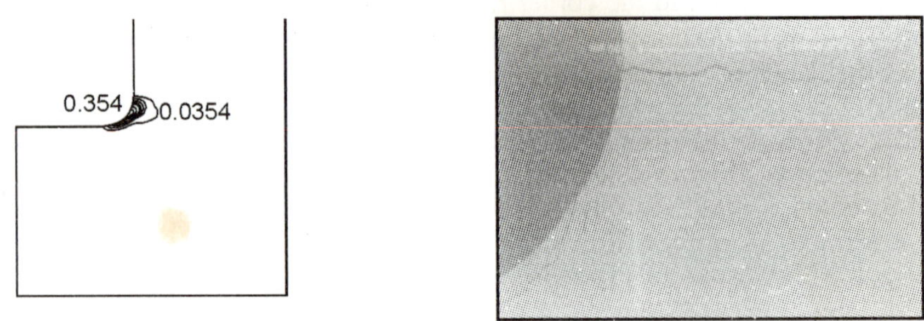

Fig. 5. Damage in flange compaction (a) FEM Simulation, (b) Experimental

Compaction of A Tool Insert

A compaction of WC-Co tool insert is simulated using the FEM model developed. Figure 5 shows the tooling assembly for compaction. Figure 6 shows the density distribution at the end of compaction in a section of the compact. One of the important information that the designer needs is about the tool stresses and deflection. Tool stresses influence the fracture and wear of the component. Estimation of deflection due to the elastic spring back of the component will help in ensuring closer tolerances and accuracy of the component. Figure 7. shows the maximum principal stresses in the tool assembly. Figure 8 shows the mean stress in the tool elements.

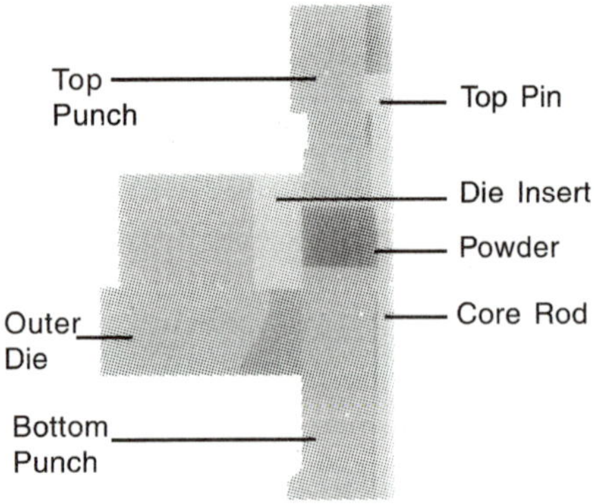

Fig. 6. Tool Setup.

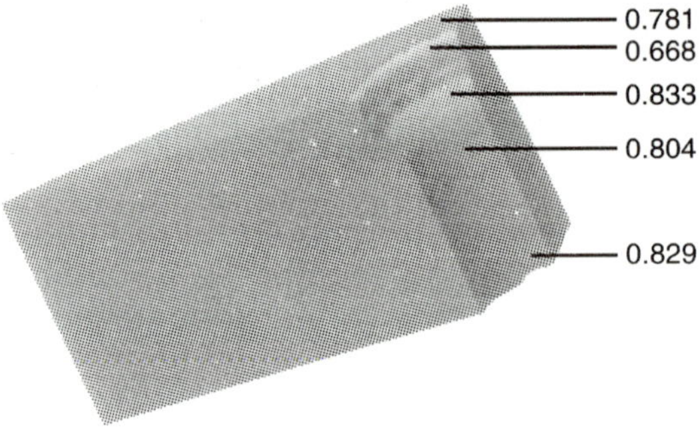

Fig. 7. Relative Density Distribution in compact.

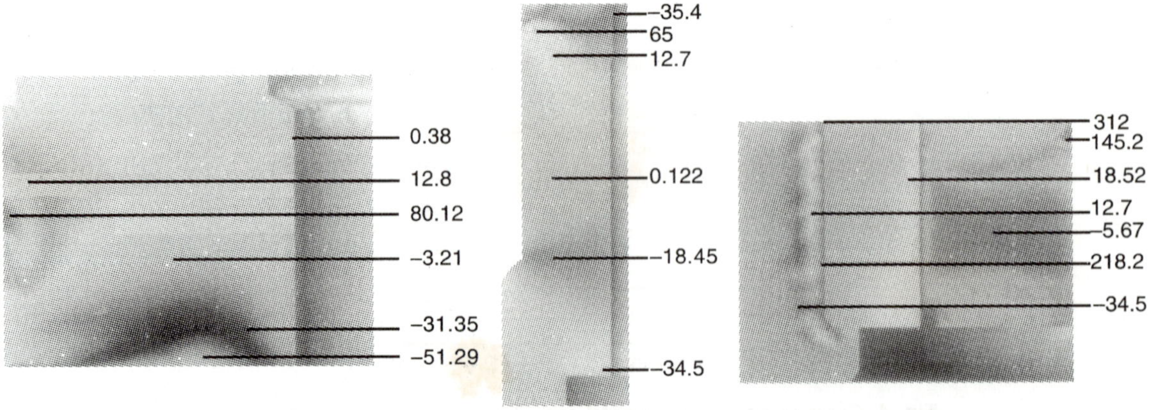

Fig. 8. Maximum Principle Stress in MPa, (a) Top Punch, (b) Bottom Punch, (c) Die.

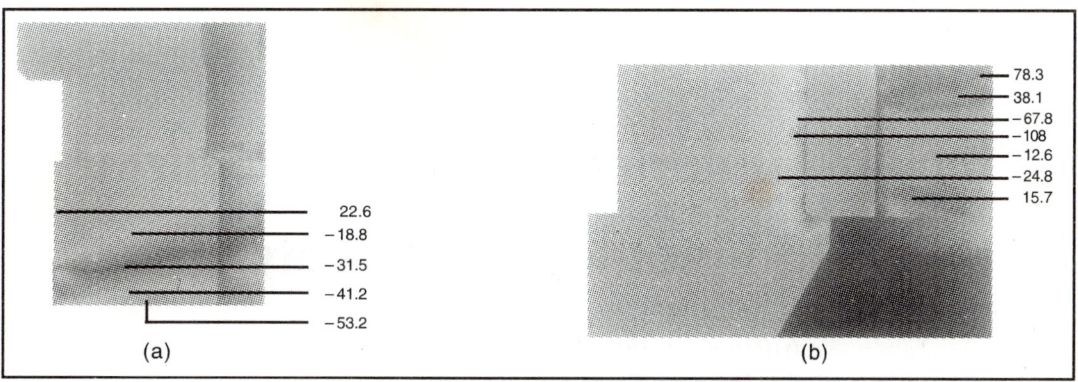

Fig. 9. Mean Stress in MPa, (a) Top Punch, (b) Die.

CONCLUSION

A comprehensive model that is suitable for the industrial p/m compaction process has been developed and implemented for the simulation of the compaction of a flanged bush and the tool insert. The model has been used to benchmark the flanged bush experiments of the European Union DIENET Programme. The model has been used to predict the density distribution and damage in the green compact. The tool stresses are also determined using this model.

REFERENCES

1. E. Ernst and R. Maassen, Practical needs for simulation of powder compaction, Proc. Euro PM 2003, EPMA, Valencia, Spain October 20-22, 2003, Vol. 3, pp 1-5.
2. Antonios, Zavaliangos, Constitutive Models for the Simulation of P/M Process, The Int. J. Pow. Met, Vol. 38, No. 2, 2002.
3. S.M. Doraivelu, H.L. Gegel and J.S. Gunasekera, J.C. Malas, J.T. Morgan and J.F. Thomas, A New Yield Function for Compressible P/M Materials, Int. J. Mech. Sci., 26, 1984, p. 527.
4. Suh, S.K. and Kuhn, H.A., In Modern developments in powder metallurgy, (Hausner, H. H. and Taubenblat, P.W., Eds.), Vol 1977.
5. S. Shima and M. Oyane, Plasticity Theory for Porous Metals, Int. J. Mech. Sci. vol. 18, 1976, pp. 285-291.
6. DEFORM User Manual, Scientific Forming Technology Corporation, Columbus, OH, 2004.
7. J. H. Tweed et al. Production And Characterisation of A Hard Metal Flanged Bushing for Compaction Modelling Study, Vol. 3, Proceedings Euro PM 2003, Valencia, Spain, Oct 20-22, 2003, pp. 57-63.

9

Efficient Sintering Operation Utilizing Non-isothermal Effects

Karthik Krishnan and Satyam S. Sahay
Tata Research Development and Design Centre
A division of Tata Consultancy Services Ltd.
54B, Hadapsar Industrial Estate, Pune-411 013, India.

ABSTRACT

Industrial sintering operations are invariably non-isothermal in nature, whereas sintering kinetics models as well as the experiments to quantify the kinetics parameters are often under isothermal condition. Application of such quasi-isothermal models to non-isothermal condition can result in erroneous model prediction. It has been observed for a number. of phase transformations in diverse material systems that during the isochronal processing, there is a functional relationship between beating rate and activation energy. Consistent with this functional relationship, it has also been observed during sintering operation that with increase in heating rate, the densification kinetics increases. These results can be effectively utilized during sintering under non-isothermal cyclic condition, where the constant temperature soaking segment can be replaced with controlled cyclic perturbations, below the required isothermal temperature. Furthermore, it will be shown that by designing and implementing cyclic treatments, as high as 15% productivity enhancement and 20% energy reduction can be achieved, while maintaining the product quality. Finally, the methodology of implementing this novel approach in the industrial sintering and other thermal processing operations, such as annealing and age hardening, will be outlined.

1. INTRODUCTION

Sintering is a critical operation having direct bearing on the product quality. The high temperature requirement as well as sluggish solid-state diffusion kinetics makes this operation energy intensive and a productivity bottleneck. Therefore, energy consumption and furnace productivity during sintering operation are the major concerns of the powder metallurgy or ceramic processing operations. Significant performance improvement can be achieved by hardware solutions, e.g. better insulation materials, fuel efficient burners, recuperator systems or modern control systems. However, implementations of all these solutions are capital intensive, which hinders their proliferation. Soft solutions such as process modeling and optimization of thermal processing operations are attractive and industrially proven solutions, which ensures the best possible process performance from the existing hardware infrastructure [1-4]. Innovation in thermal processing cycle design is another alternative by which quantum improvements can be achieved.

Design of a typical industrial sintering cycle involves selection of heating rate, sintering temperature, soaking time and cooling rate. During the cycle design, generally heating rates are perceived to reduce the furnace productivity, and therefore highest possible heating rates - at times constrained by the thermal inertia of the components or product quality - are selected. For an industrial operation, there is a tendency of designing the heat treatment cycles with minimum number of segments. Such a design philosophy was more practical in the past, when the furnaces were manually controlled and the increase in the number of segments was effected by manual set-point changes by the furnace operator, thereby increasing the risk of process variability. However, in recent times significant development has taken place in the process control technology, where low cost advanced controllers (*e.g.* multi-segment PID controllers, internet-ready controllers) and SCADA-based online control systems are available. In these modern controllers and control systems, a large number of segments can be easily programmed and the change in furnace set-points can be automatically effected. This paves way for designing and implementing efficient multi-segment process cycles where temperature or other process parameters can be dynamically changed, depending on the process requirement.

The present work proposes to utilize the advances in online control systems as well as recent innovative observations on accelerated transformation kinetics during non-isothermal (*e.g.* isochronal and cyclic) processing [5-7]. In the cyclic thermal processing, when the isothermal soaking segment was replaced with controlled cyclic perturbations, due to non-isothermal effects, the transformation kinetics were observed to get accelerated as compared to the isothermal condition [7,9]. The generality of above result has been ascertained for a number of other phase transformations and material systems. The applicability and implementation methodology of this concept in industrial operations is outlined. Furthermore, quantification of benefits - in terms of energy reduction and productivity enhancement- by designing and implementing cyclic treatments in industrial operations is presented.

2. ISOCHRONAL TRANSFORMATION KINETICS

Linear approximation to non-linear problem is necessary in many engineering disciplines. This approach simplifies the mathematical complexities and enables solutions with acceptable accuracy in most of the cases. However, in this process, one is likely to overlook the special non-linear effects, if any, present in the system. This is also true for thermal processing of materials, where most of the phase transformation models formulated for isothermal condition are used to describe non-isothermal process cycles by additively principle, *i.e.* by approximating them as infinitesimal isotherms. Recently, a number of phase transformation kinetics data in diverse materials systems were examined under isochronal or constant heating rate condition [5- 6]. These phase transformations were analyzed by the classical formulation of JMAK model, given as [6]:

$$X\ (t) = 1 - \exp(-(Kt)^n) \qquad ...(1)$$

$$K = K_0 \exp\left(-\frac{Q_c}{RT}\right) \qquad ...(2)$$

where $X(t)$ is the fraction crystallized after the time 't', and 'n' is the Avrami exponent, 'K_0,' is the pre-exponent coefficient, 'R' is the gas constant, and 'Q_c' is the overall activation energy for phase transformation.

Even though equations (1) and (2) are formulated for transformations under isothermal conditions and the kinetic coefficients are also obtained by carrying out isothermal studies, these equations are frequently utilized for describing phase transformation over non-isothermal temperature profiles by segmenting them into small isotherms and integrating the transformation kinetics over time. It must be noted that the basic assumption in such a quasi-isothermal approach is that there is no change in transformation kinetics between isothermal and non-isothermal conditions. Evidently, this methodology of evaluating non-isothermal profiles will result in erroneous predictions if the transformation kinetics changes during non-isothermal annealing due to heating rate or change in heating rate effects.

For a large number of phase transformations in varied materials systems, e.g. recrystallization in metals, crystallization of amorphous materials, crystallization from polymeric melt it was observed that the above quasi-isothermal approach results in erroneous predictions [5-6]. For example, comparison between quasi-isothermal model prediction by classical JMAK approach (broken lines) and experimental data symbols) presented in Fig. 1 (a) for isochronal crystallization of $Ge_{20}Te_{80}$ glass, shows that there is significant difference between model prediction and experimental data. Similar discrepancy between model prediction and experimental data were observed in other materials systems [5-6]. However, when an addition functional relationship (Eqn.3) between activation energy and heating rate was imposed on the classical JMAK model, as shown in Fig. 1 (a) (solid lines) the model prediction significantly improved.

$$Q_c = Q_0 + Q_1 \ln (A.\alpha) \qquad ...(3)$$

where α is the heating rate in $K.min^{-1}$ and Q_c is the total activation energy having two components, (*a*) Q_0, the heating rate independent component and (*b*) $Q_1 \ln (A.\alpha)$, the heating rate dependent component. The constant 'A' in the above equation ($A = I\ K^{-1}min$) is primarily for the dimensional consistency. The strong logarithmic relationship between activation energy and heating rate is shown for $Ge_{20}Te_{80}$ glass system has been shown in Fig. 1(b). The functional relationship given by Eqn. 3 was found to be unique for all the diverse materials systems investigated by the present authors. The functional relationship between heating rate and activation energy (Eq. 3) appears to be basic in nature. It enables in capturing the heating rate effect during isochronal kinetics by extending the validity of the quasi-isothermal model (equations 1 and 2) under non-isothermal conditions.

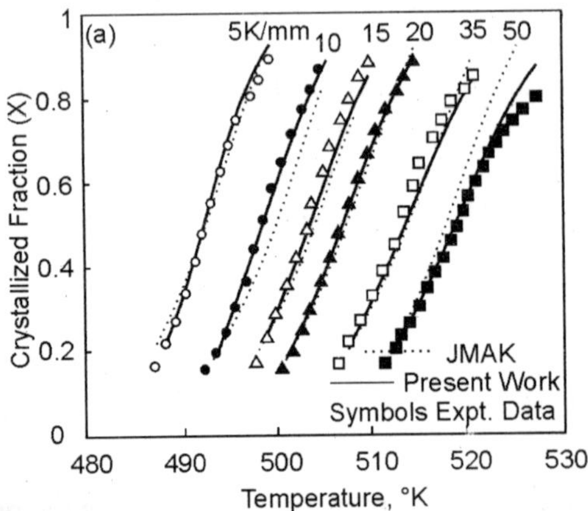

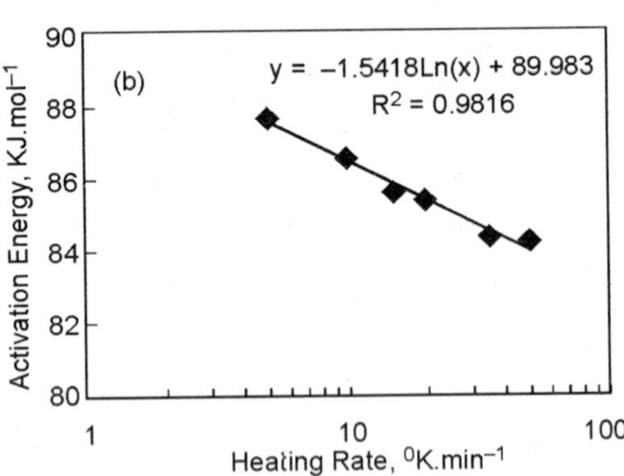

Fig. 1. (a) Comparison of model predictions with the experimental data for $Ge_{20}T_{80}$ glass system. (b) Variation of activation energy with heating rate for $Ge_{20}T_{80}$ glass.

The decrease in activation energy with increase in heating rate has been explained by the classical theory of activated complexes [6-7]. Such reductions in apparent activation energy under non-isothermal conditions have been reported in literature for several other phase transformations [5].

An important inference of the above result is that the phase transformation kinetics increases with increase in heating rate. In fact this has also been shown for sintering of ZnO powder compact [8]. In this systematic work, the effect of heating rate on sintering and coarsening of ZnO powder compacts at various heating rates between 0.50°C min-^{1} and 15°C min^{-1} were investigated. It was noted from the results on relative density vs. temperature graph at various heating rates that the relative density at any temperature showed a modest but systematic dependence on the heating rate. This is clear from the volumetric densification rate (which is equal to three times the linear densification rate) vs. heating rate plot at two different temperatures as shown in Fig. 2. In fact, it can be seen that the densification rate or sintering kinetics increased with increasing heating rate for a given temperature.

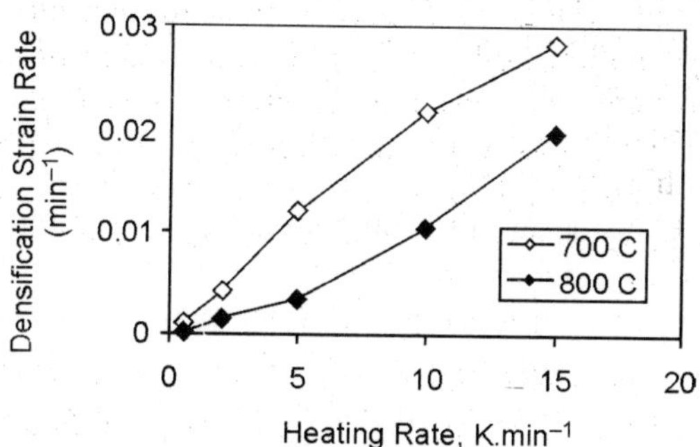

Fig. 2. Densification strain rate a as function of heating rate, during isochronal sintering of ZnO compacts. (adapted from Ref. 8)

Decrease in activation energy implies that under isochronal condition the transformation kinetics will be accelerated, which in industrial paradigm directly translates to higher productivity and lower energy consumption. However, the narrow temperature regimes available for phase transformations and slow transformation kinetics mean that there will be incomplete transformation under isochronal condition. The way out of this dilemma is to cycle between two temperatures lying in the operating regime, thereby isochronal treatment can be provided for longer period of time until the transformation is complete. In the following section, results on cyclic transformation kinetics have been presented.

3. ACCELERATED CYCLIC TRANSFORMATION KINETICS

Another implication of the quasi-isothermal approach discussed in earlier section is that the phase transformation kinetics of any non-isothermal cycle operating between two limiting temperatures T_{low} and T_{high} will be in between the isothermal kinetics at these two limiting temperatures. For example in the case of grain growth, it is expected is that for any non-isothermal cycle operating between two limiting temperatures T_{low} and T_{high}. the resulting grain size (d) will satisfy the following inequality:

$$d_{low} < d < d_{high} \qquad ...(4)$$

where d_{low} and d_{high} are the resultant grain sizes from isothermal annealing carried out at T_{low} and T_{high}, ($T_{low} < T_{high}$) respectively. It must be noted that the basic assumption in the quasi-isothermal approach is that there is no change in kinetics of grain growth between isothermal and non-isothermal conditions. Analogous inequalities can be derived for other thermally activated transformations, such as recrystallization, sintering or precipitation kinetics, by incorporating appropriate transformation formulations. For example, faction recrystalized will be examined in recrystalization kinetics, whereas relative density or grain size will be evaluated during sintering.

The results on the grain growth kinetics in AlK steel have been consolidated in Fig. 3, where the grain sizes of some of the cyclically annealed samples (filled bars) are presented along with the isothermally annealed samples (unfilled bars). A quantitative analysis of the Fig. 3(b) indicates that there is a 15% increase in the mean grain size, when the isothermal annealing temperature is increased from 650°C to 725°C. This increase in grain size with increase in annealing temperature during isothermal annealing is expected in normal grain growth due to the increased grain boundary mobility at elevated temperatures. The grain size of some of the cyclically annealed samples (between 600°C and 725°C Productivity Energy Quality for 6 h), are also shown in this figure. Considering the relationship given in Eqn.4, which was derived on the basis of quasi- isothermal grain growth kinetics, it is expected that the grain size of the cyclically annealed sample will lie between the grain sizes of the samples annealed at the two limiting isothermal temperatures. However, Fig. 3(b) indicates that the mean grain size of the cyclically annealed sample is as high as 16.5% larger than the grain size of the sample, which was isothermally annealed at 725°C. Considering the sluggish nature of the ferritic grain growth kinetics, the 16.5% increase in the grain size is significant. This is also reflected in the results from isothermal studies presented in this work, where increase in annealing temperatures from 650°C to 725°C, resulted in only 15% increase in the grain size although, due to the Arrhenius type exponential relationship (Eqn. 2), temperature is expected to have the strongest effect on the grain growth kinetics. In another case, 35% fraction recrystalized was achieved under cyclic annealing condition in the AIK steel sheet. In this case practically no recrystallization was observed when the isothermal annealing was carried out at an equivalent temperature of 503°C. In another study, acceleration of age hardening kinetics in Ti-6A1-4V has been reported [10] during cyclic age hardening treatments between 77 K and 623 K.

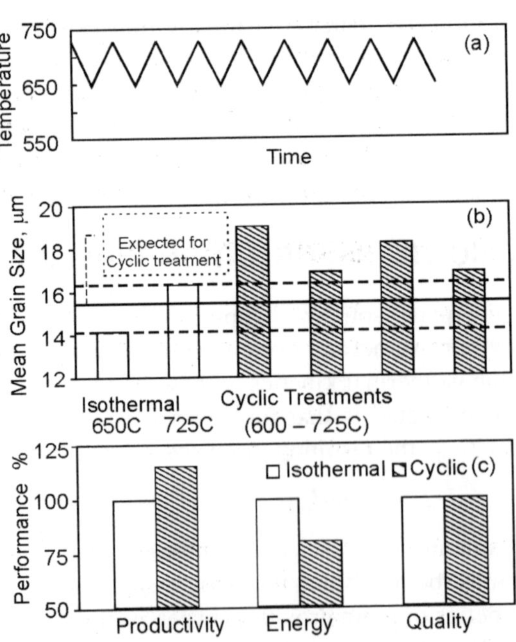

Fig. 3. (a) Thermal profile, (b) accelerated grain growth and (c) performance enhancement during cyclic annealing.

In another work [9], Schuh *et al.* have studied densification of zinc powders through thermal cycling. In their work, they have compared the relative density obtained at various time through thermal cycling between two different temperature ranges and the relative density obtained through isothermal process at the upper temperature of the thermal cycles under same pressure. In both cases it was shown that the thermally cycled densities where significantly higher than the isothermal ones at all times. Fig.4 (adapted from Ref. 9) shows the relative densities at a particular time instant. This shows that the thermal cycling densification occurs much more quickly than isothermal one at the upper temperature of the thermal cycle despite the significantly lower average temperature during thermal cycling. So, sintering via thermal cycling could be used as an alternative in lowering the sintering temperatures and enhancing sintering rates. It can therefore be concluded, that acceleration in transformation kinetics has been observed in a number of material systems during various types of phase transformations.

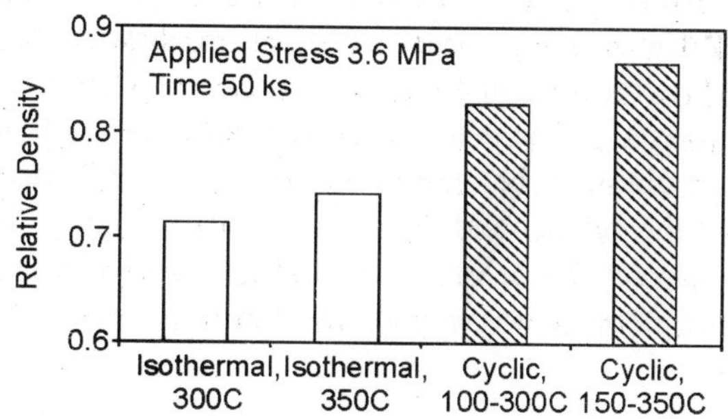

Fig. 4. Enhanced densification during cyclic sintering of Zn powder (adapted from Ref. 9).

4. INDUSTRIAL IMPLICATION AND IMPLEMENTATION

The accelerated transformation kinetics observed under cyclic condition can be utilized to enhance the performance efficiency of the industrial thermal processing, including sintering operations. The two importance performance metrics, which can be positively influenced by these observations, are productivity and energy consumption. Generally, industrial thermal processing operations are carried out to achieve various degrees of phase transformations. The acceleration in phase transformation kinetics under cyclic condition suggests that the desired degree of phase transformation can be carried out in a shorter time, while achieving the specified product quality. In an industrial scenario, such reduction in cycle time implies that the furnace productivity can be enhanced. A comparison between the cyclic annealing and isothermal annealing showed that as high as 15% enhancement in productivity can be achieved (Fig. 3(c)). Although this result is based on the laboratory trials, due to the independence of transformation kinetics on furnace size, similar gains are expected under industrial conditions. It is important to note that in the present work, the acceleration in phase transformation was achieved while maintaining the upper peak temperature of the cyclic treatment equal to the isothermal condition. As the temperature during the cyclic treatment is varied between the upper and lower peak temperatures, the effective mean temperature reduces and therefore reduction in energy consumption is expected. In addition, due to the shorter cycle time, resulting from the accelerated kinetics, there will be significant under laboratory condition due to the superposition of these two contributions. As the energy consumption in a furnace is very much scale dependent, up-scaling from laboratory to industrial furnace is expected to change this value. Nevertheless, even at the industrial scale, significant reduction in energy consumption is expected from the cyclic treatments. In addition, such industrial implementations are expected to reduce the furnace maintenance frequency due to reduction in mean temperature.

Implementation of thermal cycling in an industrial scenario will require a multi-segment controller where the heating and cooling rates during the thermal cycling can be controlled. Alternately, such profiles can be readily implemented by a SCADA based online control system. As mentioned in section 3, several thermally activated phase transformations, such as recrystallization, grain growth, spheroidization, sintering, age hardening and tempering can utilize the accelerated kinetics obtained under cyclic condition both in continuous or batch operations. In the batch operation, the cyclic profile can be achieved by temporal variation in temperature, whereas under continuous operation the spatial variation in temperature will result in cyclic condition. However, it must be noted that cyclic profiles cannot be implemented in all the industrial heat treatment conditions. A major limitation of cyclic treatment will arise during the heat treatment of large components, where due to the thermal inertia, the cyclic condition cannot be achieved in the core. As an industrial example, achieving the cyclic profile in the core of steel coils (outer diameter between 1500-2000 mm) will be impossible in the industrial batch annealing furnaces. On the other hand, the cyclic profile can be readily achieved during the continuous annealing of steel sheets. Therefore, the industrial implementation of cyclic treatment will require a judicious selection of component type and the heat treatment technology.

5. SUMMARY

In the present work it has been shown that due to non-isothermal effects, the quasi-isothermal model can result in erroneous prediction for the non-isothermal profiles. The prediction capability under isochronal condition can be significantly improved by using an additional functional relationship between heating rate and activation energy. This functional relationship has been found to be unique for large number of phase transformations in varied materials systems. Reduction in activation energy with increasing heating rate suggests acceleration in phase transformation kinetics under non-isothermal condition, which is also observed for sintering under isochronal condition. This effect can be utilized in industrial thermal processing by cyclic thermal profiles, by which isochronal profiles can be given within the required thermal regime for any duration of time. This hypothesis has been shown to be true for a number of material systems as well as for varied phase transformations, sintering, annealing and age-hardening, where accelerated transformation kinetics was observed under cyclic condition. This phenomenon can be effectively utilized in designing shorter industrial annealing cycles with resultant increase in productivity. The decrease in mean temperature for the cyclic profile as well as reduction in cycle time, results in significant decrease in the energy consumption. In an industrial furnace, this novel methodology can be implemented by using the modern multi-segment controller or a SCADA based online control system. The effective implementation of this methodology in an industrial operation requires a judicious selection of component type as well as technology.

REFERENCES

1. Satyam S. Sahay and Chetan P. Malhotra, Model Based Optimization of Heat Treatment Operations, Proc. of 7th ASM Heat Treat Show-2002 and Conference on Heat Treatment, May 2-4, Mumbai, pp. 153-160 (2002).
2. Satyam S. Sahay and Arun M. Kumar, Applications of Integrated Batch Annealing Furnace Simulator, Materials and Manufacturing Processes, Vol. 17, No. 4,439-453 (2002).
3. Satyam S. Sahay and Chetan P. Malhotra, Cost Model for Gas Carburizing, ASM Heat Treating Progress, Vol. 2, No. 2, 29-32 (2002).
4. Chetan P. Malhotra, Niranjan R. Pedanekar and Satyam S. Sahay, Cost Model for the Steel Reheating Operation, Industrial Heating-The International Journal of Thermal Technology, 67-71 (2002).
5. Satyam S. Sahay and Karthik Krishnan, Analysis of the nonisothermal crystallization kinetics in three linear aromatic polyester systems Thermochimica Acta, *in press* (2005).
6. Satyam S. Sahay and Karthik Krishnan, Modeling the Isochronal Crystallization Kinetics, Physica B 348, 310-316 (2004).
7. S.S. Sahay, C.P. Malhotra and A.M. Kolkhede, Accelerated Grain Growth Behavior during Cyclic Annealing, Acta Materialia, Vol. 51, No. 2, 339-346 (2003).
8. M.Y. Chu, M.N. Rahaman and L.C. De Jonghe, Effect of Heating Rate on Sintering and Coarsening, J. Am. Ceram. Soc., Vol. 74, No. 6, 1217-1225 (1991).
9. C.A. Schuh and D.C. Dunand, Enhanced Densification of Zinc Powders through Thermal Cycling, Acta Materialia, Vol. 50, No. 6, 1349-135s8 (2003).
10. H. Geng, S. He, and T. Lei, Thermal Cycling Behavior of as Quenched and Aged TI-6Al- 4Y Alloy, Metallurgical and Materials Transactions A, Vol. 28A, 1809-1814 (1997).

Unit–III

Properties of Powder Metallurgical Products

1

Improving the Dimensional Stability of PM Components — a Practical Approach

N. Nagarjuna
Höganäs India Limited

Mats Larsson
Höganäs AB Sweden

Ian Howe
Höganäs AB Sweden

ABSTRACT

This paper focuses on a practical approach to reducing the level of dimensional scatter in PM products and how this leads to improved quality and lower total cost. PM technology offers (a) ab abundance of powders, additives and alloying elements that can be introduced by several methods, (b) various processes and process routes and (c) varying production conditions. All of these factors have a significant influence upon the final dimensional tolerances of PM parts. This vast array of practical aspects can be manipulated to improve dimensional stability. These practical aspects range from choice of materials, advances in powder handling and consideratikons during PM part manufacturing. This paper reviews work done by Höganäs AB R & D on various powders, mix compositions, and mix systems and powder handling systems that bring about dimensional consistancy in PM components.

1. INTRODUCTION

Powder Metallurgy is the **art and science** of manufacturing metal powders, making customized mixes; converting the mixes to components by compaction in rigid dies using various types of compacting presses and sintering for metallurgical bonding at time and temperature under protective atmospheres in sintering furnaces.

There are many post-sintering operations like sizing, oil impregnation, steam treatment, heat treatment, infiltration, maching, resin impregnation and plating which are selected into the process route based on the application requirements of components.

However it is the basic powder quality and mandatory process steps of Mixing, Compaction and Sintering that play a major role in determining the dimensional stability of PM components. The size change occuring during the sintering step is a result of the powder (+ additives) itself, the compacting and the sintering conditions. Figure 1 below tries to put forward the variables occuring in Base Powder, Powder mix, Compaction and Sintering that could affect the dimensional change of sintered components.

Post sintering operations like sizing, machining, hardening and finishing can all be affected by the condition of the components coming out of the sintering furnace. It would be appropriate to state that every stage can be affected by the previous stage.

Use of consistent quality powder/mixes and highest level of controls established in the mandatory processes ensure components with dimensional stability from within-losts as well as from lot-to-lot.

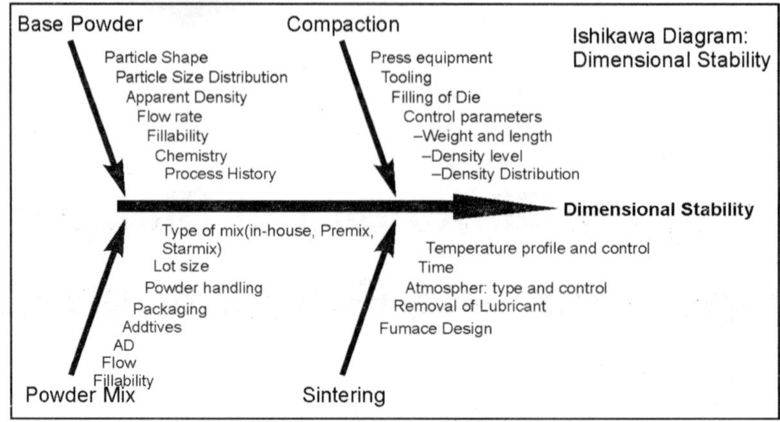

Fig.1 Ishikawa Diagram: Dimensional Stability

2. FACTORS INFLUENCING THE DIMENSIONAL STABILITY

The main factors influencing the dimensional stability of ferrous PM components are

Powder

Apparent density, compressibility, Particle morphology, grain size characteristics of the powder, fillability, alloying composition and quantities of alloying elements present.

Process

Compact density, density variation within compact, sintering temperature, sintering time and sintering atmosphere. Various methods of manufacturing metal powders lead to specific particle shapes, size distribution, particle purity and particle hardness. Even for the same type of powder these parmeters could vary leading to lot to lot variations in properties (required for the success of next operation–mixing and compacting) shape, size, A.D gms/cc (under applied pressure), flow secs/50gms, compressibility gms/cc (under zero pressure), freen strength N/mm2 (under applied pressure) and spring-back (elastic expansion as existing form compaction in rigid die). Changes in these properties causes variation in mixing and compacting output.

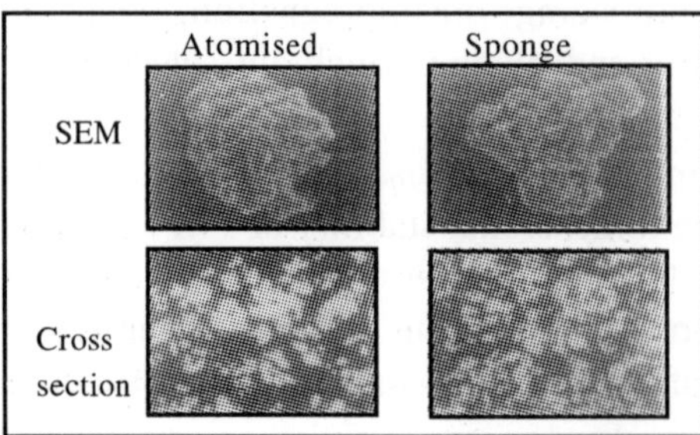

Sponge powder shows lower growth than the later. This is because some of the melting copper moves into the particle porosity in the case of sponge particles. Tooling design should consider this.

Fig. 2. Particle morphology

The behavior of sponge Iron powder (highly irregular and porous) and Atomized Iron powder (smooth-irregular and non porous) gives different dimensional results with copper.

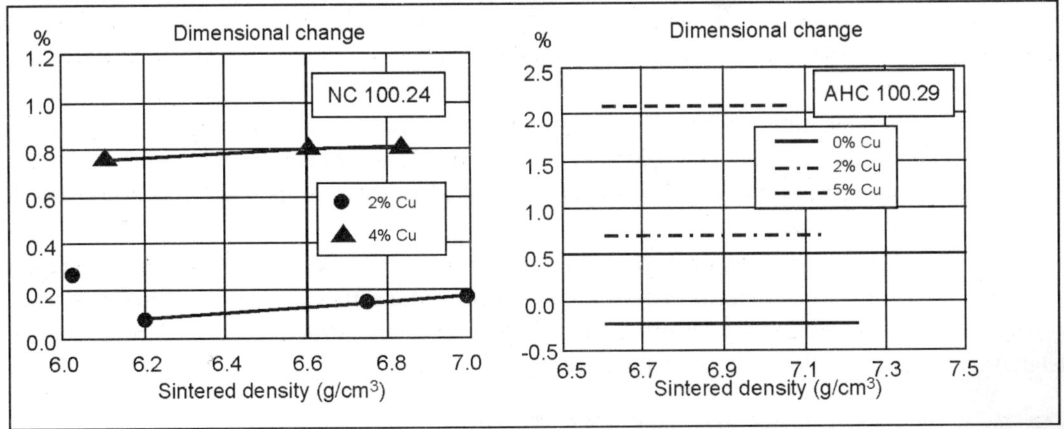

Fig. 3. Dimensional change sponge vs. atomized with copper additions

When good consistent control of tolerance and physical properties is essential the ability of the powder producer to manufacture Iron powders with high consistency in all characteristics is a key factor. At Höganäs the clean manufacturing techniques, closed loop down stream operations and practice of SPC ensures lot-to-lot consistency. Figures 4 and 5 below show as an example the exacting controls achieved at Höganäs on atomized grade powder ASC 100.29.

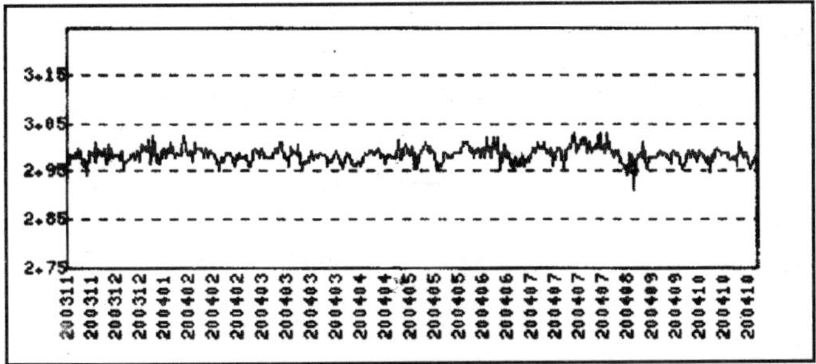

Fig. 4. ASC 100.29. AD control

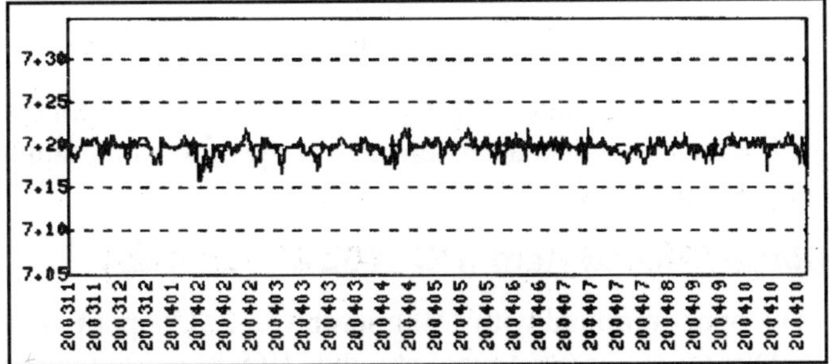

Fig. 5. ASC 100.29. G.D control

Such lot-to-lot consistency in powder properties enables the component manufacturer toi get a very high repeatability on component dimensions. The condition of alloying powders including particle sizes (fine, coarse), oxide content and type and percentage of additives like Cu, C, Ni etc is an important aspect for dimensional stability. The widely used additives for PM components are graphite and copper. Copper through liquid formation gives growth and carbon through interstitial diffusion retards growth in Fe+ Cu systems.

The type of blender, time of blending, and amount of load placed in blending, type and blend uniformity also significantly effect the size and dinsity control of compacted and sintered components due to induced variations in mix A.D., mix green strength. Then there is also the aspect of large blends vs. a number of small blends.

Höganäs has done a great deal of work in the area of ferrous metal Powders and mixing techniques whereby it is possible to produce the best of mixes and reduce the variation in sintered size. The component cost is also reduced by elimination of costly after operations to correct the parts, as size consistency is made possible in the first instance.

3. CASE STUDIES ON SOME MIX SYSTEMS

Study I: Iron Graphite mix system (NC 100.24 + 1.2% graphite)

Here two mixes were studied 1) without binder and 2) with binder bonding of graphite to Iron. Part 10 × 10 × 30 mm compacts 6.7 gms/cc were sintered at 1120 C 30 mins. in DA atmosphere. Length variation was studied.

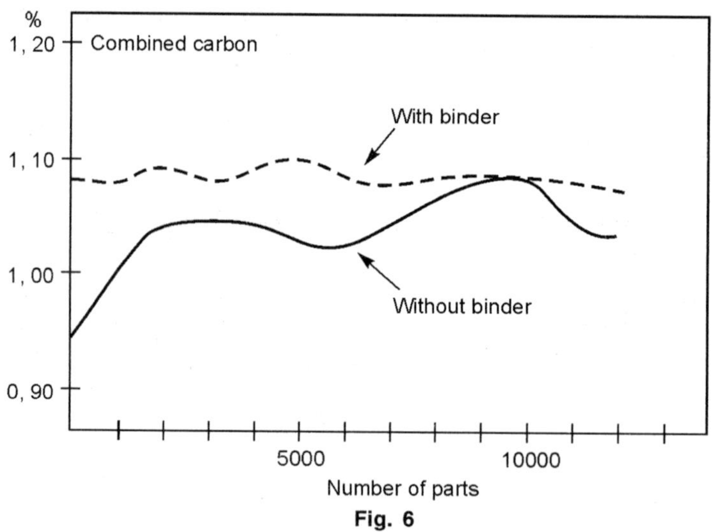

Fig. 6

Low-density graphite floats and collects on top of the iron particles giving rise to dusting and segregation. Use of binder has reduced dusting and segregation of graphite. The dimensional scatter reduced from 0.14% for mix 1 to 0.03% for mix 2.

Study 2: Iron-Copper Mix System (NC 100.24 + 3% Cu)

Two mixes 1) Copper premixed with 150 micron Cu, 2) Copper premixed with 75 micron Cu and 3) Copper added as Distaloy-Cu (30%) were studied. 10,000 parts were pressed and sintered at 1120 C in Endogas for each mix.

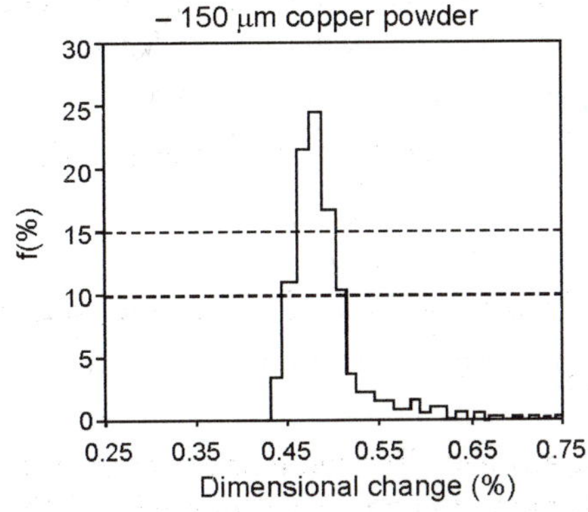

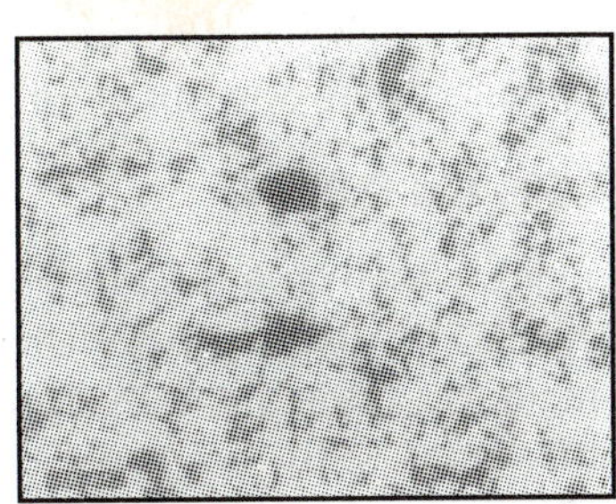

Fig. 7

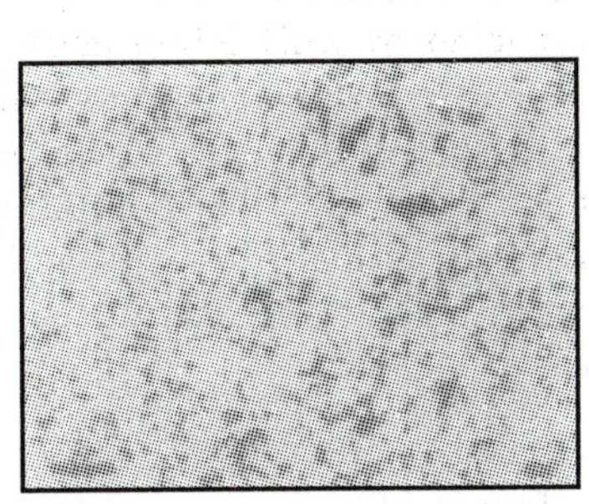

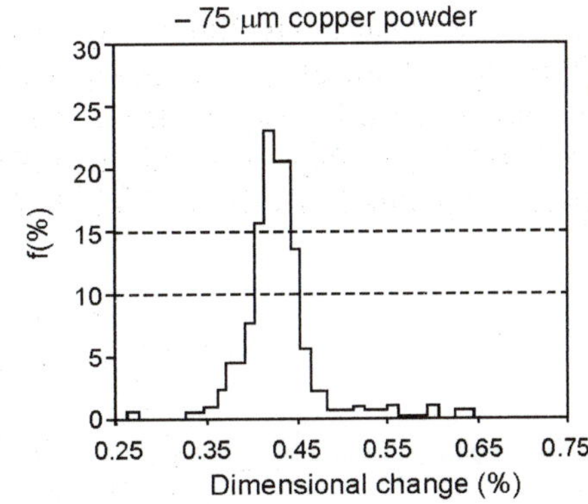

Fig. 8

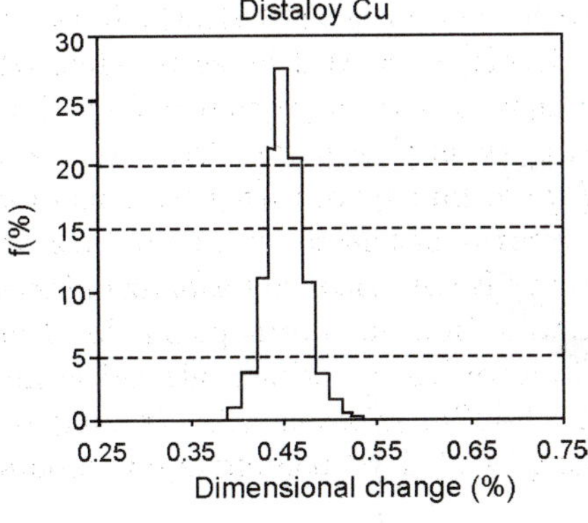

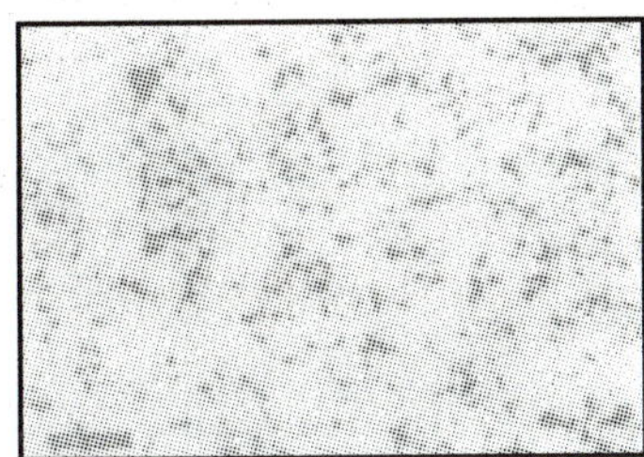

Fig. 9

Finer copper powder as compared to coarser copper powder gives better porosity morphology but continues to give dimensional scatter. It is evident that copper added in the form of Distaloy Cu (fine copper diffusion bonded to iron) gives the best results towards simensional stability and finer porosity.

Study 3: Iron-Copper-Carbon Mix Systems

Fe-Cu-C continues to be one of the most used alloying systems in the PM industry. In these type of materials however copper has a great influence on dimensional scatter followed by carbon. This influence can be reduced by appropriate mix composition and mixing technologies.

3.3 (A) Fe-2.5% Cu-0.3% C

This also means that variation of alloying content has maximum effect on scatter in dimensional changes as compared to sintering parameters like temperature and time for this system. The variation of alloying content can come about by segregation due to particle size and density of the constituents as can be seen from the following study.

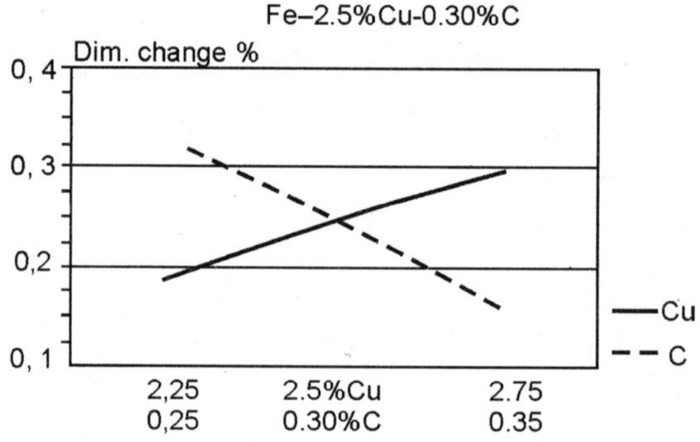

Fig. 10: This result confirms that variations in Cu and C in premix greatly induce scatter in dimensional change.

3.3 (B) Segregation Studies: NC 100.24 + 3% Cu + 0.5% Graphite + 0.8% Wax

Here the variables were copper size (100#, 200# and 325#) copper type (atomized, electrolytic) and method of alloying copper (premix and diffusion bonded).

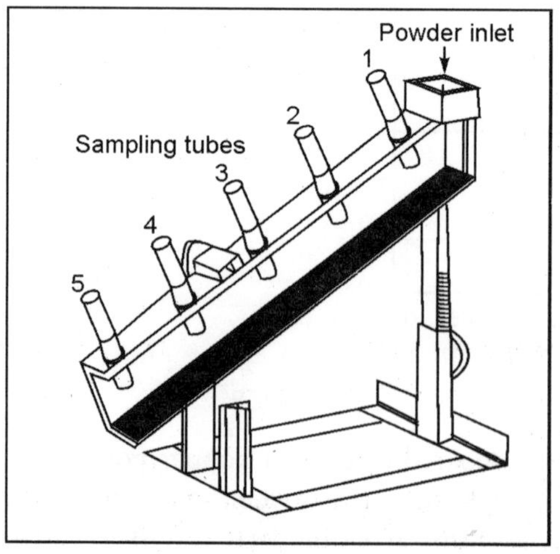

Fig. 11

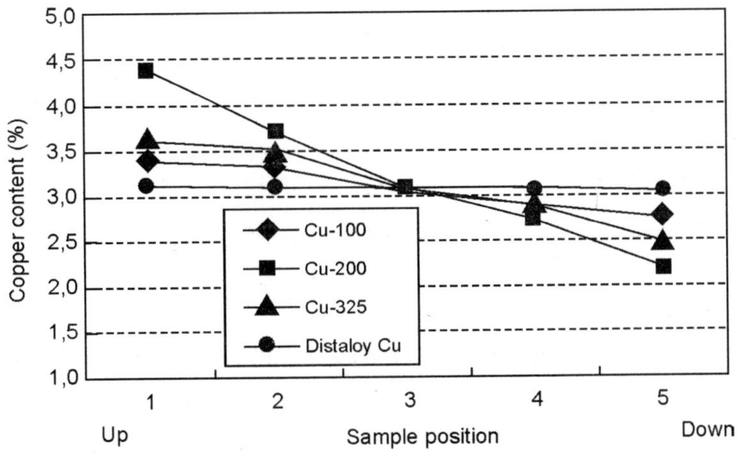

Fig. 12

All 4 mixes were passed through the segregation analyzing equipment as shown in Fig. 11. All except Distaloy Cu segregated while flowing downwards as analyzed in sample 4 and 5. With atomized copper the finer 200# segregates more than 100#. Interstingly finer 325# electrolytic copper powder shows less segregation due to particle morphology that is more irregular as compared to atomized copper powder. The net effect of both particle shape and size is a result of particle morphology (smooth, jagged or rough) and the basic shape. Copper added as Distaloy Cu shows very good dimensional stability.

3.3 (c) Optimum mixes evalution: Iron-Copper-Carbon Mix Systems

Mix A = Premix of (ASC 100.29 + 3% Cu (–200#) + 0.5% C (ultra Fine) + 0.9% wax)
Mix B = Premix of (ASC 100.29 + 3% Cu (–100#) + 0.5% C (ultra Fine) + 0.9% wax)
Mix C = Starmix of (ASC 100.29 + 30% Distaloy A–Cu + 0.5% C (ultra Fine) + 0.9% wax)
Mix D = Starmix of (ASC 100.29 + 20% Distaloy A–Cu + 0.5% C (ultra Fine) + 0.9% wax)
The variables studied through design of experiments were

Variable			
Copper	100#	200#	Dist A-Cu
Copper content	3%	2%	
Carbon content	0.5%	0.8%	
Mixing method	Premix	Starmix	

The contribution to dimensional change for these parameters is as shown in figures 11 to 14 below for the 4 types of mixes under consideration.

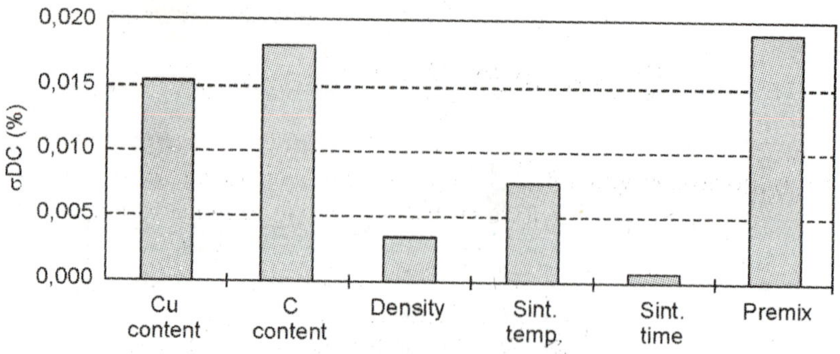

Fig. 13. Mix A Premix Cu–200#

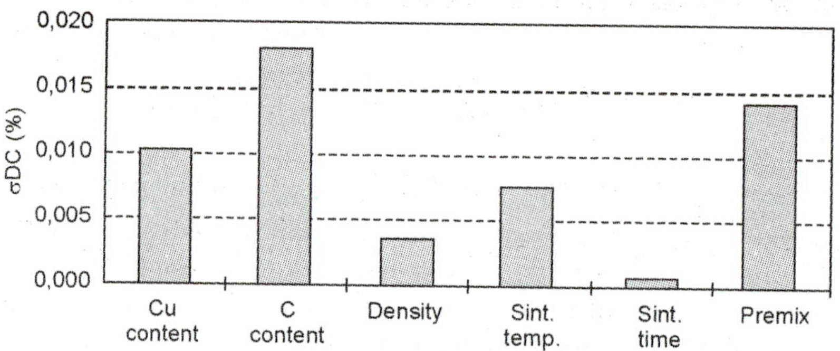

Fig. 14. Mix B Premix Cu–100#

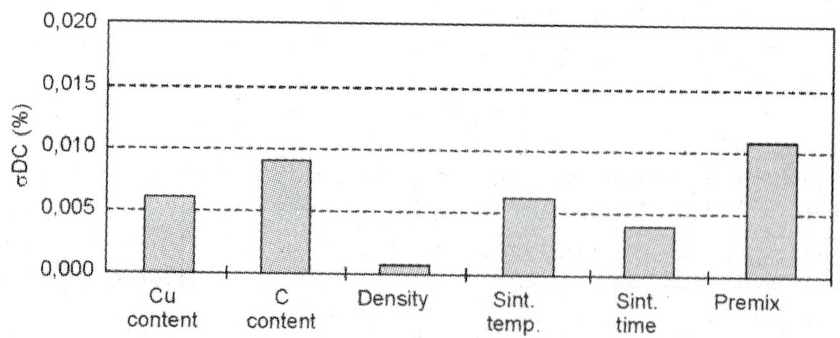

Fig. 15. Mix C Starmix Distaloy Cu–30%

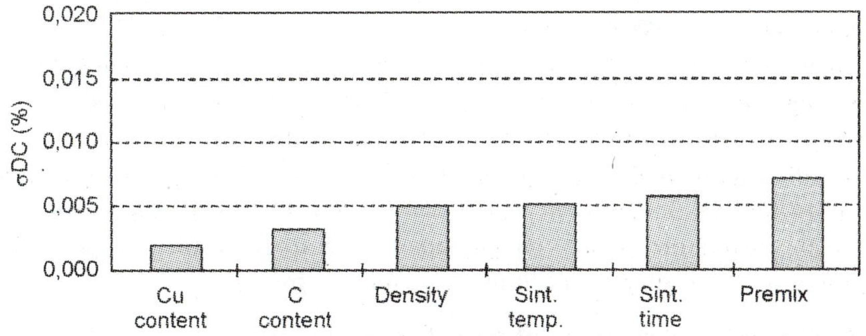

Fig. 16. Mix D Starmix Cu–20%

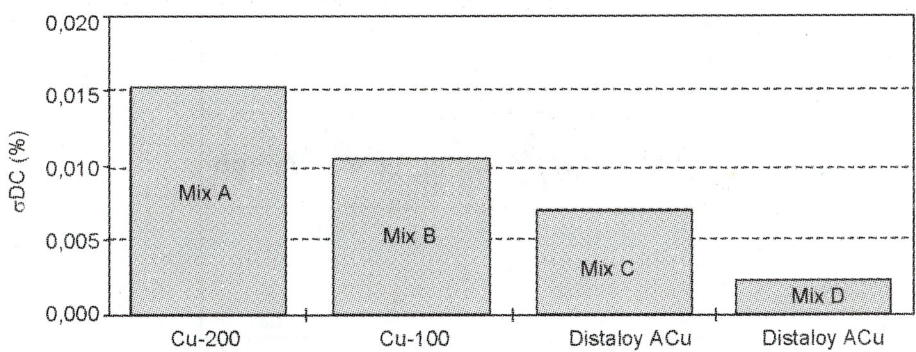

Fig. 17. Scatter due to Copper.

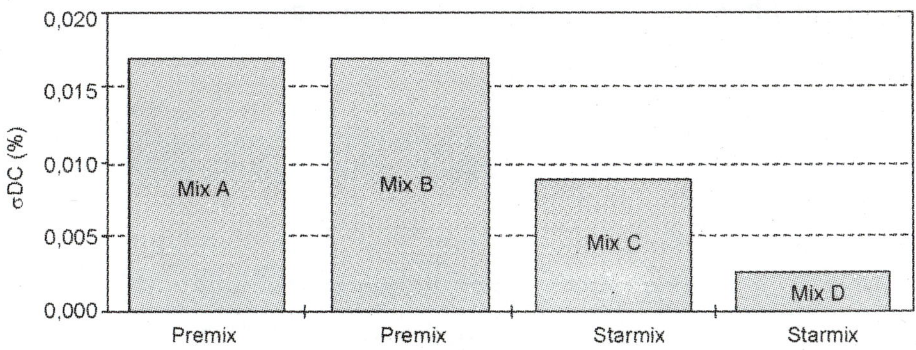

Fig. 18. Scatter due to Carbon

In premix A scatter due to Cu is higher than in premix B due to finer Cu giving rise to segrregation. In Starmixes C and D the influence of Cu and C has been appreciable reduced. This is due to use of Distaloy Acu and bonding of graphite to iorn.

It is seen that finer copper gives the highest DC scatter as it has a tendency to segregate in mixing, mix handling and filling of the tool die. Use of Distaloy Acu is a practical method to reduce the scatter. Carbon too has a strong influence on dimensional scatter.

The fineness of graphite makes it segregate to large extent when mixed with–100# Iron powder. The best method to eliminate dimensional scatter due to ultra fine graphite is to bond the mix.

By virtue of this technique organic binders bond the graphite to the iron particle. It can be seen that Starmix reduces the % scatter by 1/7th as compared to premix.

3.3 (D) Optimizing mix composition Iron-Copper-Carbon

Use of SC 100.26 with appropriate content of copper and carbon allows good dimensional stability. Höganäs AB has developed premix grades PM 0205 and PM 0208 with A.D 2.65 gms/cc with control in +/–0.04

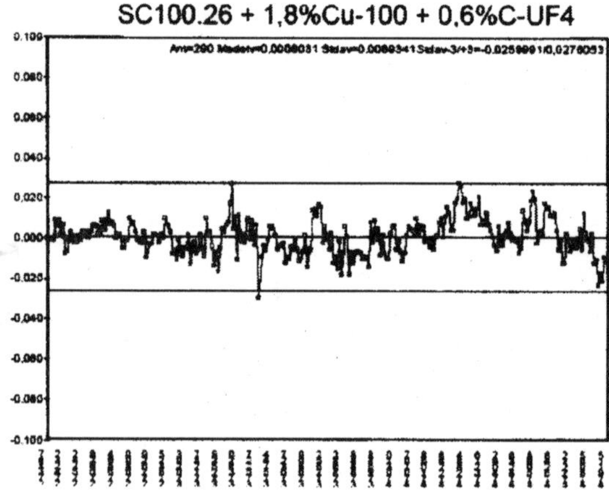

Fig. 19

3.3 (E) Iron-Copper-Phosphorous-Carbon mix systems

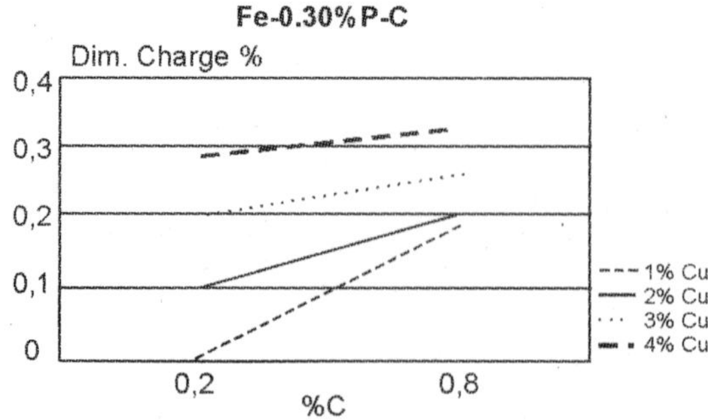

Fig. 20

This system can be viewed upon as another practical way to reduce the sensitivity towards dimensional scatter of Fe-Cu-C systems by additions of about 0.3% P.

This system with 4% Cu has a near stable dimenstional change with graphite additions between 0.2 to 0.8% combined carbon. This means dimensional control can be improved by using copper addition as Distaloy Acu.

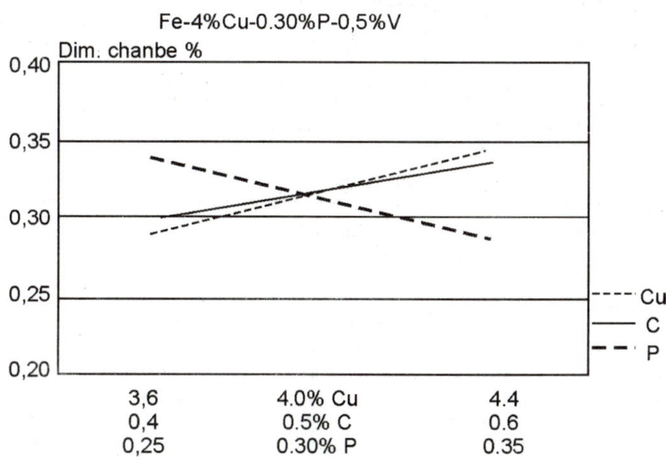

Fig. 21

3.3 (F) Iron-Copper-Nickel-Molybdenum-Carbon mix systems

In this system adjusting Cu and Ni percentages can control dimensional change effectively. Using Distaloys (wherein the additives are diffusiion bonded to iron particles) is the best approach. Distaloys hardly very at all.

Distaloy DC is very good for steady dimensions. 2% Ni bonded to prealloyed Astaloy Mo is very stable in sintering at different part densities. With carbon levels there is a difference in dimensional change but remains stable at range of part density.

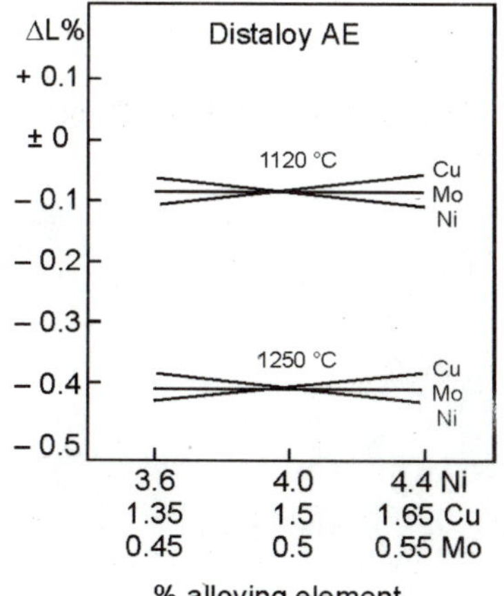

Fig. 22

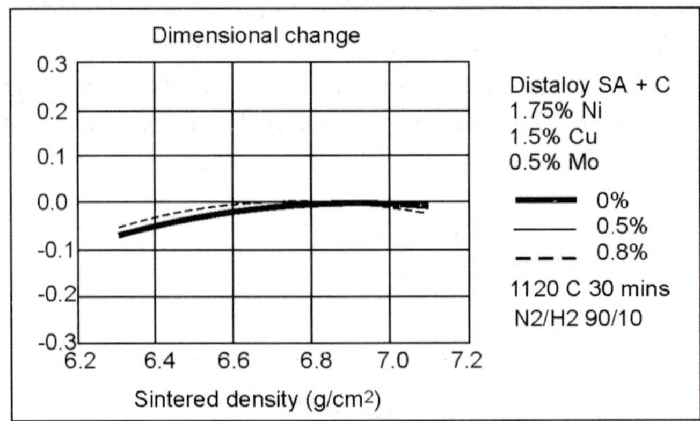

Fig. 23

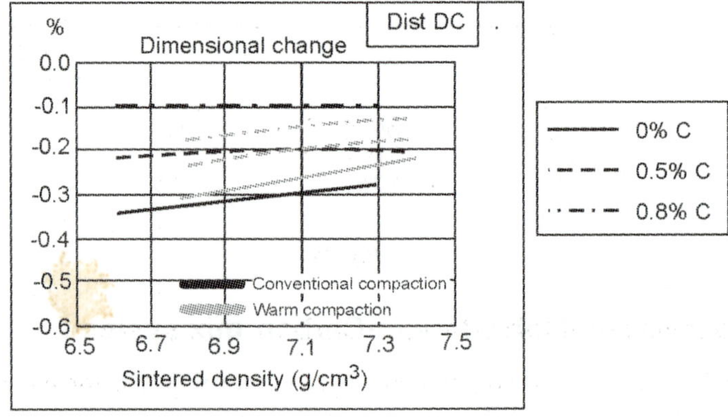

Fig. 24

4. REDUCTION IN WEIGHT SCATTER AS A METHOD FOR DIMENSIONAL CONSISTENCY

(A) Hopper Design

Three designs with angles 90°, 60° and 30° for hoppers were used for trials as shown in Fig 25.

Fig. 25

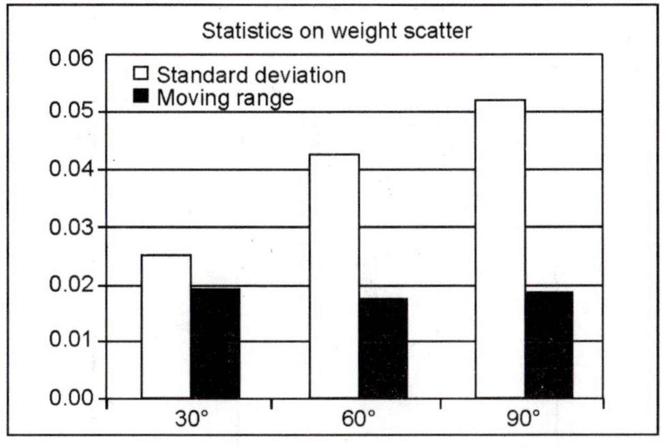

Fig. 26

With properly designed powder feeding hopper the particle segregation is avoided and weight scatter (Fig. 26) is significantly improved thus facilitating dimensional stability.

(B) Flex Bag System vs. Auger System (Part and composition are customer specific)

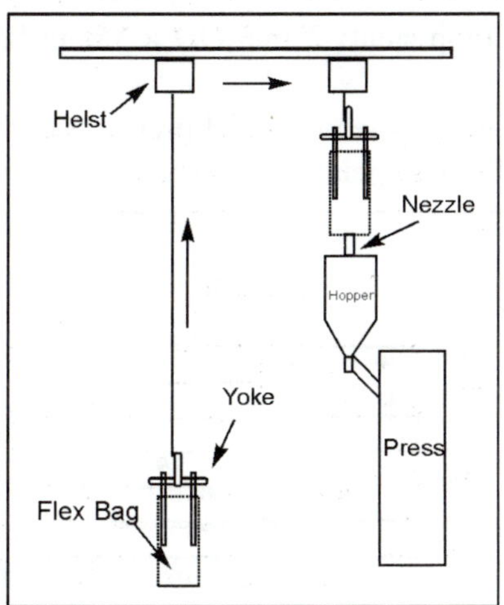

Fig. 27.

		Standard Deviation	
		Flex bag	*Auger*
Height	mm	0.020	0.027
Weight	Gms	0.4	0.7
OD	mm	0.022	0.038
C	%	0.03	0.045
Cu	%	0.03	0.11
Adjustment/ton	nos.	2	4.8

The results in general show that feeding material from flex bags (Fig. 27) directly to the press is a good practice that minimizes handling of powder mix. This decreasese scatter in weight, dimensions and chemistry as well as minimizes press adjustments.

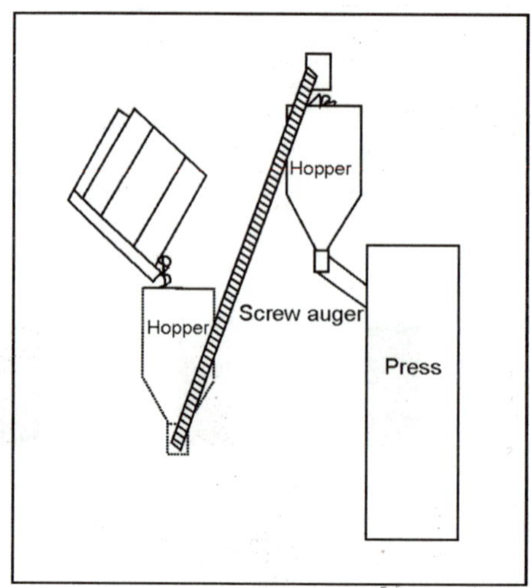

Fig. 28.

(C) Fillability of mix as a tool for better controls in processing PM parts

Case Study: Astaloy Mo + 0.5% P + 0.30% MnS + 0.75% was

Type: Premix and Starmix; Component: 75mm OD × 55mm ID × height 11 mm; Press: 200 ton Dorst mechanical press.

The Starmix had a hall flow of 23.5 seconds and premix had flow of 27.9 seconds. The feeding of the mix into the die cavity can be manipulated to bring out the best results for a given press.

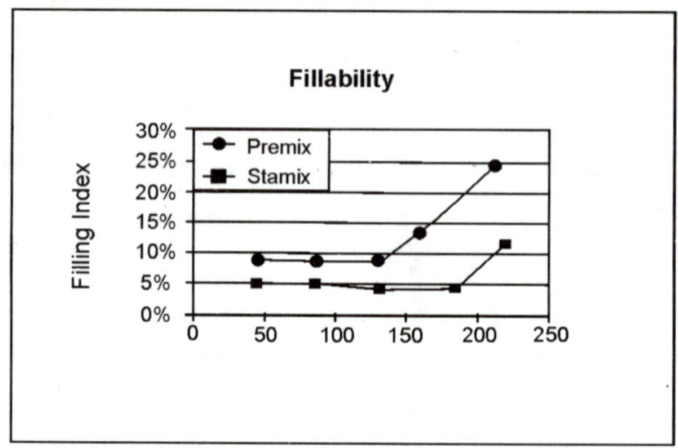

Fig. 29

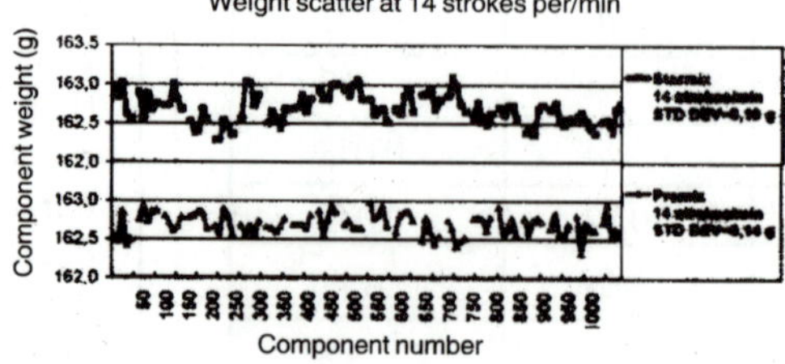

Fig. 30

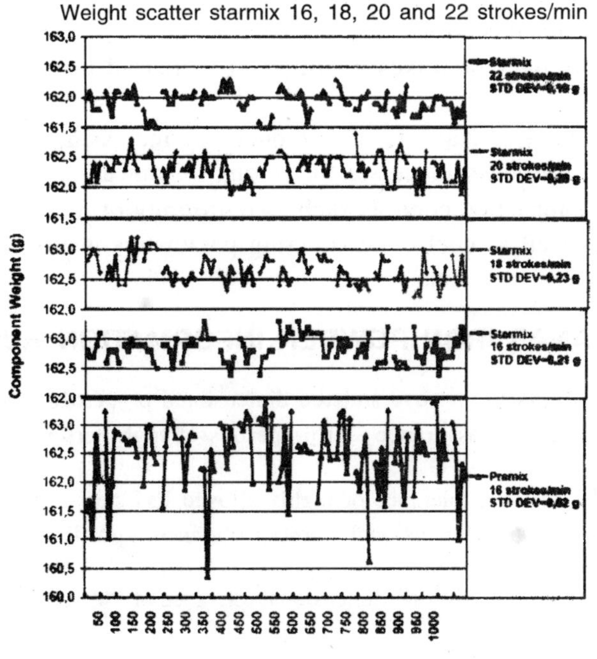

Fig. 31

Starmix gave better consistency in the parts in terms of compact accuracy and sintered dimensional consistency. (Fig. 31)

Dusting in the middle of trial 1, 16 strokes/min

Premix

Starmix

Cumulative value

Premix 120,0

Starmix 54,6

Fig. 32

Premix flow/fillability (Fig. 29) could not work at higher speed i.e. >16 strokes inducing more weight scatter (Fig. 30)

Starmix increased productivity by 40%. Also dusting was less with Starmix (Fig. 32)

CONCLUSIONS

1. The most widely used Iron–Copper–Carbon mix system is most sensitive to scatter in dimensions with fluctuations in alloying content, segregation and dusting. However selection of appropriate alloying content, optimization of mix composition, use of good mixing technology and bonded mixes achieves good dimensional stability, a reason why Höganäs "ready to us mixes" are increasingly being sought after by PM component manufacturers.
2. Distaloy mixes are very stable for dimensional control.
3. Good hopper design reduces weight scatter and controls dimensions better.

4. Feeding material directly from flexbags to press reduces problems associated with mix handling and gives better dimensional control
5. Starmix shows better fillability over premixes resulting in increased productivity and dimensional control.
6. From the above advantages it is clear that total cost reduces by increasing productivity, eliminating extra operations or corrective processes for getting tolerances.
7. It is ofcourse imperative that compaction controls such as uniformity of press conditions, accuracy of die dimensions as well as control of sintering parameters is ensured by the component manufacturers.

REFERENCES FROM HOGANAS IRON POWDER INFORMATION HUB

1. Ulf Engstrom, Goran Wastenson, Ways to improve sintered dimensional properties, PM 84.
2. Jan Tengzelius, Mats Larsson, Tolerance assurance and properties of high performance steels, PM 93.
3. Jan Tengzelius, Properties and dimensional tolerance of highly alloyed PM steels, PM 89.
4. Mats Larsson, Mechanical properties and dimensional stability obtained with different alloying techniques.
5. Jan Tengzelius, Ulf Engstrom, Means to improve the dimensional tolerance of PM steel components, PM 84.
6. Caroline L, K Ishii, T Tsuboi, Material with stable dimensional change independent internal density variations PM 93.
7. Eva J, Mats L, Paul S, Improved-mixing technique PM 94.
8. Mats L. Higher tolerances by improved materials and processes.
9. Ulf E, Caroline L, Efficient PM part production through bonded mixes having unique powder properties PM 02.
10. Hilmar V et al Efficient PM part production through bonded mixes having unique powder properties PM 02.
11. Olle Thornblad, Dimensional consistency of powder compacts PM 01.
12. Mats L Olle T, Close dimensional tolerances through material selection PM 96.

2

Effect of Friction Additive and Solid Lubricant on the Friction and Wear Properties of an Iron Base Sintered Aircraft Brake Friction Material

D. Dutta

Foundry & Forge Division, HAL, Bangalore, India.

ABSTRACT

Friction characteristics of an aircraft brake friction material containing different amounts of natural graphite and silicon carbide were investigated by using a sub-scale brake dynamometer rig. Run-down stop mode of braking was employed to study the change of friction coefficient, torque variation and friction force oscillation. Results showed that the relative amount of Graphite and SiC played important roles in determining the friction characteristics. Graphite in the friction material improved the friction stability and SiC aggravated the torque variation during brake applications.

1. INTRODUCTION

A basic attribute of powder metallurgy is the ability to combine materials in powder form that are otherwise immiscible. This unique advantage allows the production of complex metal-ceramic friction material composites in which copper or iron is used as the matrix material and into which special particulate non-metallic and ceramic additives, *i.e.* solid lubricants, oxides and other compounds, are added for imparting special frictional, anti-seizure, wear and thermal properties that are desired. These special additives enable the resultant friction material to operate at higher operating loads and temperatures compared to resin-asbestos based organic friction materials.

P/M friction materials are used in aircraft brakes. For aircraft braking applications matrix powders having high green strength are essential. Such powders characteristically also have high internal porosity, low apparent density and irregular shapes. The morphology, shape and size and physical characteristics of the special ceramic and non-metallic additives also have a profound influence on the ultimate performance of the friction material and need to be carefully and judiciously selected and optimized for each application through elaborate experimentation and evaluation.

Generally, the major portion of the matrix is either copper or iron with about 5-15% low melting metal such as tin; 5-25% lubricant which may be lead, litharge, graphite, Molybdenum disulphide, Boron Nitride or galena; up to 20% friction additives such as silica, alumina, magnetite, silicon carbide, silicon nitride or mullite; up to 10% wear resistant materials such as cast iron grit or shot, steel wool and finally friction and wear stabilizers such as calcium and barium sulphates.

This paper describes the influence of the various additives, in particular friction and solid lubricant additives, on the tribological and frictional properties for a high-energy aircraft brake fitted with sintered metal-ceramic brake pad composite material. As an example, the effects of the relative amounts of an abrasive, Silicon carbide, and a solid lubricant, natural graphite, on the friction and wear characteristics were investigated. A sub-scale brake dynamometer rig was used to investigate the friction characteristics in run-down braking mode (decreasing speed). The change of friction coefficient, friction force torque variation, and the amount of wear were carefully examined to investigate the roles of SiC and Graphite on brake friction characteristics.

2. EXPERIMENTS

2.1. Friction Materials

Friction materials investigated in this study were non-asbestos inorganic sintered type materials containing six different ingredients. The ingredients in the friction material comprise matrix metal, reinforcing ceramic phase, friction modifiers/stabilsers and antiseizure additives. The composition of the friction material studied in this work is shown in Table I. Nine friction materials containing different amounts of Natural Graphite and SiC were produced. The amounts of SiC were changed as 3, 5 and 7 wt.%, and that of graphite as 7, 9 and 11 wt%, respectively. The friction material specimens were named as *S-X-Y* according to the weight percent of graphite (*X*) and SiC (*Y*) in the friction material (Table 2) . The amount of ingredients except graphite and SiC was fixed with only the major matrix constituent, iron, varying to make up the balance. Friction material specimens were produced by a conventional procedure for dry formulation following dry mixing, compaction, pressure sintering, post-pressing, and grinding. The size of the friction material brake pad specimens made was approximately 28 sq cm and had two riveting holes 3 cm apart in the vertical central axis of the pad. The surface hardness values of the specimens were 108 to 134 BHN ($P/d^2 = 10$) and average porosity was approximately 8%.

Table 1. The composition of the friction materials studied in this study

Ingredient	*Weight%*
Copper	10%
Barium Sulphate	6 %
High temperature lubricant	3 %
Silicon Carbide	3-7%
Natural Graphite	7-11%
Iron	Balance to make up 100%

Table 2. The 9 Friction Material Specimens

Solid Lubricant : Natural Graphite (wt%)		*Silicon carbide Wt (%)*		
		3%	5%	7%
	7%	S-7-3	S-7-5	S-7-7
	9%	S-9-3	S-9-5	S-9-7
	11%	S-11-3	S-11-5	S- 11-7

2.2. Brake Rotor and Braking Arrangement

The rotor mating part used in this study was made of an alloy gray cast iron. The diameter of the brake rotor was 40 cm and the thickness of the rotor was approximately 2.6 cm. The microstructure of the cast iron rotor exhibits a mix of A-type and B-type graphite with predominantly B-type graphite of rosette morphology. A floating caliper type braking was used in this experiment that had two pistons on the inboard side of a rotor.

2.3. Sub-scale Dynamometer Test

Friction tests for this study were carried out using a single end sub-scale brake dynamometer of Dynaspede make. A schematic diagram of the brake dynamometer is shown in Fig. 1. The dynamometer test procedure consists of a bedding procedure followed by 50 stop tests. The wheel load (normal force on the rotor) for this test was set at 460 kg and the detailed conditions for the dynamometer procedure are listed in Table 3. This test provides a change of friction coefficient and the amount of friction force oscillation during run - down mode of brake applications. Brake temperature was measured from a K-type thermocouple embedded on the pad holders at 1 mm below the brake pad working surfaces.

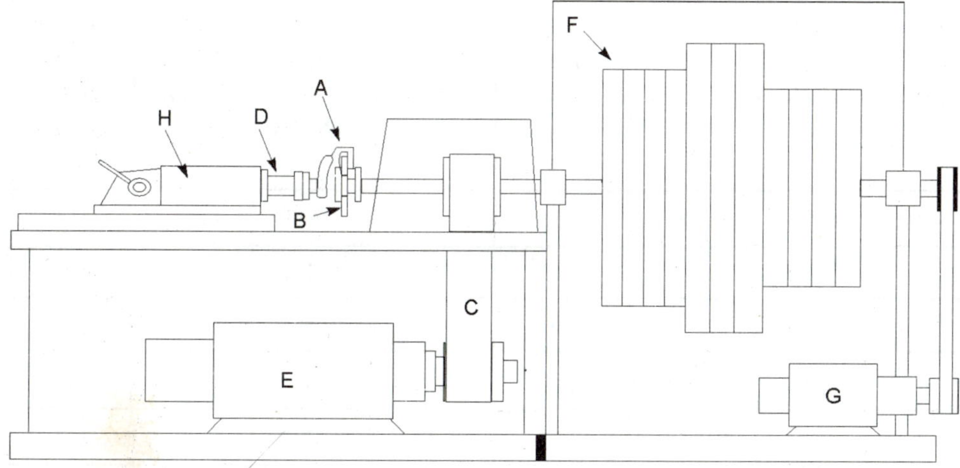

Fig 1. A schematic diagram of the single end brake dynamometer. A Caliper assembly, B brake rotor, C driving belt, D torque transducer, E maindriving motor, F inertia weight.

Table 3 : Dynamometer Test Parameters Maintained

Sl. no.	*Conditions maintained For dynamometer test*	*Value of the parameters*
1	Inertia of gyrating mass	30.1 kgm^2
2	Braking Load(Normal Force)	460 Kgf
3	Friction Radius (Torque arm)	180 mm
4	Brake Application speed	100 kmph (1462 RPM)
5	Brake Energy Simulated	35900 Kgfm

3. RESULTS AND DISCUSSION

3. 1. Friction characteristics during stop tests

Friction materials for an aircraft brake system require a stable friction force regardless of temperature, applied pressure/normal load, and sliding speed. This is because a pilot expects the same level of friction force at various stop conditions. In this experiment, friction coefficients under run-down stop conditions were measured to study the effect of the Graphite and SiC on friction coefficient, friction stability, brake force/ drag / brake torque and temperature rise at the end of 50 stops. During these tests the rotor speed continuously falls with the progress of the test. An average friction coefficient for each friction material was obtained from the 50 friction coefficient values obtained in each test. The stability of the friction coefficient is derived from the standard deviation obtained from these 50 values.

Table IV gives the summary of the friction test results of the 9 different friction materials possessing different combinations of Natural graphite and SiC.

Table 4. Summary of Friction Test Results

Sample identification	*Avg Coeff of friction* μ	*Std Devia of* μ	*Avg Friction Force (kgf)*	*Avg Torque (kgfm)*	*Avg Peak torque (Kgfm)*	*Peak to Mean torque ratio*	*Maxm Brake Temp Rise (°C)*	*Wear wt loss (gms)*
S-7-3	0.284	0.018	131	47.2	60.3	1.278	212	5
S-7-5	0.315	0.022	145	52.2	67.9	1.300	233	6
S-7-7	0.346	0.023	159	57.2	78.4	1.37	249	7
S-9-3	0.256	0.016	118	42.5	51.2	1.223	198	4
S-9-5	0.284	0.016	131	47.1	57.7	1.225	217	5
S-9-7	0.322	0.018	148	53.3	68.5	1.285	237	5.5
S-11-3	0.242	0.0074	111	40.0	46.3	1.158	188	3.5
S-11-5	0.268	0.0076	123	44.3	51.4	1.160	202	4
S-11-7	0.305	0.0081	140	50.4	61.2	1.214	225	4

Fig. 2 shows the level of the average coefficient of friction for the nine friction materials. As expected from the roles of natural graphite and SiC, the coefficient of friction was highest in the case of using the most aggressive friction material S-7-7 and lowest in the case of using the most mild friction material S-11-3 . Friction stability was obtained by calculating the standard deviation of the friction coefficients from 50 stop of each test. Fig. 3 shows that friction stability is improved by adding more solid lubricant, *i.e.* natural graphite, to the friction material, graphite helping establish a stable friction film is built up on the friction interface in the case of using friction materials containing 11 wt.% of graphite. This result suggests that the solid lubricant, natural graphite, maintains its lubricating capability at elevated temperature by forming stable films at the friction interface. The stable friction film formed on sample no S-11-5 is shown in Fig. 4a as compared to the lack of film formation on sample no S-7-7 (Fig. 4b) which exhibited high wear and excessive friction and temperature rise.

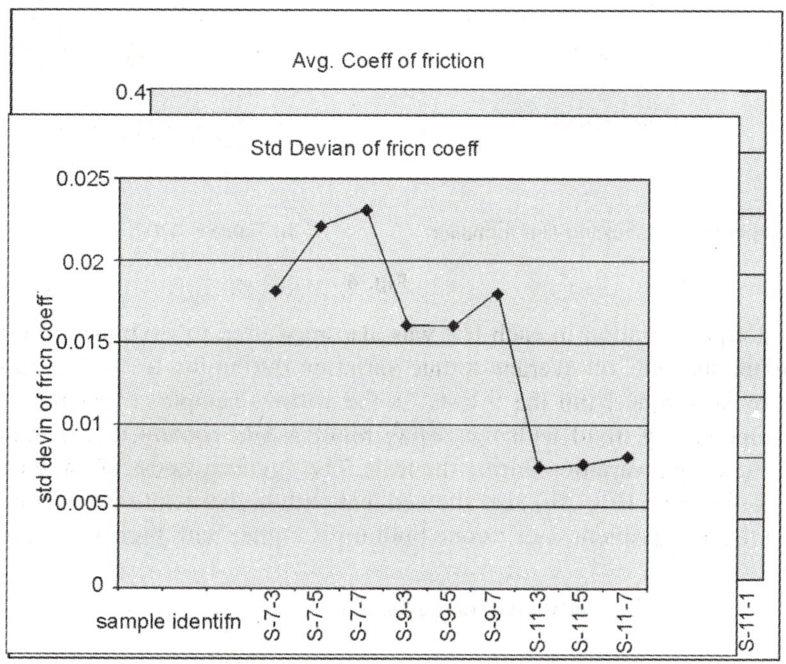

Fig. 2. Variation of Average Friction coefficient.

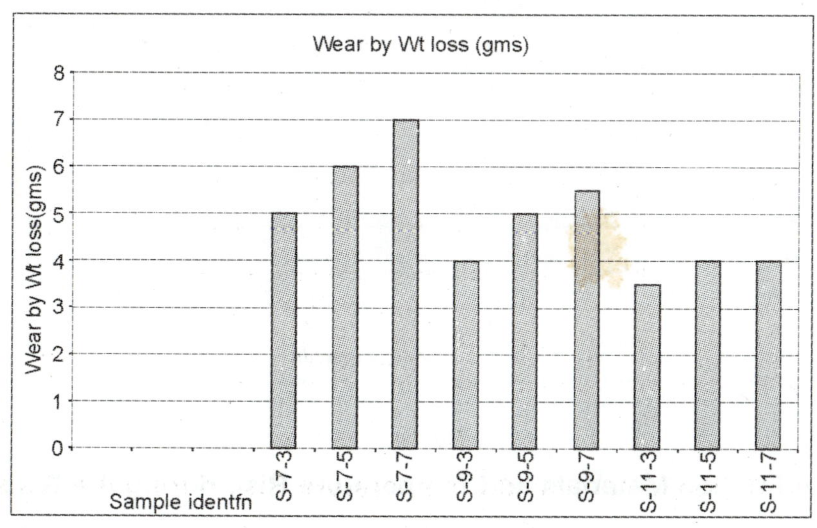

Fig. 3

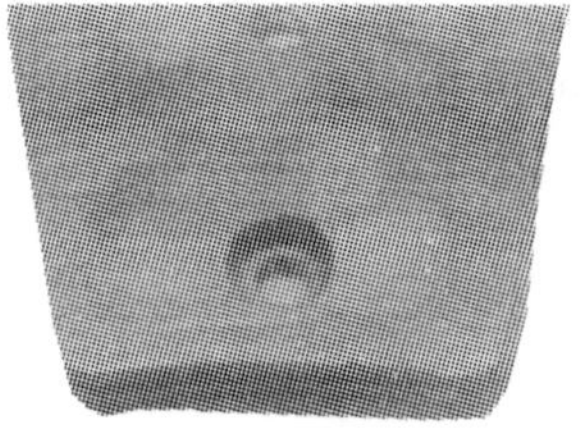

(a) Sample S-11-5 showing film formation

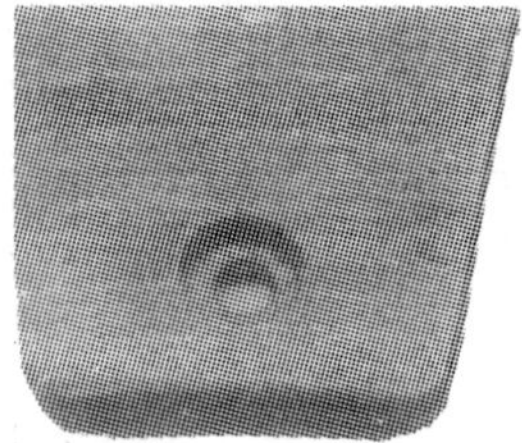

(b) Sample S-7-7 showing almost no film

Fig. 4

Maximum torque variation in each test was also measured to examine the effect of the relative amounts of graphite and SiC on average torque variation during the tests. Fig. 5a shows the average mean and peak brake torque from the 9 tests on the various samples of each friction material. The figure shows an increasing trend with increasing relative SiC content that demonstrated the strong influence of SiC on torque variation during the tests. The aggressiveness of the torque depicted by the peak to average torque ratio (Fig. 5b) also showed that with higher relative SiC content compared with Graphite content the aggressiveness of torque built up is higher with higher relative SiC content, e.g., in S-7-7.

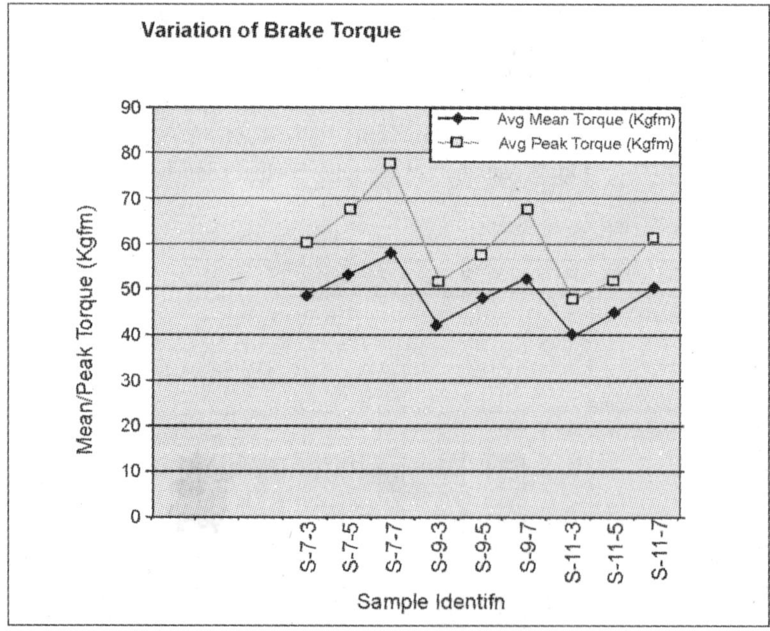

Fig. 5a

3.2. Wear of the Friction Materials and Temperature Rise during the Tests

The wear amounts of friction material samples were measured from the weight change of the samples during the 50-stop dynamometer test. The wear of friction material ranged from 3.5 to 7 gms and the variation of the friction material wear is depicted in Fig. 3. The figure shows that the relative amounts of graphite and SiC strongly affect the wear of the friction material. A reason for the large amount of wear of the aggressive friction materials containing less graphite and more SiC appears due to the severe abrasive action of SiC and insufficient lubrication, resulting in high friction temperatures. Lower wear and the stability of wear values with materials containing higher graphite content of 11% appears to be due to the formation of a stable surface film that controls friction and wear and also limits the metal to metal rubbing contact and abrasion and thus lowering frictional temperature rise. The variation of maximum brake temperature rise of the various samples is also depicted in Fig. 6 which clearly brings out the fact that higher the SiC content relative to the graphite content, higher is the friction coefficient and wear and also higher is the maximum brake temperature rise. Wear of the mating rotor also was found to be lower in the case of samples containing 11% graphite.

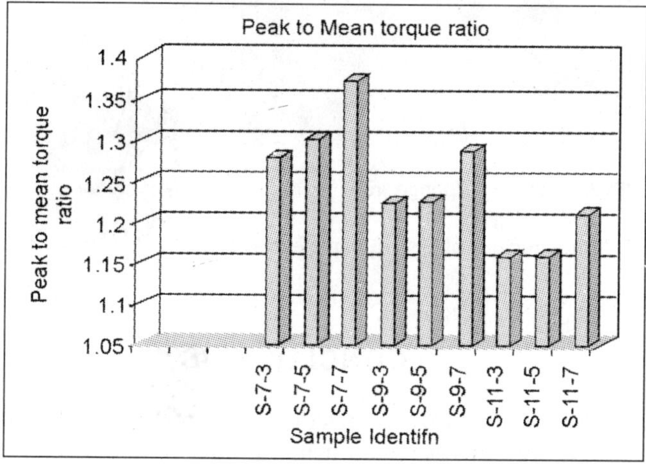

Fig. 5b

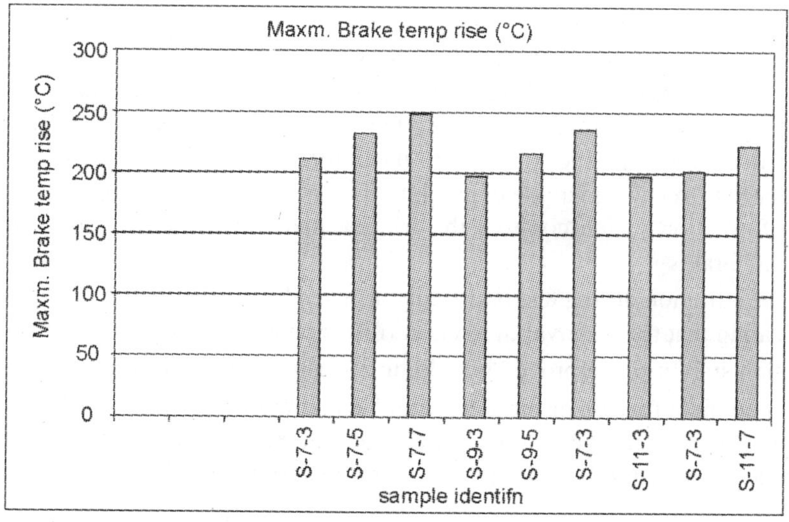

Fig. 6

Fig. 7 shows back plates, Ni plated diffusion layer, stronge iron, friction materials constituents along with the microstructure of S-11-5 sample, iron based friction material brake pad.

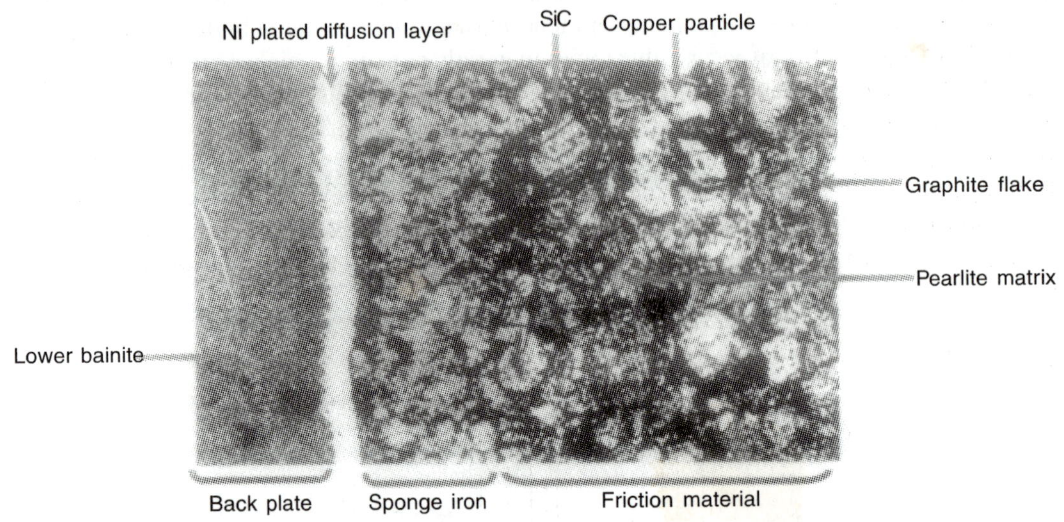

Fig. 7. Microstructure of S-11-5 sample iron based friction material brake pad.

CONCLUSIONS

Friction characteristics of nine different friction materials containing different amounts of natural graphite and silicon carbide were investigated using a brake dynamometer. Run-down mode of brake application was applied during the dynamometer test to examine the friction performance. The results showed that the relative amount of natural graphite and SiC had strong influence on the level of friction coefficient, friction stability, torque, wear and temperature rise variation. The important results from this study are summarized as follows:

1. The average friction coefficient increases with relative increase in the abrasive ingredient, viz., SiC. Increasing solid lubricant content, *i.e.* natural graphite has a tendency to moderate the friction coefficient and also keep the same under better control. Torque, aggressiveness of torque developed (peak torque to mean torque ratio), temperature rise and wear also get regulated by relative increase of graphite. This appears to be due to a self-sustaining tribological film formation which controls the friction and wear characteristics during high - energy aircraft braking.
2. Torque variation during test is strongly affected by the amount of SiC in the friction material, indicating that the removal of friction film is undesirable in reducing torque variation.
3. Friction stability is improved by adding more graphite in the friction material.
4. The amount of friction material wear and temperature rise during this dynamometer procedure shows that the relative amounts of SiC and graphite in the friction material plays a crucial role in determining the amount of friction material wear during the service.

3

Fatigue Mechanism in Low Alloy Sintered Steels

H. Khorsand
Assistant Professor of Mechanical Engineering Department- Materials Group of Khajenasir University of Technology, Research Center of SAPCO, Tehran, Iran.

S. M. Habibi
Department of Materials Science and Engg., Shiraz University, Pajouhesh Tarrahan Pars Co., Tehran, Iran

ABSTRACT

The powder metallurgy process is ideally suited to the mass production of structural parts of complicated shape and with tight tolerances without appreciable loss of material. Becausee of its unique attributes such as reducing weight and production cost, powder metallurgy has made very significant in roads as a substitution technology in automotive parts, such as cams and gears in this process, we can improve the mechanical properties such as fatigue resistance of PM materials by changing of manufacturing parameters, for example, density, sintering temperature, sintering time and ... . In this research, the fatigue behaviour of sintered steel (Fe-1.5 Mo-3 Mn-0.7 C) was investigated with varing of density and sintering temperature. The results showed that in this steel, increase of porosity plays an important role in the fatigue limit of specimens also with sintering at high temperature which results more spherical pores, the fatigue strength was improved. In this research, also SEM fractography was used to determine mechanism of fatigue fracture.

INTRODUCTION

Powder metallurgy, especially its ferrous branch, has traditionally been closely linked to the automotive industry; today the fraction is rather between 70 and 80% [1]. Recently the automotive industry consumes approximately 27 Ibs of P/M parts per vehicle. In 1977, this figure was 15 Ibs per vehicle. Ford motor company predicts that they have over 50 Ibs of P/M product in its vehicles by the year 2002 [2]. The mechanical properties of P/M steels are related directly to their microstructure and the level of porosity. In addition, the alloying mode has a significant effect on the microstructure and pore morphology in sintered steels [3].

Fatigue is becoming a more and more important field of application for powder metallurgy. And the many researchers have worked about fatigue of conventional ferrous P/M alloys like Fe-P, Fe-Cu. Fe-Cu -C, Fe-Cu-Ni within the past ten years. The ability of powder metallurgy to achieve parts with high qualities for special requirements opens up broad possibilities for a development or fatigue loaded P/M parts as well as for a substitution of conventional materials. The range of fatigue strength where stresses are in the order of the yield stress and consequently the amount of endurable cycles is relatively small ($N<5*10^4$ cycles) is called low- cycle fatigue (LCF). In the finite fatigue life range ($5*10^4 < N < 10^6$ cycles), the stress levels are lower than the yield stress and higher than the co-called endurance limit. The fatigue behaviour is described by the falling branch of the S-N curve, which is normally obtained from load controlled constant amplitude tests. Metallic parts of security belts and door hinges of a car are designed for this fatigue life area. Fatigue data for bending are higher than for axial loading. For a better understanding of the fatigue response of P/M materials. It is necessary to consider their specific microstructure [3- 9]. The typical microstructure of single phase P/M material is characterised mainly by the presence of pores. The pores may appear isolated, interconnected, or a mixture. In 1970, Haynes published a summary of the fatigue properties of a variety of sintered metals and alloys, Most of the data were obtained from plane bending mid rotating tests. Since then, only short reviews concerning fatigue properties of P/M materials have been published as part of general information on powder metallurgy [10, 11]. Sonsino and Esper recently issued a manual for fatigue design of P/M components [12]. From the results presented in papers, it turned out that pore length, pore curvature, and pore spacing, are the most prominent microstructural features, and it may be deduced how each of these parameters influences the endurance limit. These findings enabled some modeling of the microstructure by a statistical approach, so that σ_e could be given by the following relationship:

$$\sigma_e = g(H^4/V_p^2\beta)^{1/6} \qquad ...(1)$$

where g = constant, H = pore curvature, V_p = volume fraction of porosity, and β = average pore spacing. The use of equation (1) is restricted to porosities larger than 8%. Similar results were reported by Piotrowski et al. [13], Lindqvist [14] and Sonsino et al. [15]. Lindqvist [16] studied the fatigue strength behaviour of Ni, Cu, Mo, and C steels with varying Ni content from 1.75 to 8%. and reported an increase due to matrix strengthening and enclosing pores in the austenite phase. Almost no effect of the initial powder size on the fatigue behaviour was detected. The role of carbon is similar to that in conventional steels: higher content results in higher strength. Obviously the reduction of porosity increases the fatigue strength. On the other hand, due to open porosity, gaseous and liquid media may penetrate into the interior, which may influence the liquid phase sintering process [17]. Ledoux and Prioul determined σ_e, of pure iron, sintered for two different types of powder: ASC (water atomised iron powder), and NC (reduced spong iron powder), using push-pull and bending fatigue testing at R = –1. as already known from tensile properties, specimens prepared from NC powder show better values of fatigue strength compared with specimens produced from ASC powder [18]. Detailed investigations yielded that density/porosity alone can not satisfactorily describe the fatigue behaviour of such P/M materials. Lower sintering temperature associated with higher compacting pressures and high sintering time leads to higher σ_e values. However, sintering times exceeding 2 h do not improve the fatigue strength [19]. Christian and German examined the influence of initial particle size of the starting powder on fatigue strength. The authors reported that for low densities, small particles are preferable to achieve high σ_e values, whereas for higher densities, larger particles seem to he advantageous. Thus for low porosity, rounder pores and a larger pore spacing proved favourable for high values of σ_e. In porous materials, especially in high densities, for increasing of strength, we can use a heat treatment process such as hardened and tempered. H. Danninger and D. Spoljaric in 1996 showed the effect of heat treatment on the mechanical properties of Mo sintered steels [20]. Bertilsson demonstrated that the sintered density primarily affects fatigue resistance and elongation with a lesser effect on tensile strength [21]. A fatigue strength of 38% of the static tensile strength is often an overestimate [22]. Porosity significantly affects fatigue performance, although the role of porosity in fatigue is somewhat different than that in tension Many investigations [3, 4. 10, 11-13] indicate the preferred sites for crack initiation are pores or pore clusters located at near the surface of the specimen. Holmes and Queeney [23] proposed that the relatively high stress concentration at pores, particularly surface pores, is responsible for localized slip leading to crack initiation. In general, an angular pore creates a higher stress concentration and stress intensity factor than a round pore [24]. Christian and German [25] showed that total porosity, pore size, pore shape, and pore separation are important factors that control the fatigue behaviour of P/M materials. Pores have also been proposed to act as linkage sites for crack propagation through interpore ligaments [16].

EXPERIMENTAL PROCEDURE

The low alloy steel (Fe – 1. 5% Mo – 3% Mn) with 0. 7%C Was selected for experiments. To obtain this analysis, the prealloyed steel powder astaloy Mo which is iron powder with 1.5% Mo was initially supplied. Carbon as a fine natural graphite (UF4) and manganese powder with high purity were added and 0.75% zinc acetearate was added to the mix as lubricant. The powders were blended in a tumbling mixer and then compacted in a pressing machine with a floating die to form rectangular green specimens in dimensions of 90×10×10 mm^3. Higher density in the specimens was achieved through double pressing and double sintering technique. The compact samples were sintered at 1120°C and 1250°C in an atmosphere of cracked amonia and high vacuum (10^{-4} torr) respectively. (After annealing at 700°C for double pressing - double sintering samples). The cooling rate of specimens after sintering in industrial furnace and vacuum furnace was 30 and 20°c/min respectively. Some of the specimens were heat treated by austenitizing at 800-850°C for 30 minutes under controlled atmosphere and then quenched in oil and temperad at 300-350°C for 60 minutes. The density of samples was measured in accordance with recommendation of ISO 2738/1987 standard. Tensile and fatigue samples were prepared according to recommendation of ISO 3928(1977) and ASTM - G99. Fatigue tests were carried out by rotate - bending method to achieve a fatigue limit of 10^7 cycles. The rotate rate was 2850 RPM and the mean stress was zero. Fracture surfaces were studied by electron microscopy.

RESULTS AND DISCUSSION

The influence of total porosity and sintering temperature on the fatigue strength of specimens have been shown in Table 1.

It can be seen that fatigue limit is decreased with increasing of porosity and it is increased with increasing of sintering temperature. It this research work, S - N curve (a diagram with the nominal stress amplitude as a function of cycles to failure), that is one of the most important graphs to clarify the fatigue behaviour of materials, was prepared. In Figure 1 and Figure 2, the S-N curves of this steel under different manufacturing conditions are drawn.

According to A. Hadrboletz, B. Weiss and several researchers' results, It can be seen that the curves have not become horizontal even after 10^{+7}cycles. D. Spoljaric and H. Danninger [26] apparently showed that fatigue limit of PM steels are obtained in much higher cycles (around 10^9). Of course,presence of internal pores is the most important reason because they culminate in changing of fatigue fracture mechanism. Figure 3 represents the correlation between fatigue strength and porosity percentage at the two different sintering temperatures. In this figure, It is showed that we can increase fatigue strength with correct controlling of manufacturing conditions such as sintering temperature and density. The pore morphology was quantified by image analysis and its results have been shown in Table 2.

Table 1. Physical and Mechanical properties of sintered steel Fe-3 Mn- 1. 5 Mo-0. 7C

Conditions	*Sintering Temperature* (°C)	*Density* (g/cm^3)	*Fatigue Strength* (*MPa*)	*Tensile Strength* (*MPa*)	*Hardness* (*HV30*)
Sintered	1120	6.5	80	210	150
		6.8	135	320	165
		7.2	272	555	191
	1250	6.5	165	315	160
		6.8	215	405	172
		7.2	335	610	202
Sintered Heat treated	1120	6.5	–	–	–
		6.8	–	–	–
		7.2	–	–	–
	1250	6.5	205	400	220
		6.8	–	–	–
		7.2	460	815	310

Table 2. Physical and Mechanical properties of sintered steel-Fe-3Mn-1.5Mo-0.7C

Density (g/cm^3)	*Sintering Temperature* (°*C*)	*Shape Factor*	*Ave. Pore Length* (μ*m*)
6.5	1120	0.32	29
6.5	1250	0.51	21
6.8	1120	0.41	21
6.8	1250	0.63	15
7.2	1120	0.58	17
7.2	1250	0.86	13

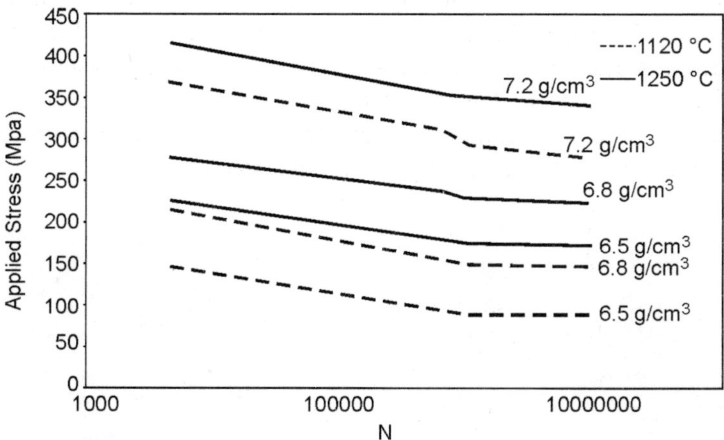

Fig. 1. The effect production conditions on S-N curves.

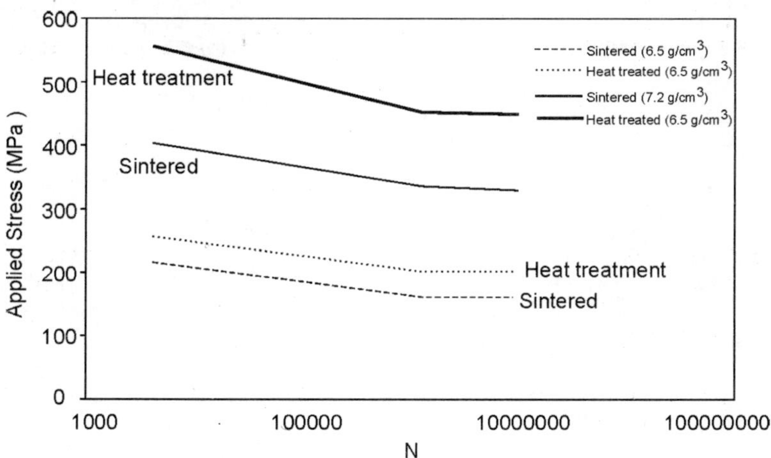

Fig. 2. The effect of heat treatment on S-N curves.

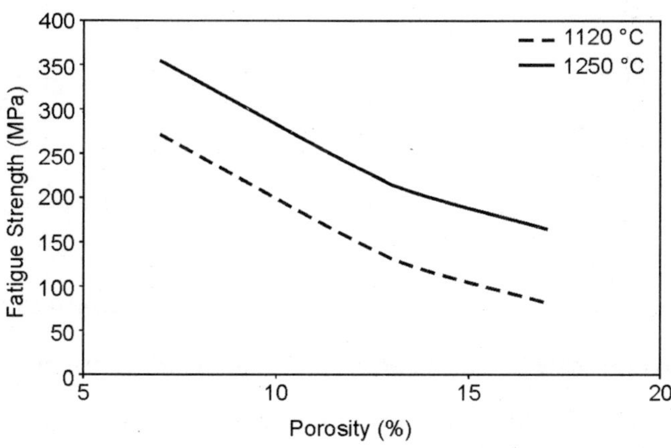

Fig. 3. Fatigue Strength against porosity.

The fatigue strength is markedly raised by increasing of shape factor. The shape factor is close to 0 if a pore is very irregular and close to 1 if it is more or less circular. However the effect of shape factor on mechanical properties of sintered steels is very important. The results show that average pore length is decreased with increasing of density and sintering temperature. of course, these results interpret that density is the more important than sintering temperature and density affects obviously the mechanical properties of these PM steels with decreasing of average pore length.

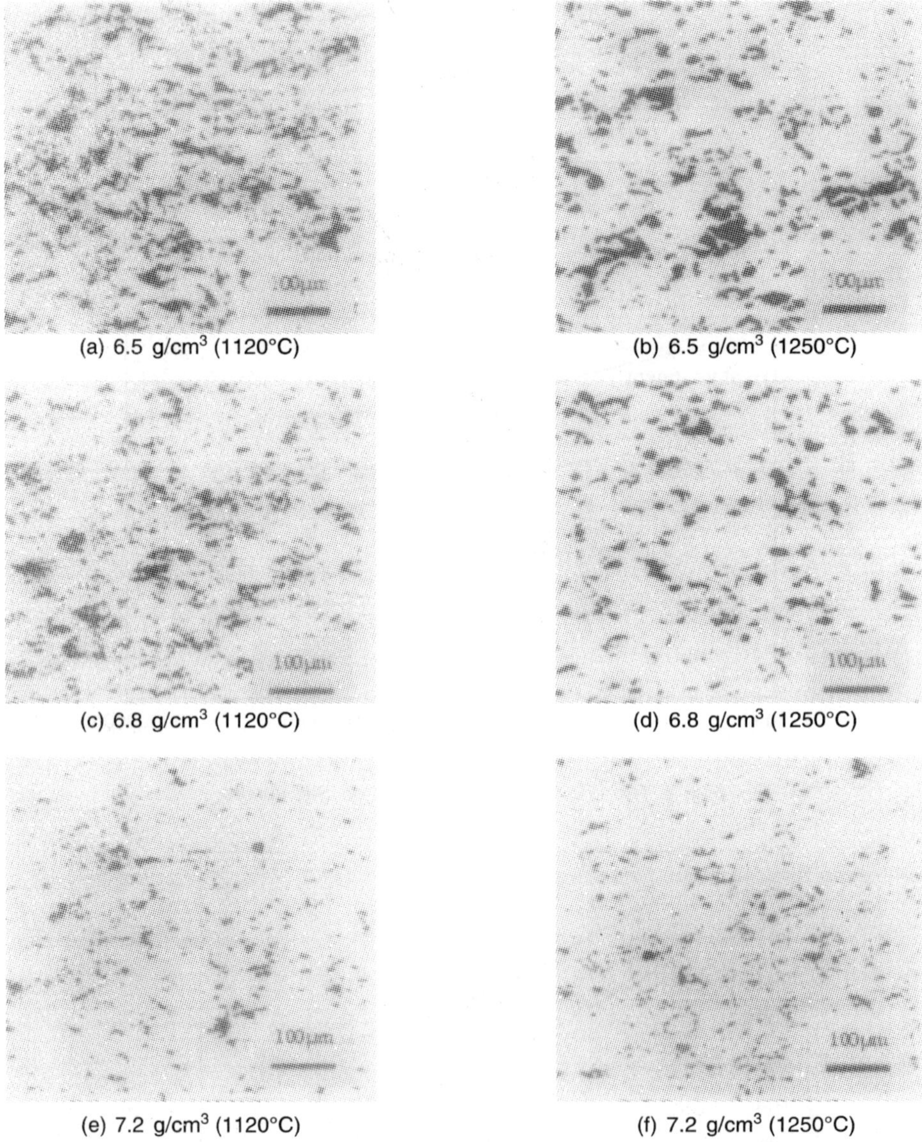

(a) 6.5 g/cm^3 (1120°C) (b) 6.5 g/cm^3 (1250°C)

(c) 6.8 g/cm^3 (1120°C) (d) 6.8 g/cm^3 (1250°C)

(e) 7.2 g/cm^3 (1120°C) (f) 7.2 g/cm^3 (1250°C)

Fig. 4. The microstructure of' specimens with different densities and sintering temperature prior to etching.

It was turned out that pore length and pore curvature are the most prominent microstructural features and it was deduced how each of these parameters influences the endurance limit. Pores act as stress raisers especially with the low shape factors and pores with the high shape factors (near to 1) can decrease the rate of crack propagation and they, play an important role in increasing of fatigue and tensile strength. In this steel, increase of sintering temperature is the most important parameter for increasing of shape factor (table 2) and pores which have high shape factor have a delaying effect on crack propagation due to blunting of crack tip. However, Shape factor was increased with increascant of density as it has been presented in table 2 that these results are agreement with Esper's results about steels (Fe-Cu-Ni-C) and (Fe-Cu-C) [15]. Of course, effect of density is less than effect of sintering temperature on the shape factor. The microstructure of this steel before the etching under the different manufacturing conditions have been shown in Figure 4.

Of course, from the results presented in this paper, It shows that primary steps of increasing in shape factor is more effective on fatigue strength than the following increase of shape factor. It is very interesting that we don't need the microstructures with completely circular pores. Because blunted of sharp tips of irregular pores improves intensively fatigue strength.

It is well known that the static properties of sintered materials are dependent on total porosity and in this research was seen that especially in low densities, the mechanical properties almost independent of the matrix structure. According to other researchs' results [19] that the fatigue behaviour of PM steels at a low density ($p < 6.75$ g/cm^3) independent of the chemical composition. In this research at 6.5 g/cm^3 density which pores have govering role, increscent of matrix strength by heat treatment did not have outstanding effect on the fatigue strength. But in the high densities (especially 7.2 g/cm^3), all of the obtained results showed that increase of matrix strength is more effective than changing of pore morphology on the mechanical properties. Pearlitic microstructure of this steel after the etching has been clearly shown in Figure 5 and tempered martensitic microstructure also can be seen in Figure 6 (for heat treated samples).

Fig. 5. Pearlitic microstructure of specimens.

Fig. 6. Martensitic microstructure of specimens.

Obtained results interpret that the strength of these PM steels under periodic loading are affecting more by heat treatment and porosity percentage in compare with static behaviour. (Figure 7).

In this research. SEM fractography was used to find features of the fracture surface. Fracture surface of tensile samples shows two kinds of fracture: Ductile rupture, due to microvoid coalescence occurred locally and brittle failure generally occurred in large pearlitic grains (Figure 8). Microvoid coalescence can be clearly seen at sintering necks in Figure 9 and in Figure 10, brittle failure of a pearlitic grain has bean presented.

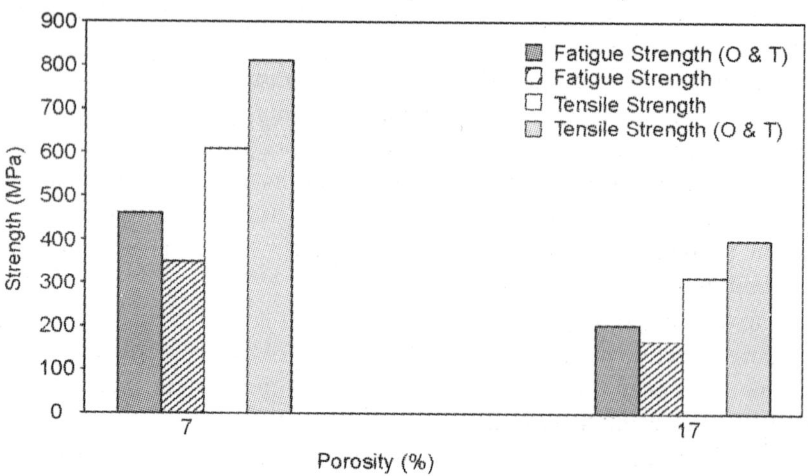

Fig. 7. The effect of heat treatment and Porosity on Strength.

The investigation presented here is focused on the fatigue crack initiation and crack formation in relation to the microstructure. Fatigue cracks tend to initiate at pore clusters or inclusions near the surface because of the higher localized stress intensity of these areas (Figure 11). after the nucleation of a fatigue crack. It propagates through the interpore ligaments (sintering necks) especially for high porosity specimens. Although an irregular shaped pore could act as a nucleation site, a smooth pore could arrest the growth of a fatigue crack. A. Berg marks, L. Alzati and U, Persson [27] showed that in PM steel (Fe-Mo-Cu-Ni-C), The crack propagation through the interface between a ductile and a very hard phase have resulted in striations. H. Drar [28] was also successful that showed striations in fatigue fracture surface. In this research, we could observe striations as they have been shown in figure 12. Also in powder metallurgy steel (Fe-Mo-Cu-Ni-C) at 7 g/cm^3 density, striations were observed by w. chawala etal.

In fatigue fracture surface of this steel especially for low porosity specimens, brittle fracture of large pearlitic grains was observed similar to tensile fracture surfaces.

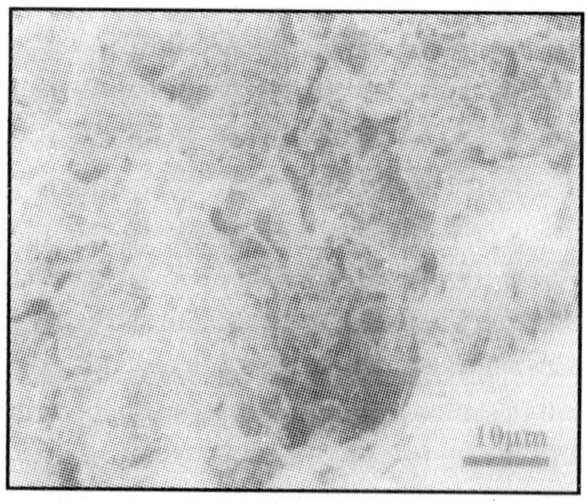

Fig. 8. Fracture surface of tensile samples.

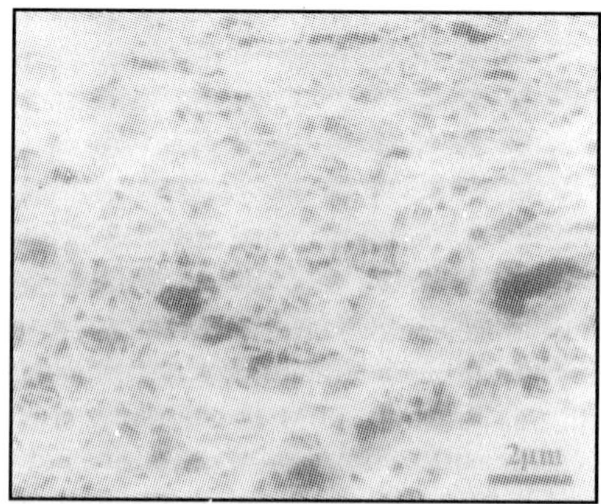

Fig. 9. Ductile rupture of tensile samples.

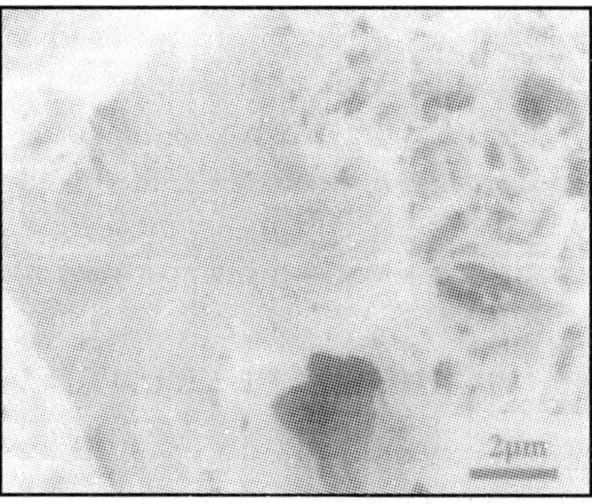

Fig. 10. Brittle failure of a pearlitic grain.

Fig. 11. Fatigue crack initiating site.

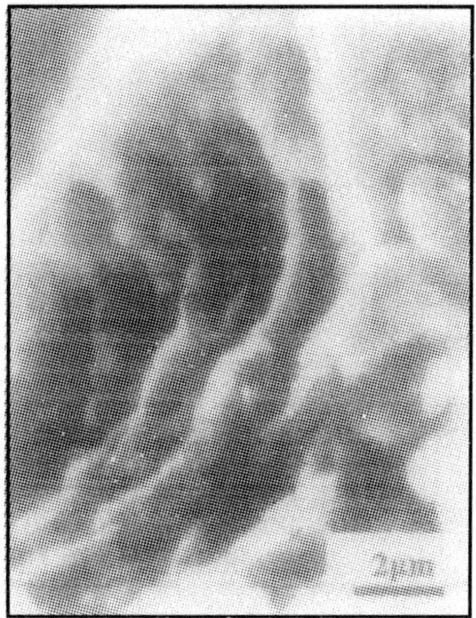

Fig. 12. Striations in fatigue fracture surface.

CONCLUSIONS

1. Pore length and pore curvature are the most prominent microstructural features that influence the endurance limit.
2. In the low densities, increase of sintering temperature and shape factor is the most important parameter for increasing of fatigue strength.
3. In the high densities, increase of matrix strength by heat treatment is more effective than changing of pore morphology on the fatigue strength.

4. The effect of density is less than effect of sintering temperature on the shape factore.
5. Pores with the lowe shape factors act as stress raisers and pores wich have high shape factor have a delaying effect on crack propagation due to blunting of crack tip.
6. Fatigue crack tends to initiate at pore clusters or inclusions near the surface because of the higher localized stress intensity of these areas and then it propagates through the interpore ligaments (especially for high porosity samples).In the fatigue fracture surface, brittle fracture of pearlitic grains especially for low porosity specimens was observed.

ACKNOWLEDGEMENTS

The authores wish to thank the Mashad Powder Metallurgy Co. and Iran Powder Metallurgy Complex for making samples and SAPCO. For the financial support for this work.

REFERENCES

1. A. Salak, Ferrous Powder Metallurgy, Cambridge International Science Publishing, Cambridge, 1995.
2. B. James, Recent Developments In Ferrous Powder Metallurgy Alloys, Metal Powder Report, Vol. 46, No. 9, 2001, pp. 26-30.
3. H. Danninger, G. Jangg B. Weiss, and Stickler, Microstructure and Mechanical Properties of Sintered Iron Part I, Powder Metallurgy, International, Vol. 25, No. 3,1993, pp. 111- 117.
4. S.M. Habibi, K. Janghorban, H Khorsand, H. Yoozbashizade, 2002 The Effect of Microstructure on Mechanical Properties of Ferrous PIM Materials, PM Auto'02, Isfahan, Iran, April 21-25 2002.
5. H. Danninger, G. Jangg, B. Weiss, mid'Stickler, Microstructure and Mechanical Properties of Sintered Iron Part II, Powder Metallurgy International, Vol. 25, No. 4,1993. pp. 170-173.
6. H. Khorsand, S. M. Habibi, H. Yoozbashizade, K. Janghorban & 5.M. S. Reihani, Investigation of Microstructure and Fatigue Properties of Low Alloy Sint-ered Steels By Heat Treatment, PIM Science & Technology Briefs, Vol. 4, No. 2. 2002, pp. 18-22.
7. A. E. Segall, T Cimino, T. Murphy, and T. Haberberger, The Influence Of Microstructure and Carbide Particles on The Creep Behaviors of P/M Alloys, $PM^2TEC2001$, International Conference on Powder Metallurgy and Particulate Materials, New Orleans Louisiana USA, May 13 - 17, 2001.
8. Z.Y. Liu, N.H. Loh *, K.A. Khor, S.B. Tor, Microstructure Evolution During Sintering of Injection Molded M2 High Speed Steel, Materials Science and Engineering A, Vol. 293, No. 1, 2000, pp. 46-55.
9. Alan B. Davala, Amie H. Graham, Robert J. Causton,Application of High Performance Materials and Processes-Alloy Systems, PM^2TEC '98, International Conference on Powder Metallurgy and Particulate Materials, Las Vegas, Nevada USA, May 31-June 4,1998.
10. R. Haynes, Fatigue Behavior of Sinter-ed Metals And Alloy, Powder Metallurgy, Vol. 13, No. 26, pp. 465-510.

11. W. Brian James and Robert C. OtBrien High Performance Ferrous Pim Materials: The Effect of Alloying Method on Dynamic Properties, The Annual Powder Metallurgy Conference, San Diego, California. USA, June 11-14,1989.

12. C. M. Sonsino, Methods to Determine Relevant Material Properties for the Fatigue Design of Powder Metallurgy Parts, Powder Metallurgy International, Vol. 16, No. 1, 1984, pp. 34-38.

13. A. Piotrowski, and G. Biallas, Influence of Sintering Temperature on Pore Morphology, Microstructure and Fatigue Behaviour of MoNiCu Alloyed Sintered Steel, Powder Metallurgy, Vol. 41, No. 2,1998, pp. 109-114.

14. Bjorn Lindqvist, Koki Kanno, Considerations when Sintering Oxidation-sensitive PM Steels Introduction, PM2TEC2002, International Conference Powder Metallurgy & Particulate Materials, Orlando, Florida USA, June 16-21 2002.

15. E. I. Esper and C. M. Sonsino, Fatigue Design for P/M Components. The European Powder Metallurgy association, Shrewbury, 1994.

16. Bjorn Lindqvist, Koki Kanno, Voshinobu Takeda, Susumu Takahashi, Kiichi Kanda, Sintering of Prealloy 3cr-0.5mo Steel Powder in a Carbon/Carbon Composite Mesh Belt Furnace" PM2TEC2002, International Conference on Powder Metallurgy & Particulate Materials, Orlando, Florida USA June 16 - 21, 2002.

17. N. Chawla. S. Polasik, K. S. Narasimhan,. M. Koopman, and K. K. Chawla, Fatigue Behavior of Binder-Treated Powder Metallurgy Steels, The International Journal of Powder Metallurgy, Vol. 37, No. 3, 200 1. pp. 49-5 7.

18. L. L4edoux and C. Prioul. Fatigue and Fracture of Porous Steels and Cu-infiltrated Porous Steels, PM2 TEC98, International Conference on Powder Metallurgy & Particulate Materials, Las Vegas, Nevada USA, May 31-June 4. 1998.

19. A. Hadrboletz A, B. Weiss, Fatigue Behavior of Iron Based Sintered Material., Intl Mater Rev, vol. 42, No. 1, 1997, pp. 144.

20. H. Danninger, D. Spoljaric, and B. Weiss, Microstructural Features Limiting the Performance of P/M Steels, The International Journal of Powder Metallurgy, Vol. 33, No. 4,1997, pp. 43-53.

21. 1. Bertilsson and B. Karlsson, Dynamic Properties of Sintered Steel, Powder Metallurgy, Vol. 30, No. 3,19871, pp. 183-188.

22. Robert C. O'Brien, Impact and Fatigue Characterization of Selected Ferrous P/M Materials, The Annual Powder Metallurgy Conference, Dallas, Texas USA, May 17-20,1987.

23. J. Holmes, and R.A. Queeney, Fatigue Crack Initiation in a Porous Steel, Powder Metallurgy, Vol. 28, No. 4,1995, pp. 231-235.

24. H. Khorsand, H. Yoozbashizadeh, S. M. Habibi, K. Jonghorban and S. M. S. Rahmani, Carbon, Porosity and Fatigue in Sintered Steel, Metal Powder Report, Vol. 5 7, No. 4, 2002, pp. 32-36.

25. K.D. Christian and R.M. German, 'Relation between Pore Structure and Fatigue Behavior in Sintered Iron-Copper-Carbon', The International Journal of Powder Metallurgy, Vol. 31, No. 1, 1995, pp. 51-61.

26. D. Spoljaric, H. Danninger, B. Weiss, and Rostickler, Fatigue Behavior of Mo-Alloyed Sintered Steels and its Consequences for Application, Proc. PM94. Paris,Vol. 2,1994, pp. 827-832.

27. Anders Bergmark, Luigi Alzati and Ulrika Persson, Fatigue Crack Initiation in PM Steel, PM2TEC2002, International Conference on Powder Metallurgy and Particulate Materials, Orlando, Florida USA, June 16-21, 2002.

28. H. Drar, Metallographic and Fractographic Examination of Fatigue Loaded PM-steel with and without MNS Additive, Materials Characterization, Vol. 45, No. 4,2000, pp. 211-220.

4

Fatigue Performance of PM Steels in a Sintered State

Luigi Alzati, Anders Bergmark, Jessica Andersson.
Höganäs AB, SE-26383 Höganäs, Sweden.

ABSTRACT

An overview of the plane bending fatigue performance of PM steels in as-sintered state is presented. Pre-alloyed Molybdenum and Chromium-Molybdenum materials, diffusion-bonded Ni-Cu-Mo and Fe-2% Cu pre-mixes are reviewed. The effect of varying density level and Carbon content upon fatigue performance is analyzed separately. Diffusionbonded materials exhibit heterogeneous microstructures with high performance metallographic phases distributed along the sinter-necks and around the pores. Optimized carbon content in diffusion-alloyed materials is characterized by a well-distributed martensitic microstructure. Furthermore, fatigue crack initiation is found in the low yield stress phase in the base powder core. PM steels based on pre-alloyed powders exhibit one phase or random distribution of two or more phases with no systematic orders in relation to the pores, hence fatigue crack initiation is related to sinter-neck failure. Influence of density is more pronounced for pre-alloyed materials than compared to diffusion-bonded materials.

INTRODUCTION

Microstructure and Density are the most relevand parameters influencing PM steels fatigue performance. The comprehension of materials' microstructure allows the interpretation of fatigue properties [1, 2, 3].

The role of microstructure is analyzed in this paper by the comparison of different alloying systems and different alloying techniques: pre-alloyed, premixed and diffusion-alloying system and different alloying techniques; pre-alloyed, premixed and diffusion-bonded powders at density 7, 1 g/cm^3 are reviewed. two carbon levels have been considered for each material.

The role of density is shown for materials based on one-pre-alloyed and two diffusion bonded powders. Various compaction techniques hav e been used. all the materials hav e been processed by the same sintering route: 1120°C/30' 0,8°C/s the cooling rate. One example of sinter-hardened material with 2,5°C/s cooling rate has also been included.

MATERIALS

Table 1 lists all the investigated materials. Powders have been mixed with 0.8% Amidwax, cold compacted at 7.1-7. 15 g/cm^3, sintered in belt furnace at 1120°c in 90/10 N_2/H_2 atmosphere for 30', with 0.8°C/sec cooling rate. One diffusion-bonded material has also been tested in sinter-hardened conditions with 2,5°C/s cooling rate.

Table 1. Chemical composition and alloying technique for all the investigated iron-based powders

Group	*Material*	*Pre-alloyed elements*	*Diffusion Bonded elements*	*Premixed elements*
Pre-alloyed	Astaloy85Mo	0,85%Mo	–	–
	Astaloy Mo	1,5%Mo	–	–
	AstaloyCrL	1,5%Cr, 0,2%Mo	–	–
Premixed	ASC100.29	–	–	2%/Cu(-100mesh)
	Astaloy85Mo	0,85%Mo	–	2%Cu(-325mesh)
	Astaloy85Mo	0,85%Mo	–	2%Ni(Inco123)
Diffusion-	DistaloyDC	1,5%Mo	2%Ni	–
	DistaloyAE	–	4%Ni, 1,5%Cu, 0,5%Mo	– –
	DistaloyHP	1,5%Mo	4%Ni, 2%Cu	–
	DistaloyDH	1,5%Mo	2%Cu	–

EXPERIMENTAL PROCEDURES

Four point bending fatigue tests have been performed for as-sintered, un-notched ISO 3928 test bars with modified cross section [1]. The following test parameters have been used: load ratio R= – 1 (i.e. mean stress equal to zero); frequency: 25-29 Hz; broken specimens stopped for 1.5% increased compliance; run out at 2 millions cycles; fatigue endurance [$S_{A,50\%}$] and standard deviation were determined by the staircase method, according to MPIF Standard No. 56, 2001. The staircase step has in all cases been 10MPa. Ordinary Light Optical Microscopy (LOM) observations have b een made on polished and etched samples in order to analyze materials' microstructures.

Scanning electron microscopy (SEM) fractography observations have been made on specimens tested at stress levels less than 5% higher than the $S_{A,50\%}$ value of the respective material.

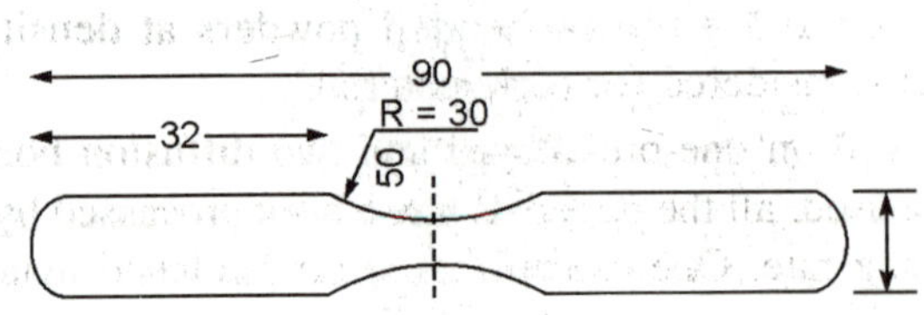

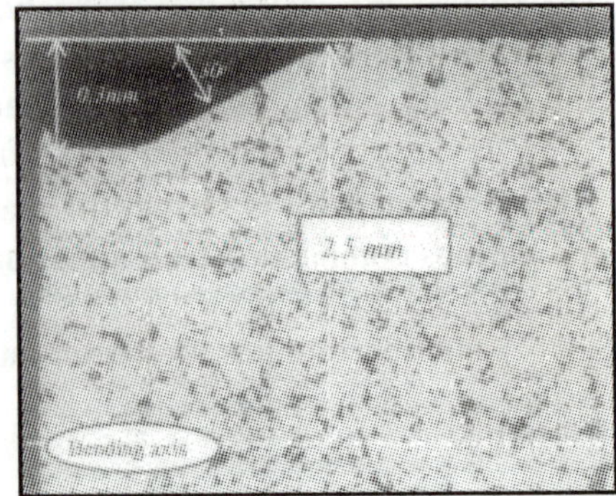

Fig. 1. Modified ISO 3928 test specimen with chamfered edges.

FATIGUE TESTS RESULTS

Density close to 7.1 g/cm^3

All the results obtained for materials cold compacted at density close to 7.1 g/cm^3 are summarized in Table. 2.

Table 2. Bending fatigue endurance evaluation at 2 million cycles, according to MPIF Standard No. 56 (2001) for as-sintered materials at 7.1 g/cm^3 density. Sintering performed in belt furnace at 1120°C in 90/10 N_2/H_2 atmosphere for 30, 0.8°C/sec cooling rate, except where noted.

Group	*Material*	*Carbon [%]*	*S 50% [MPa]*	*Std Dev [MPa]*
Pre-alloyed	Astaloy 85Mo	0.48	180	11
	Astaloy 85Mo	0.76	196	8
	Astaloy Mo	0.42	201	<5
	Astaloy Mo	0.82	210	<5
	Astaloy CrL	0.55	236	8.5
	Astaloy CrL	0.74	252	<5
Premixed	ASC100.29 +2%Cu (-100)	0.49	190	<5
	ASC100.29 +2%Cu (-100)	0.77	220	<5
	Astaloy 85Mo +2%Ni	0.5	201	16
	Astaloy 85Mo +2% Ni	0.78	210	<5
	Astaloy 85Mo + 2%Cu (-325)	0.49	211	15
	Astaloy 85 Mo + 2%cu (-325	0.78	233	<5
Diffusion-Bonded	Distaloy AE	0.46	224	7
	Distaloy AE	0.81	274	8.6
	Distaloy DC	0.47	230	<5
	Distaloy DC	0.83	239	<5
	Distaloy HP	0.48	251	13
	Distaloy HP	0.85	326	<5
	Distaloy DH	0.49	209	15
	Distaloy DH	0.76	236	22
	Distaloy DH (2,5?C/s)	0.47	236	13
	Distaloy DH (2,5?C/s)	0.73	271	17

Densities higher than 7.1 g/cm^3

Bending fatigue tests have been performed for Diffusion-bonded distaloy AE and Distaloy DC and pre-alloyed Astaloy CrL at different density levels. One carbon lev el has been set for each material, in order to keep microstructure as constant parameter. Cold compaction, warm compaction, double pressing double sintering and powder forging have been performed to obtain the four density levels shown in Fig. 2.

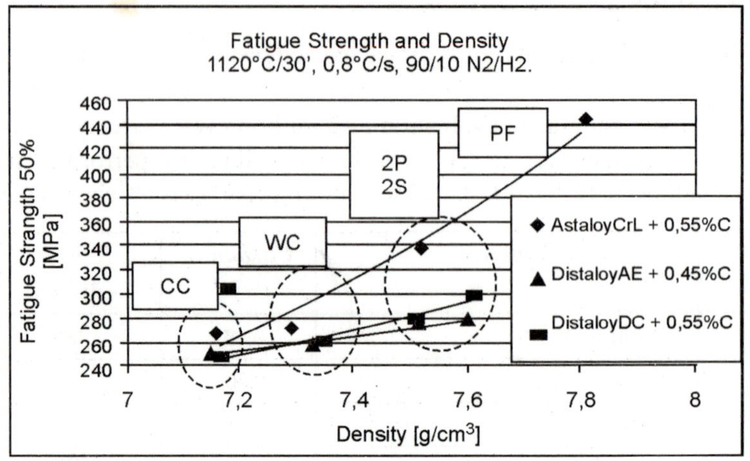

Fig. 2

Bending fatigue endurance evaluation at 2 million cycles, according to MPIf standard No. 56 (2001) for as-sintered materials at different densities. sintering performed in belt furnace at 1120°C in 90/10 N_2/H_2 atmosphere for 30°, 0.8°C/Sec cooling rate. Four levels of density have been achieved by Cold Compaction (CC), Warm Compaction (WC), Double pressing double sintering (2P2S), Powder Forging (PF).

METALLOGRAPHY

Pre-alloyed Materials

Astaloy 85Mo shows upper bainitic structure at 0,48°C; at 0,76°C spots of very dense upper bainite can also be detected. astaloy Mo shows upper bainitic strucιure at 0.42°C, while at 0,82°C also spots of lower bainite can be detected [3].

Astaloy CrL shows upper bainitic and pearlitic structure at 0.55% and 0.74°C respectively [2]. Upper bainite is denser in astaloy CrL than in Astaloy Mo and even more than in Astaloy85Mo, at similar Carbon levels; i.e., the ferritic areas among cementite lamellas are the smallest for Astaloy CrL.

Premixed Materials

Standard Fe+2% Cu mixes show ferritic-pearlitic structure. Cu-rich areas surround the base powder particles.

Astaloy 95Mo +2%Ni (Inco 123) +0,5%C shows upper bainitc base powder particles and austenitic Ni-rich islands partially surrounded by spots of martens ite. At 0.78°C, the base powder particles show denser upper bainite; still the austenitic islands do not show a continuous martensitic shell.

Astaloy85Mo +2%Cu(-325) shows fully upper bainitic structure at 0.49°C;spots of denser upper b ainite [?30+30?M^2] can be detectecd in correspondence of the Cu-rich areas. At 0.78°C the base powder particles show denser upper b ainite and spots of lower bainite/martensite [?50+50?m] are present in the Cu-rich areas.

Diffusion-bonded Materials

DistaloyAE shows ferritic/pearlitic and fully pearlitic base powder particles at 0.46° and 0.81°C respectively. A Cu-Ni-rich network surrounding the base powder particles is consisting of lath martensite/ lower bainite with austenitic islands and plate martensite with austenitic islands at 0.46° and 0.81°C respectively.

Both Distaloy DC and Distaloy HP show upper bainitic base powder particles at Carbon levels close to 0.5% and 0.8%. Ni-rich austenitic islands in Distaloy DC show thicker martensitic shells for higher carbon level. Cu-Ni-rich network in Distaloy HP shows the same characteristics that are seen for Distaloy ae [3] (see Figure 3).

As-sintered distaloy dH shows fully upper bainitic structure at 0.49°C, while at 0.76°C lower bainitic/martensitic spots can be detected in correspondence of some sinter-necks. it must be pointed out that no continuous martensitic network is established. Sinter-hardened Distaloy DH shows at 0.47°C upper bainitic base powder particles and lower bainitic/martensitic Cu-rich network; at 0.73°, the base powder particles show mainly martensitic structure, with bainitic core. The Cu-rich network is martensitic (see Figure 3).

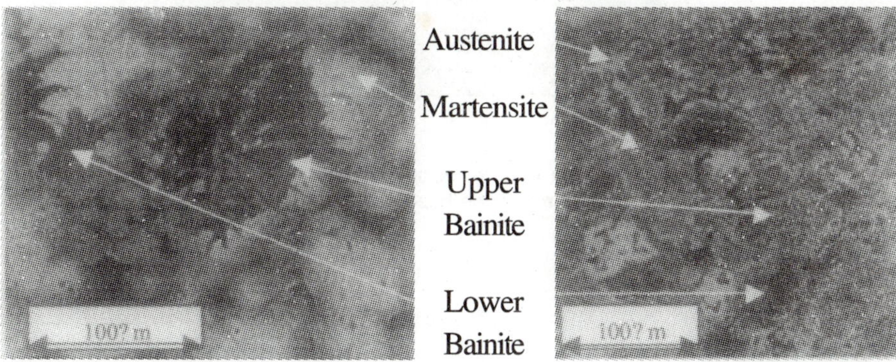

Fig. 3. To the left: Microstructure of distaloy dH (2.5°C/s) +0.73°C. To the right: microstructure of Distaloy HP+0.85°C. At density close to 7.1 g/cm^3, these microstructures allowed? 50% equal to 271 MPa (326 MPa) and standard deviations equal to 17 Mpa (< 5 MPa) for Distaloy DH (Distaloy HP).

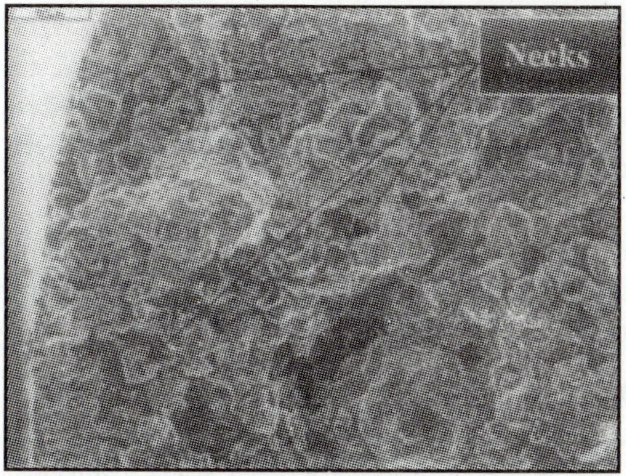

Fig. 4. Crack initiation and early propagation area for astaloy Mo +0.84%C. mainly inter-particles fracture, i.e. through the sinter-necks.

FRACTOGRAPHY

SEM fractography reveals different behavior of diffusion-bonded materials, compared to pre-alloyed materials. The latter show crack early propagation mainly developed by inter-particles modality, i.e., through the sinter-necks (see Figure 4). In crack initial propagation area distaloy materials show mainly trans-particles fracture, i.e., through the base powder particles (see Figure 5).

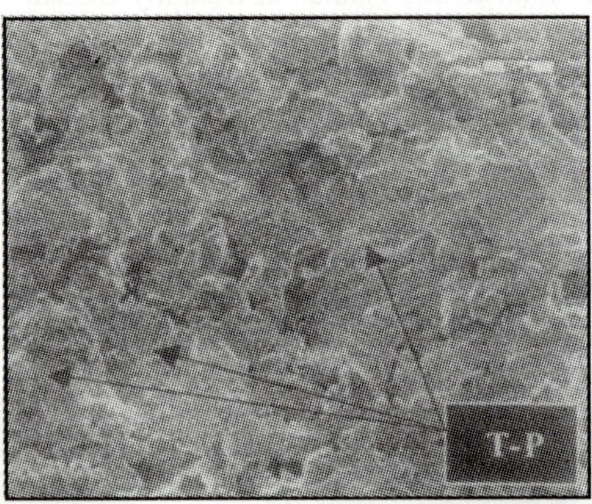

Fig. 5. Crack initiation and early propagation area for distaloy HP + 0.65%C. Mainly trans-particles fracture ("t\T-P"), i.e. through the base powder particles.

DISCUSSION

Fatigue Tests Results

Pre-alloyed materials

Different levels of density of upper bainite (i.e. amount of ferrite among the cementite lamellas) give reason for the different performance of Astaloy CrL, Astaloy Mo and Astaloy 85Mo. Tests results for the high Carbon level series show beter performance of Astaloy CrL pearlitic structure, compared to upper bainitic Astaloy 85Mo and Astaloy Mo.

Premixed materials

2% Ni and 2% Cu additions increased the performance of Astaloy 85Mo by 10% and 15% respectively.

Due to the better performance of upper b ainitic compared to ferritic/pearlitic base powder particles, Astaloy 85Mo +2% Cu(325) shows 7% to 10% higher fatigue strength than ASC100.29 +2%Cu (–100). Smaller Copper particles size might also have played a beneficial role.

Diffusion-bonded materials

DistaloyAe shows microstructural modificatio n level both in the base powder particles and in the Cu-Ni-rich network with increasing Carbon level from 0.46% to 0.81%: the reduction of the ferritic portion in the ferritic/pearlitic base powder particles and the transformation to fully plate martensite of the Cu-Ni-network has beneficial effect on the fatigue strength.

Increased carbon level from 0.47 to 0.83% has limited effect on the microstructure of Distaloy DC; consequently, the fatigue strength shows only slight difference (3.5%).

The beneficial role of optimized network microstructure is highlighted by distaloy HP: from 0.48% to 0.85°C no significant difference is observed in the bainitic base powder particles; therefore, the 30% increase in fatigue strength is to be considered due to the formation of fully plate-martensitic Cu-Ni-network.

As-sintered distaloydH shows similar microstructure to Astaloy 85Mo +2% Cu(325), at comparable carb on levels. The fatigue performance is therefore similar. It has to be noticed that fatigue strength values are slightly lower than what obtained for Distaloy DC: the latter presents formation of martensite in small areas in correspondence of the Ni-rich sinter-necks.

Sinter-hardened Distaloy DH +0.47°C shows mostly-martensitic/lower bainitic Cu-rich network. The amount of lower bainite is so high that the same performance of as-sintered Distaloy DH +0.76°C is achieved.

Sinter-hardened distaloy DH +0.73° shows 15% higher fatigue strength, compared to what found at 0.47°C. This result is remarkably lower than what shown by Distaloy HP. The reason has to be found in the microstructural modification due to increased carbon level: at 0.73°C martensite is present not only in the network, but also in large portions of base powder particles. This structure is less favorable for fatigue compared to the thinner, fully plate-martensitic and thinner network shown by Distaloy HP +0.85°C (see Figure 3).

In Distaloy materials the effect of 2% Ni is higher than the one of 2% Cu (compare distaloy DC to as-sintered distaloy DH); the opposite result was found in premixes of Astaloy 85Mo. the reason is the pronounced martensitic transformation occurring in Ni-rich areas in the diffusion-bonded distaloy DC, due to better diffusion of Nickel, combined with higher Molybdenum level in the base powder.

Scatter in Fatigue Strength Results

Pearlitic and bainitic pre-alloyed materials (Astaloy CrL, Astaloy 85Mo and Astaloy Mo) show scatter in fatigue strength (standard deviation) lower than 5%. This result is comparable with what found for bainitic dense steels [5]. Slightly higher value has been found for Astaloy 85Mo +0.48°C, probably due to the fact that large areas of ferrite among cementite lamellas acting as weak phase.

ASC100.29 +2% Cu(100 mesh) shows very low standard deviation n (<3%): the heterogeneous ferritic/pearlitic structure of base powder particles is surrounded by a reinforced Cu-rich shell. The low scatter indicates that these phases have similar yield strength.

In premixes of Astaloy 85Mo with Copper or nickel very low scatter is shown at Carbon levels close to 0.8%. Standard deviations of respectively 7% and 8% are generated at Carbon level close to 0.5%, probably due to the presence of inhomogeneous distribution of the metallographic phases (austenitic/martensitic islands and dense upper bainitic areas in mixes with 2% Ni and 2% Cu respectively).

In Distaloy materials scatter is generated when a continuous high yield phase network is not formed (as-sintered Distaloy DH) or is not fully consisting of martensite (Distaloy HP +0.48°C, sinter-hardened Distaloy DH +0.47°C). In these cases the scatter is between 5 to 10%.

For sinter-hardened Distaloy DH +0.73°C upper bainitic islands act as weak points in the extended portion of material that is transformed into martensite and are responsible for the 6% standard deviation.

Fatigue Fracture Behaviour and Response to High Density

Homogeneous PM steels are characterized by equal phase compositon in the powder particles and in the sinter-necks. Due to a phenomenon of stress conc entration, crack initiation is related to sinter-necks failure, i.e. the material is sensitive to porosity level. Therefore, the increase of density is beneficial for homogeneous PM steels (Astaloys).

Heterogeneous PM steels with sinter necks effectively reinforced by the presence of a continuous martensitic network (Distaloy Ae, Distaloy DC, Distaloy HP,

Astaloy 85Mo+2% Cu, sinter-hardened Distaloy DH) present crack initiation mostly related to base powder particles failure, whose microstructure (ferrite/pearlite or bainite) has lower yield threshold compared to martensite [4, 6]. With regards to sensitivity to porosity, the sinter-necks are optimized by teh presence of martensite. Increase of desnity will therefore have less significant effect compared to homogeneous PM steels, with regards to fatigue endurance.

CONCLUSIONS

1. Microstructure analysis gives reason for PM steels fatigue strength values and the relative scatter. Fractography is a useful method to reveal materials' sensitivity to porosity.
2. Sinter necks are effectively reinforced once that a continuous high yield microstructure network is formed among the base powder particles.
3. Increasing desnsity is beneficial on the fatigue performance of homo geneous PM steels (Astaloys) and not effective on materials characterized by network-like microstructures (Distaloys).

REFERENCES

1. A. bergmark, L. Alzati, U. Persson, Fatigue Crack Initiation in PM steel, PM2TEC Powder Metallurgy World Congress 2002, MPIF, Orlando (USA).
2. A. Bergmark, O. Bergman, L. Alzati, Pre-alloyed Chromium materials for highly fatigue loaded parts, PM2TEC Powder Metallurgy world congress 2004, MPIF, Chicago (USA).
3. L. Alzati, A. Bergmark, Diffusion alloyed powders with high fatigue performance in the as-sintered state", EPMA world conference 2004, Vienna (Austria).
4. Matsunaga H., Mars O., Johannesson B. and Murakami Y., Statistical distribution of powder sizes and fatigue strength of powder metal, 1999, pp. 71-79.
5. Murakami y., Effect of Small Defects and Nonmetallic Inclusions, Elsevier 2002.
6. Mars O., Bengtsson S., Bergmark A., Study of the defect-size effect on the endurance limit for sintered steel. Proceedings of the International Conference df PM'99, "Deformation and Fracture in Structural PM Materials", Editor: L'. Parilak, H. Danninger, Piest' any, September 19-22, 1999, IMR SAS Kosice, Slovak Republic, pp. 52-58.

5

Studies on Strength and Fracture Behaviour of Al-SiC Particulate Composites

R.K. Pandey
Professor, Applied Mechanics
Department, Indian Institute of Technology,
Hauz Khas, New Delhi, 110016, India

D.K. Sehgal
Professor, Applied Mechanics
Department, Indian Institute of Technology, Hauz
Khas, New Delhi, 110016, India

Z. Hasan
Research Scholar, Applied Mechanics Department,
Indian Institute of Technology,
Hauz Khas, New Delhi, 110016, India

ABSTRACT

Studies have been conducted in Al-SiC particulate composites with addition of 10% and 20% SiC particulates (diameter 23 μm) by weight, in a 2000 series Al-alloy used as matrix. The strength, ductility and CTOD fracture toughness were investigated in the base alloy and in the composites. The results indicated that the tensile strength and ductility get reduced in the composites due to presence of the particulate phase. On the other hand, the yield strength of the 20 % composite remains higher than the base alloy. There does not appear to be any degradation in the fracture toughness of the 10 % SiC composite as compared to the base alloy.

1. INTRODUCTION

Due to their light weight and good corrosion resistance, Al-alloys are proposed for various applications relating to automotive industries such as engine pistons, connecting rods, etc. The other desired properties for these mechanical components are fracture strength, fatigue resistance, wear resistance etc. A particulate composite based on an Al-alloy with uniform distribution of hard ceramic particulate phase is considered to be an appropriate material to meet most of the desired requirements for the above applications [1, 2, 1]. However, no much information is available on the subject. The present investigation has been conducted in an Al-alloy based particulate composites using SiC as the particulate phase. The focus of this work is on strength and fracture behaviour of' the composites.

2. MATERIALS

The Al-alloy used in the present work belongs to the 2000 series and its composition (in wt %) is given in Table 1.

Table 1. Chemical Composition of the Base Al-Alloy

Element	*Cu*	*Zn*	*Mn*	*Fe*	*Mg*	*SI*	*TI*	*Sn*	*Pb*	*Al*
wt%	5. 10	1. 13	0.59	0.56	0.45	4.09	0. 11	0.01	0.04	rest

The Al- alloy was initially cast in the form of a plate of thickness 40 mm. It was further forged to a thickness of 17.3 mm resulting in a reduction of about 57%. Two types of particulate composites were prepared by using liquid metallurgy technique. These are containing 10% SiC particulates and 20% SiC particulates by weight. These composites were forged subsequently causing a reduction in thickness of about 52% and the final plate thickness obtained was 19.2 mm. The average particulate size employed in this work is 23 μm. No significant grain orientation or directionality was noticed in the base alloy and the composite plates.

3. EVALUATION OF TENSILE PROPERTIES

The tensile properties of the base alloy and the composites were obtained by conducting tensile test on cylindrical tensile specimens. For this purpose the tensile specimens were machined with a gauge length of 20 mm and gauge diameter of 9.00 mm as per the ASTM standard [4]. A typical tensile specimen is shown is Fig. 1.

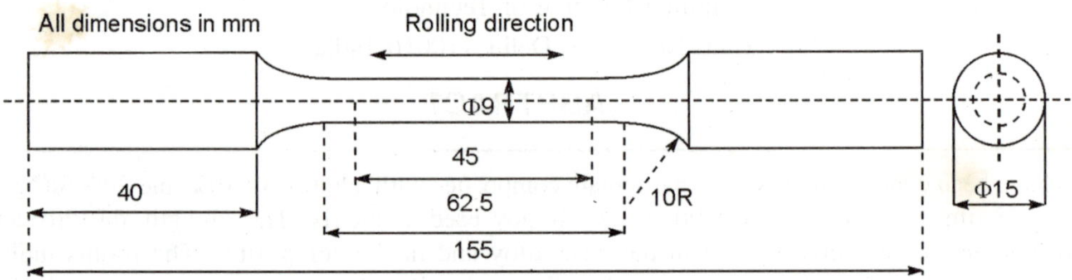

Fig. 1. Typical Tensile Specimen.

The tensile test was conducted on a Zwick machine of 250 kN capacity at a test speed of 1 mm/minute. At least three specimens were tested for each material. The stress-strain diagrams were obtained and were used to compute the proof stress, tensile strength (ultimate tensile stress) and the % elongation. Typical stress-strain diagram for the base alloy as well as the composites are presented in Fig. 2. The tensile properties are reported in Table 2.

4. FRACTURE MECHANICS TEST AND EVALUATION OF FRACTURE TOUGHNESS

The fracture toughness of the base alloy and the composite were evaluated by conducting standard fracture toughness test based on fracture mechanics approach. For this purpose standard compact tension (CT) specimens were machined from the plates keeping thickness (B) of test specimen as 12 mm and width (W) as 56mm. A typical CT specimen is shown in Fig. 3. The CT specimens were precracked as per ASTM standard [5] to generate a precrack length of 2.0 mm at the notch tip. The fracture toughness test was conducted by using an MTS machine (Model 858 Table top system) of 25 Ton capacity by providing a standard clip gauge at the mouth of the notch. The load-crack mouth opening displacement (CMOD) diagrams were obtained and analyzed to determine the fracture toughness. Typical Load-CMOD diagrams for the base Al-alloy and the 10% SiC composite are shown in Fig. 4 and 5 respectively. Since it was not possible to obtain valid K_{IC} toughness in the alloys, the elastic-plastic toughness, CTOD (δ) was evaluated in the present case.

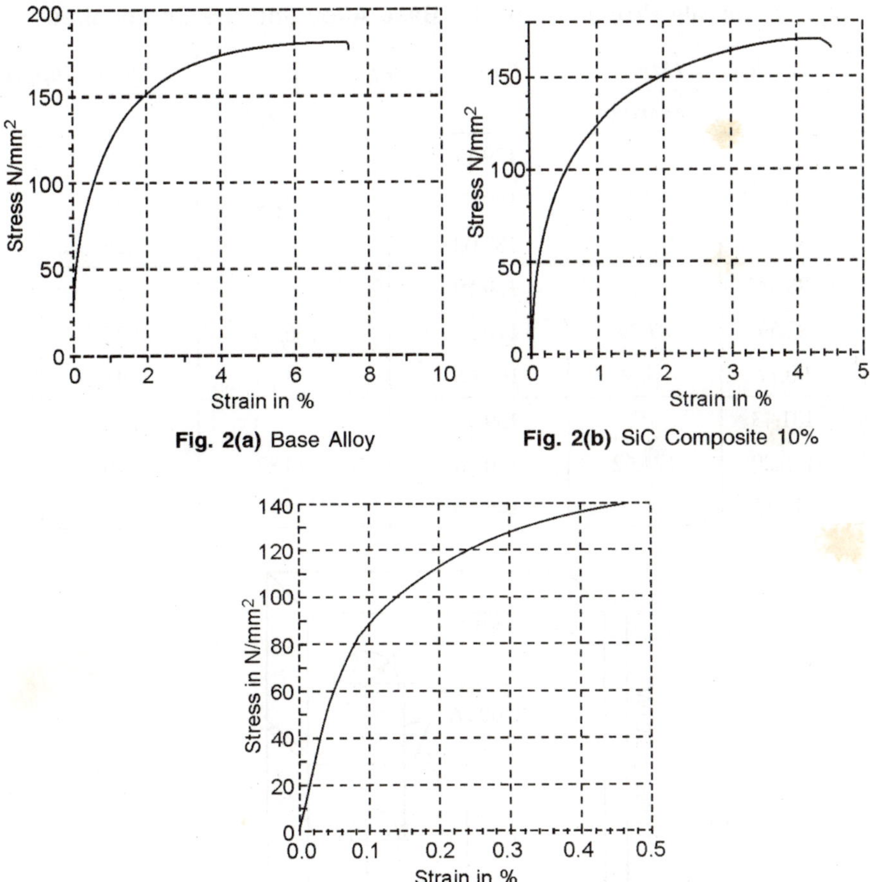

Fig. 2(a) Base Alloy

Fig. 2(b) SiC Composite 10%

Fig. 2(c) Al-SiC Composite 20%

Fig. 2 Typical Stress Strain Diagrams for the Base Alloy and the Composites.

The following equations were employed to determine the CTOD toughness [6]

$$\delta = \frac{K_1^2\left(1-\nu^2\right)}{2E\sigma_{ys}} + \frac{0.46(W-a_0)V_p}{0.46W+0.54a_0+C-W)+Z} \quad ...(1)$$

where

$$K_I = \left[\frac{F}{BW^{0.5}} f'\left(\frac{a_0}{W}\right)\right] \quad ...(2)$$

Here K_1 is the stress intensity factor corresponding to the critical load, W-Specimen width, a_0– initial crack length (including precrack length), Vp-Plastic component of CMOD, E- elastic modulus, σ_{ys}-yield strength of material, *C*-total width of specimen, *Z*-thickness of the knife edge, $f(a_0/W)$ a polynomial function in terms of (a_0/W), *F*-Critical load.

The force CMOD diagrams were analyzed to obtain the critical load and the values of various parameters in equations (1) and (2). The critical CTOD (δ_c) was determined which represents the fracture toughness. The values of CTOD (δ_c) toughness are presented in Table 3.

Table 2. Tensile Properties of the Base Alloy and the Composites

Material	*Proof Stress (Mpa)*		*Tensile Strength (MPa)*		*% Elongation*	
		Average		*Average*		*Average*
Base Alloy	86.16		179.17		6.72	
	78.58	88.58	169.89	177.00	7.87	7.28
	101.01		182.02		7.25	
10 % sic	88.34		166.59		4.14	
Composite	82.59	87.29	170.26	166.50	4.35	3.97
	90.77		162.66		3.41	
20% SiC	131.43		139.94		0.46	
Composite	118.29	124.62	150.96	145.87	0.76	0.60
	124.15		146.70		0-59	

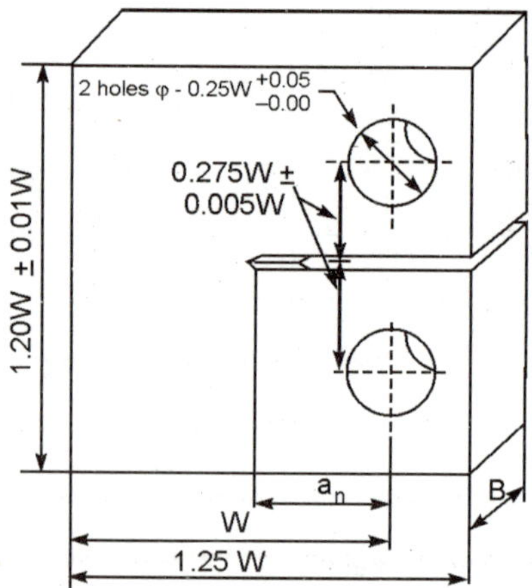

Fig. 3. Standard Compact Tension Specimen for Crack Propagation Studies.

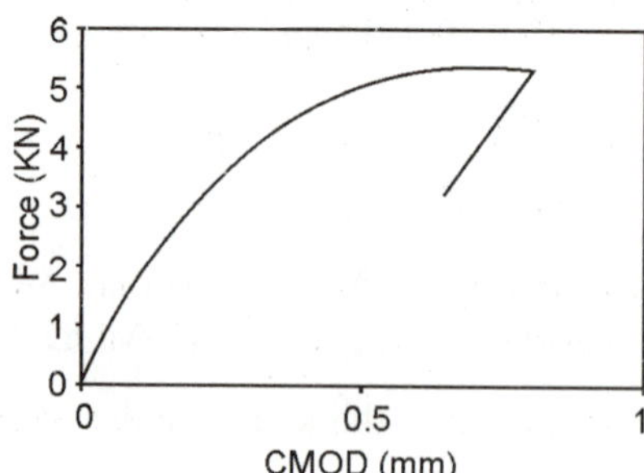

Fig. 4. Typical Force Vs CMOD Diagram for the Base Alloy.

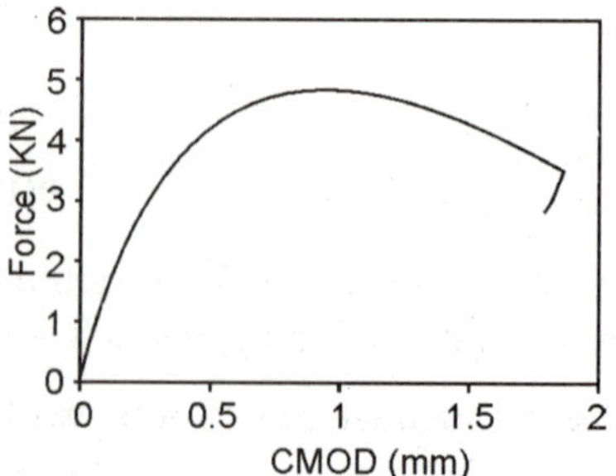

Fig. 5. Typical Force Vs CMOD Diagram for Al-SIC 10%.

Table 3. CTOD (δ_c) toughness of Materials

Material	*Yield strength MPa*	*Plastic CMOD (V_P) mm*	*K_I MPa $\sqrt{m}$*	*CTOD (δ_c) mm*	*Equivalent K_{IC} MPa $\sqrt{m}$*
Base Alloy	88.58	0.585	20.30	0.1574	44.18
Al-SiC 10%	87.29	0.66	18.48	0.1680	45.00

5. DISCUSSION

5.1 Effect of Particulate Phase on Tensile Properties of the Composites

From the results presented in Table 2, it is noticed that the yield strength of the base alloy is 88.58 MPa which remains practically unaffected by addition of 10% of SiC particulates. However, with further increase in particulate phase to 20 wt% the yield strength increases by a factor of 1.4. On the other hand, the tensile strength of the composite decreases with addition of the particulate phase consistently. The tensile strength of the base alloy decreases from 177 MPa to about 146 MPa in the 20% SiC composites. Obviously, the tendency for void formation on the SiC particulates reduces the tensile strength of the composite as compared to the base alloy.

The effect of particulate phase on the ductility of composites is quite substantial. Whereas the elongation at fracture in the base alloy is 7.28%, it decreases to a level of 3.97% in the 10% SiC composite and it becomes 0.60% in the 20% SiC composite. Thus, the ductility is a very sensitive function of the particulate phase in these composites.

5.2 Effect of Particulate Phase on the Fracture Toughness of the Composites

The effect of particulate phase on the fracture toughness of composites is shown in Table 3. The CTOD (δ_c) toughness of the base alloy is 0.157 mm and the 10% SiC composite also exhibits δ_c toughness in the same range (*i.e.* 0.168 mm). Therefore, the toughness of the 10% composite does not seem to be affected by the particulate phase. The equivalent K_{Ic}toughness was computed [7] using the relationship

$$K_{Ic}(\delta) = \sqrt{2E\sigma_{ys}\delta_c} \qquad ...(3)$$

The K_{Ic} toughness value for the base alloy and the 10% composite is about 44 MPa √m which is considered to be a reasonably high value for the composite. The fracture toughness of the 20 % SiC composite could not be evaluated.

Thus, it is noticed that though the ductility of composite decreases due to presence of SiC particulates (10 wt %), the fracture toughness does not get impaired. The composite is likely to possess superior wear resistance as compared to the base Al-alloy. Hence, it could be possible to design an appropriate particulate composite with SiC as particulate and Al- alloy as matrix to provide an optimum combination of strength, ductility and fracture toughness along with the abrasive wear resistance property.

6. SUMMARY

Based on the studies conducted in particulate composites employing a 2000 series Al-alloy (Al-Cu-Zn-Si) as the matrix with SiC particulates following conclusions can be made.

1. The addition of 10% particulate phase in the Al-alloy does not affect its original yield strength (of 88 MPa) however, the yield strength increases substantially by a 20% addition of the particulate phase.
2. The tensile strength and the ductility decrease continuously with the addition of the particulate phase in both the composites investigated. The ductility gets seriously affected due to the presence of particulate phase and it decreases from the initial 7% to 0.5% (approx) in the composite.
3. The CTOD fracture toughness of the 10% SiC composite remains practically unchanged at the original value of about 0.155 mm ($\cong 44\,\text{MPa}\sqrt{\text{m}}$) of the base alloy.

ACKNOWLEDGEMENT

The authors are grateful to the authorities of the Regional Research laboratory, Thiruvananthapuram for the supply of material for the present investigation.

REFERENCES

1. S.V. Nair, J.K. Tien and R.C. Bates, SiC-Reinforced Aluminium Metal Matrix Composites, lnt. J. Met. Review, 30(6), 275 (1985).
2. M.K. Surappa and P.K. Rohatgi, Preparation and Properties of Cast Al-Ceramic Particulate Composite, J. Mats Science, 16, 983 (1981).
3. J. Yang, C. Cady, M.S. Hu, F. Zok, R. Meharabian, and A.G. Evans, Effects of Damage on the Flow Strength and Ductility of a Ductile Al-alloy Reinforced with SiC Particulates, Acta Metall., Mater., 38, 2613-2619, (1990).
4. E8M-03, Standard Test Methods for Tension Testing of Metallic Materials, Annual Book of ASTM Standards, (2003).
5. El 820-01, Standard Test Methods for Measurement of Fracture Toughness, Annual Book of ASTM Standards, (2001).
6. BS 7448-1, Fracture Mechanics Toughness Test, British Standard Institution, (1991).
7. T.L. Anderson, Fracture Mechanics- Fundamentals & Applications, CRC Press London, (1995).

6

Effect of Solid Lubricant Addition on the Tribological Behaviour of Sintered Steels

S. Dhanasekaran and R. Gnanamoorthy
Department of Mechanical Engineering, Indian Institute of Technology,
Chennai, India

ABSTRACT

Sintered materials find increasing application as bearings and gears in many automotive and other engineering industries. The mechanical and tribological properties of the sintered steels are influenced by the presence of pores. The material composition and operating conditions affect the friction and wear characteristics of the sintered materials. Sintered steels with different compositions containing solid lubricant molybdenum disulphide are developed. The tribological properties of the sintered steels are evaluated using a pin-on-disc type friction and wear testing machine under unlubricated testing conditions. The coefficient of friction was measured during the tests at different normal loads. The wear characteristics were studied using mass loss and dimensional changes that happen during testing. Addition Of MOS_2 decreased the steady state coefficient of friction and also contributed in reducing the wear loss when slid against the hardened and tempered En 31 steel. A smoother wear tracks were observed in the solid lubricant added samples.

1. INTRODUCTION

Gears, bearings, connecting rods, and many other complex shaped machine parts are fabricated in an economical way using powder metallurgy processing technique. The performance of these machine elements not only depend on the strength and ductility but also on the friction and wear characteristics [1]. Many new material compositions are under development to improve the strength, ductility, hardness, toughness and also the friction and wear characteristics. In the sintered steels, unlike the conventional materials, the presence of pores affects the physical and mechanical properties of the material [2]. Inherent porosity decreases the strength, ductility and modulus but expected to contribute positively under lubricated contact applications. Lubricants stored in the pores reduce the coefficient of friction and wear loss [3].

The molybdenum di sulphide, is a widely used solid lubricant. As such it might also be a useful addition in the development of self lubricating parts which are used in different anti wear applications [4]. During operation the solid lubricants from the pores is expected flow to the metal surface provide lubrication [3]. Many parts used in automobiles, in particular those with high strength, contain carbon, nickel and molybdenum in their composition [5].

An attempt is made to develop sintered steel compacts with solid lubricant for tribo applications. Sintered compacts were prepared from elemental metal powders and solid lubricant powder molybdenum di sulphide. Effects of addition of molybdenum di sulphide on the microstructure, friction and wear characteristics and wear scar morphology are discussed.

2. EXPERIMENTAL PROCEDURE

The Fe-0.6% C composition in as sintered condition with and without 3% molybdenum di sulphide is considered for development in the current studies. The fine high compressibility iron powder of particle size less than 212 μm [ASI 100.29] was mixed with 0.6% carbon of particle size less than 150 μm and 0.7% of zinc stearate were used. The powder size of MoS_2 is approximately 5 μm. All the powder compositions were mixed using a double cone mixture for 30 min. The rotating speed of double cone mixture was maintained at 150 rpm. Cylindrical pins of dimensions 5mm diameter and 15 mm length were compacted in a hydraulic press. The pressed compacts were sintered in nitrogen at 1348 K. The theoretical density of the sintered steel pin was $\pm$6.8 g/cc. The actual density of the sintered steel was measured as per ASTM standard [6]. The metallographic samples were prepared by grinding, alumina polishing and etching with Nital 5 % for 5 secs.

The friction and wear behavior of the sintered materials was evaluated using a pin on disc tribometer. The details for the tribometer are described elsewhere [7]. The disc is made of hardened and tempered En-31 steel with 90 mm diameter and 10 mm thickness. The disc was heat treated at a temperature of 1118 K and oil quenched and tempered at 423 K for 1 h. The hardness of the disc varies from 60 to 63 HRc. The sliding wear tests were carried out at a constant sliding speed of 0.3 m/s upto a sliding distance of 4000 m. Tests were conducted at different normal applied load ranged from 5 N to 20 N. During experiments the friction force was continuously measured. The wear rate was evaluated from the mass loss and dimensional measurements.

3. RESULTS AND DISCUSSION

The pore distribution and microstructure of Fe-C and Fe-C-MoS_2 compositions are shown in Figure 1. The pores are isolated in Fe-C alloy. This individual isolated pore volume is higher than the interconnected pores. In samples containing MoS_2, the pores are mostly interconnected [8]. These interconnected open pores were observed on the surface of the pins.

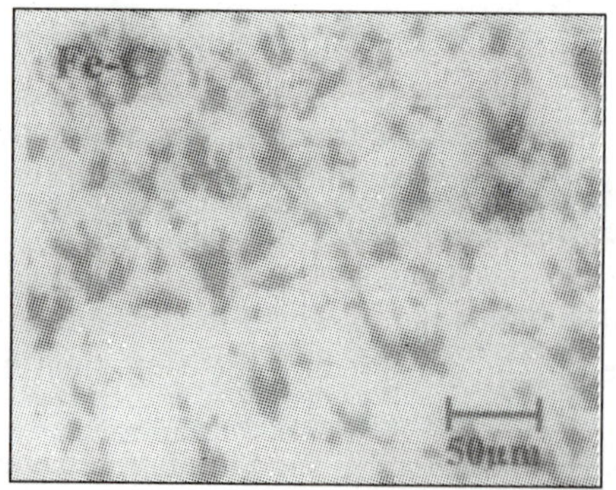

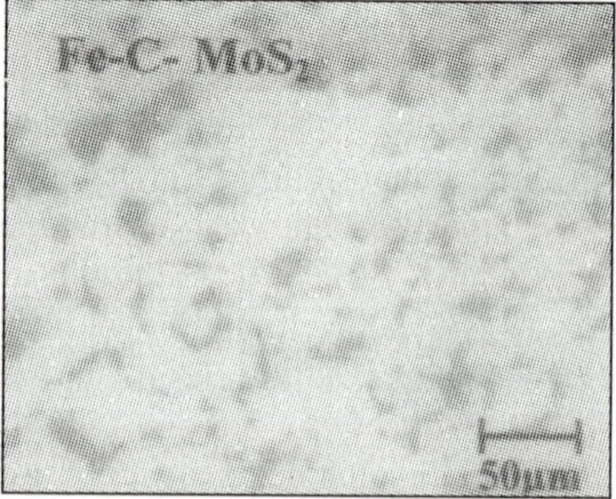

Fig. 1. Pore distribution of as sintered steel.

Figure 2 shows the coefficient of friction evolution of test materials at 10 N and 15 N applied normal loads. The coefficient of friction reaches a steady state after initial running in period. The steady state coefficient of friction is lower for the material containing solid lubricant at all normal loads investigated. Presence of the solid lubricant particles in the compacts are released due to sliding action and contribute to the reduced coefficient of friction. Samples containing MoS_2 exhibited a lower coefficient of friction compared with the materials without addition Of MoS_2. At low loads the difference in the coefficient of friction was very low. At low applied normal loads the coefficient of friction in both samples are lower ranging between 0.27 and 0.31. The effect of the solid lubricant MoS_2 was found increased at high normal loads investigated.

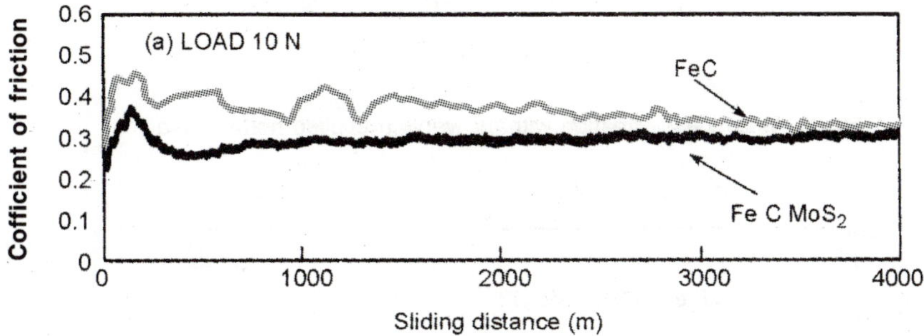

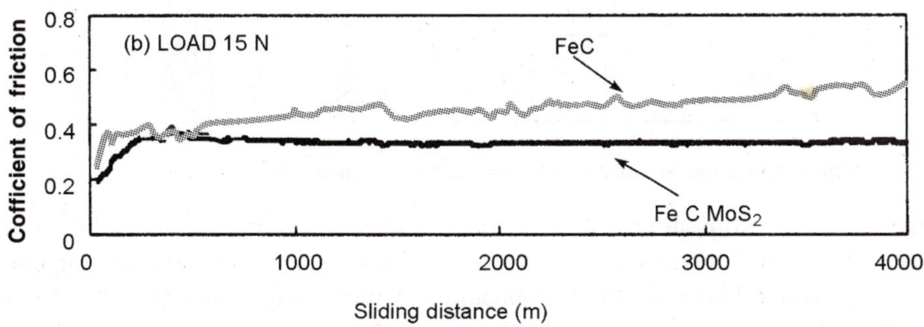

Fig. 2. Variation of friction coefficient at the normal load of (a) 10 N and (b) 15 N.

The effect of the MoS_2 solid lubricant powder on the wear resistance was evaluated from the mass loss measurements carried out. The wear of the sintered steel pin with solid lubricant containing composition is lower at all normal loads (Fig.3). The effect of solid lubricant was pronounced at high applied normal loads. At low load the loss of materials very low, because the oxide formed on the surface of the pin and there is no metal to metal contact exists. So the wear loss very less and the smooth wear track were observed on contact surface of the Fe-C pins. Fe-C-MoS_2 compositions the stagnated solid lubricants transferred from the surface pores and form a white layer on the contact surface. When increasing the loads the deformation of the materials high and leads to the release of solid lubricants on the surface. The thickness of the lubricant film formation also depends upon the sliding speed. Wear rate variation with different applied normal load is given in Fig. 4. The wear rate is calculated as volume of worn out material divided by the sliding distance. When comparing the wear rates of two compositions, Fe-C alloy exhibits severe wear rate.

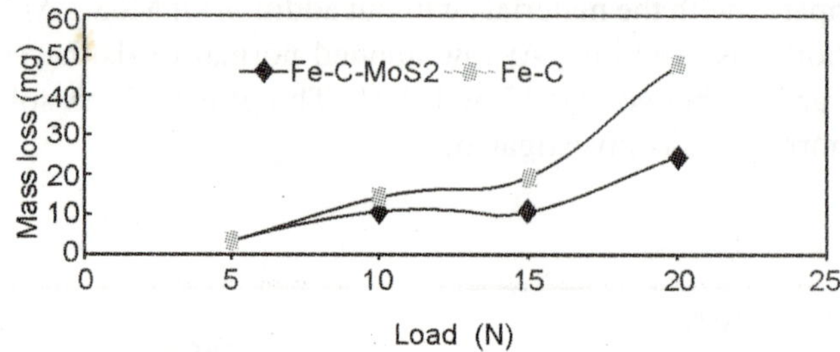

Fig. 3. Mass loess variation with the varying applied normal loads.

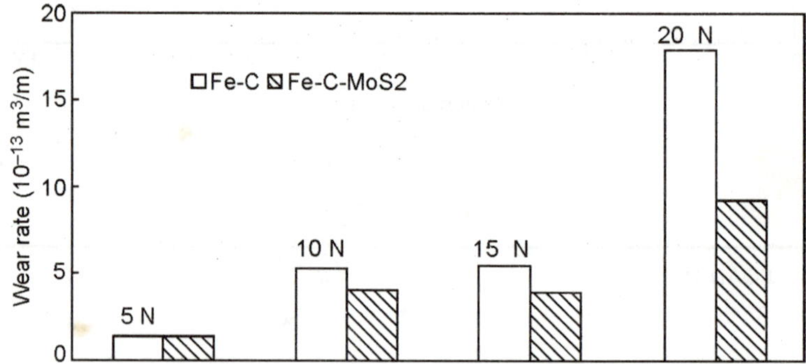

Fig. 4. Wear rate of test samples measured at different normal loads.

Optical micrograph of worn surface of Fe-C and Fe-C-MoS_2 compositions is shown in Fig. 5. The worn surface of Fe-C alloy composition clearly shows the severe deformation during sliding. But at low loads the wear scar exhibits a smooth appearance. Increasing the normal loads the sliding surfaces continuously highly stressed and the deformation in the contact area increased. The depth of deformation of surface depends upon the load and alloying addition. The depth of grooves formed is higher in Fe-C alloys at high normal loads. At 20 N loads the worn surface indicates deeper grooves formed in the pin. The measured surface roughness (centre line average) after tests for Fe-C alloy composition tested at 20 N load is 0.371 μm. The worn surface of Fe-C-MoS_2 composition showed no plows on the surface. In this composition smooth wear tracks formed resulting in surface roughness of 0.185 μm. The worn surface becomes comparatively fine then Fe-C alloy composition and this indicated that mild wear existed in Fe-C-MoS_2 composition.

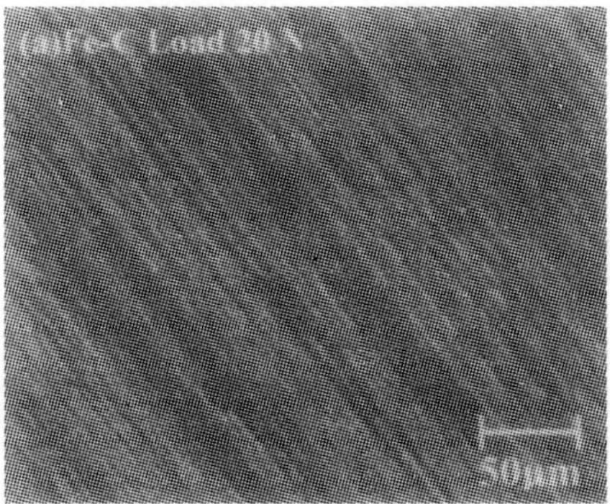

Fig. 5. Worn surface of test pins.

4. CONCLUSIONS

Solid lubricant MoS_2 added Fe-C compositions were prepared from elemental powders by compaction and sintering. The tribological properties of solid lubricant added specimens are found improved. There is a reduction in the coefficient of friction at all normal loads for the solid lubricant added compositions. The wear resistance is improved by the addition of solid lubricant. A smooth wear tracks were observed.

REFERENCES

1. Maitra G M, Hand book of gear design,Tata Mcgraw Hill Pub Co., New Delhi, 1994.
2. Bocchini G, Int. J. Powder Metal. 1986, Vol 22, pp 185-202.
3. Rapoport L,Leshchinsky V, Lvovsky M, et al., Superior Tribological properties of powder materials with solid lubricant nanoparticles, Wear, Vol. 255, pp. 794-800 (2003).
4. Sustarsic B, Kosec L, Jenko M, Leskovsek V, Vacuum sintering of water atomised HSS powders with MoS_2 additions, -pVacuum, Vol. 61, pp. 471-477 (2001).
5. Herbert Danninger, Gerald Frauendienst, Klaus-Dieter Streb, Raimund Ratzi, Dissolution of different graphite grades during sintering of PM steels, Materials chemistry and physics,Vol. 67, pp. 72-77(2001).
6. ASTM Int (2002) (Designation B 328-96)108.
7. Srinath G, Gnanamoorthy R,Friction and wear characteristics of nylon-clay nanocomposites under dry sliding conditions, 1st Int Conf on Advanced Tribology, pp. B-13 Dec 2004.
8. Rapoport L, Lvovsky M, Lapsker 1, Leshchinsky V, Friction and wear of bronze powder composites including fukkerene like WS_2 nanoparticles, Wear, Vol. 249, pp. 150-157(2001).

7

Wear Behaviour of Sintered Ferrous Alloys

Mohammed Hamiuddin
Department of Mechanical Engineering,
AMU, Aligarh - 202002, India.

ABSTRACT

This investigation deals with wear behaviour of sintered iron alloys compacted at 589 and 700 MPa pressure and sintered at 1423 K for 1.8 ks under hydrogen atmosphere using heating and cooling rates of 16 and 32 K per minute to and from sintering temperature respectively. Some of the samples were carburized at 1173 K for 5.4 ks.

Wear tests were carried out using a pin-on-disc type rig with 1550 RMP speed and 250 mm square grinding area with loads varying between 0.5 to 2.0 kg. Wear rate was inversely proportional to sintered density. Heat treated samples showed higher wear resistance.

Results have been discussed on the basis of amount and morphology of pores present in the sintered and heat treated samples.

1. INTRODUCTION

Power Metallurgy (PM), a technique for fabricating metal components by heating compacted metal powders, has existed for over 80 years. However, only in the last three decades has PM become widely recognized as a superior way of producing high quality components for a variety of important and demanding applications. The PM industry manufactures semi-dense (porous) and fully dense products using several different technologies, including pressing and sintering, injection moulding, hot isostatic pressing, forging etc. The end application dictates the properties desired, and thus the processing route adopted. Producing near-net shape components with tight dimensional tolerance at moderate temperatures provides PM with a distinct advantage over other metal forming processes such as forging, metal casting or machining. PM minimizes finishing operations and energy use while adding design flexibility and lowering cost to end-use customers. These advantages have enabled the PM industry to grow at an average annual rate of around 10% over the last two decades and capture impressive share of key markets such as automotive (Table 1). There is considerable overlap in this table and many components were developed and put to applications concurrently.

Table 1. Evolution of PM — A Road Map

Production of PM Items	*Years*
(*a*) Low density with controlled shapes and sizes of pores (not compacted but molded for filtration of gases and liquids) (*b*) Low density-compacted to around 60-70% of theoretical density for self lubricated bearing (*c*) Sintered carbide tools, generally 90 W-10 Co	Before 1960
Bronze and Ferrous bearings, shock absorber parts — simple shapes	1960-70
Gerotors, Fuel Injection parts, Spur gears, Seat reclining, bronze structural parts, Parts for two-wheelers	1970-80
Mechanism-parts, Refrigeration compression pistons, Timing belt pulleys, Bearing caps, Valve guides Compressor valve plates, Oil pump gears	1980-90
Valve seat inserts, Synchrohubs, Rocker arm tips	1990-2000
Primary driven gears, Sinterhardenable parts, Warm compaction	2000+

While there are various other applications of PM components e.g. in business machines, house hold equipments, electrical contacts etc, but its growth has mainly been due to application of PM components in automotive industries. If one compares the application of PM components in modest developing countries like India (4-5 kg per passenger car) with the most developed country of the world, *i.e.* USA (15-16 kg/car), potential for growth of PM technology in a global society becomes all the more prominent [1].

Automotive or any other parts have very rarely been developed (and sustained the development) by PM processes until and unless their performance and economics have been justified and established as well. Better cost competitiveness has been achieved due to production of complex shapes avoiding expensive machining operations despite the fact that powder cost is much higher than the equivalent wrought steel, for example (around 3-4 times more). While complexity alone could sustain growth in the last decade, the limits to complexity are now being challenged and hence future growth due to complexity will bring only incremental growth. Major growth of the market will come through development of parts with superior mechanical properties. PM parts will start replacing wrought parts that undergo dynamic and impact loading conditions such as transmission gears. To achieve this, a through understanding of the tribological properties of P/M materials is needed. There are various factors that influence engineering decisions on the use of P/M components. For example, manufacturing advantages including material savings, lower energy consumption and environmental freindliness. These are points in favor of the additional use of P/M. These economic drivers have historically provided immediate incentives to use P/M. From the service point of view, improved mechanical properties including tribological behavior are valued. These improvements do not provide the same short term impact but are increasingly more important to the engineering acceptance of P/M in highly loaded component applications. Thus, the present investigation is just a beginning of systematic investigation based on past casual work of the author [2-4].

2. WEAR MECHANISMS

The three most common types of failures requiring replacement of components are fatigue, corrosion and wear in decreasing order of catastrophic nature of failures. Wear increases power losses and oil consumption and decreases operating efficiency, ultimately necesseciating part replacement.

Friction and wear are not intrinsic material properties but are characteristic of engineering system. There is no simple universal relationship between friction and wear. In general, however low friction will result in low wear and changes in friction are likely to denote changes in wear rate and/or mechanisms. Any change in load, speed or environmental conditions may cause catastrophic changes in the wear rate of one or both of the surfaces in contact. Fig. 1 shows some of the many variables which contribute to friction and wear in engineering systems and it may be possible to obtain a better solution to a particular wear problem by altering, for example, one of the wear surfaces [5].

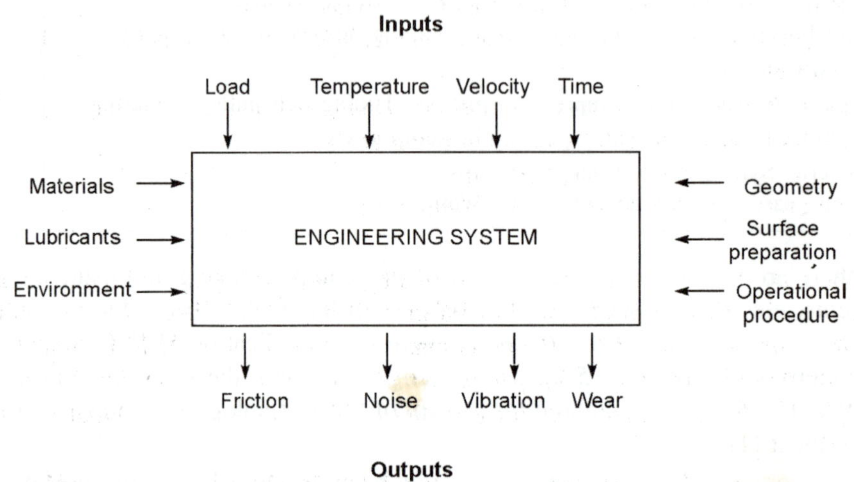

Fig. 1. System Approach in Tribology.

To date, there is still no adequate definition of wear of solid components. However, wear is generally defined as removal of materials from solid surfaces as a result of mechanical and/or chemical action due to disturbance of frictional bounds. Wear occurs by surface interactions at asperities, and components may need replacement after a relatively small amount of material has been removed or if the surface roughness is unduly increased.

The generation and circulation of a wear debris, particularly in systems where the clearances are small, may be more of a problem than the actual amount of wear. The debris may cause oil blockage or seizure or both decreasing the system's operating efficiency.

Wear in industrial situations have been classified as (*a*) Abrasive ~ 40% (*b*) Adhesive ~ 30% (*c*) Erosive 10% (*d*) fretting ~ 10% (*e*) Chemical 10%.

Sintered parts have some intrinsic advantages over other wrought, cast, or plated parts since the presence of porosity acts as a reservoir for lubricants and wear debris. Additional microstructural variables in porous parts include isolated and rounded or inter-connected pores and isotropicity due to the very nature of processing technique starting with powders. The type, size, shape, orientation and distribution of pores are dependent on processing parameters during mixing, compacition, sintering and post sintering stages which makes study of wear quite complex. In an earlier study [3] of wear behaviour of sintered phosphorus containing ternary iron alloy powder compacts, it was possible to achieve materials with a low coefficient of friction of 0.25 – 0.45 and wear rate of less than $14 \times 10^{-6}\ m^{-1}$ with phosphorus addition of up to 0.6%.

In the present work an attempt to limit the number of complex alloying elements was made. Alloying included only 0.6 mass % P and an addition of 0.6 % graphite for an average sintered C-content of 0.5 % to compensate for P induced sintering shrinkage. The C addition also improves the hardness and strength of the sintered samples to a suitable level for use in friction and wear applications.

3. EXPERIMENTAL PROCEDURE

Commercial grade PASC 60 powder was used for 0.6 mass % P iron while fine graphite of average particle size 4 μm was used as a source of carbon. The chemisry, particle size distribution and detailed characteristics of this powder are reported elsewhere [2,3]. Powder was mixed manually for 1 hr using a small amount of acetone. No lubricant was added during mixing. Cylindrical green compacts of approximately 18 mm diameter and 15 mm length were made in a single acting hydraulic press at 589 and 700 MPa pressure using die-wall lubrication with a commercial lubricant Kenolube. Green densities of 6.85 ± 0.05 and 7.05 ± 0.05 Mgm^{-3} were achieved at the two compacting pressures respectively. The compacts were sintered for 1.8 ks at 1423 K under hydrogen atmosphere (dew point 233 K) in a laboratory type tube furnace using heating and cooling rates of 16 and 32 K per minute respectively.

After measuring sintered density and hardness, some of the samples were carburized at 1173 K for 5.4 ks in order to increase the surface carbon content from 0.5 to 0.8% up to a case depth of 3 mm followed by quenching and tempering. There was loss of carbon content during sintering and as sintered carbon contents were 0.5 ± 0.05 m%. Sintered densities were 7.1 to 7.25 Mgm^{-3} with apparent hardness ranging from 180 to 240 HV 10. Heat treatment increased the hardness to 450 HV10.

All the samples were polished using 4/0 grade emery paper, cleaned in acetone and subjected to wear testing on a pin-on-disc type of test rig. All the tests were run against a heat treated, ground and polished tool steel disc (En32) having a hardness of HRC 64. Before conducting each test, the surfaces of the disc were cleaned with acetone to remove any dirt or grease. A digital electronic Mettler balance having an accuracy of 10^{-4} g was used to measure the weight loss of the pin after each run to estimate the volume loss during wear. Initially, the weight loss was measured after every 5 minutes of sliding up to 20 minutes and thereafter, at an interval of 10 minutes for a total sliding period of 2 h. The tests were conducted at relatively low loads of 0.5 to 2.0 kg at a constant sliding velocity of about 1 m/s. The ambient temperature was 303 - 308 K and humidity was 60-70%.

4. RESULT

Most of the pores in the samples were isolated and rounded. The wear volume loss was found to increase with increasing load. However, a load of up to 1 kg does not seem to be adequate in producing significant wear for the couples of pin and disc studied in this investigation (Fig. 2). The data points were observed to lie primarily on two segments. The first segment appears to be the case of primary severe wear and the second one, indicating a linear relationship for steady state wear. The rate of wear volume loss vs. time of sliding (*i.e.* sliding distance) decreases after 10-20 minutes. The wear rates of the samples at different loads were determined from slopes of these segments. Although the first segment is not very prominent, an effort has been made to estimate the wear rate from the tendency exhibited by the data (Fig. 3). It is observed that the second segment has a lower wear rate as compared to the first segment for the same load. The wear rates in the second segment show an almost linear behaviour with increasing load even with low loads ranges.

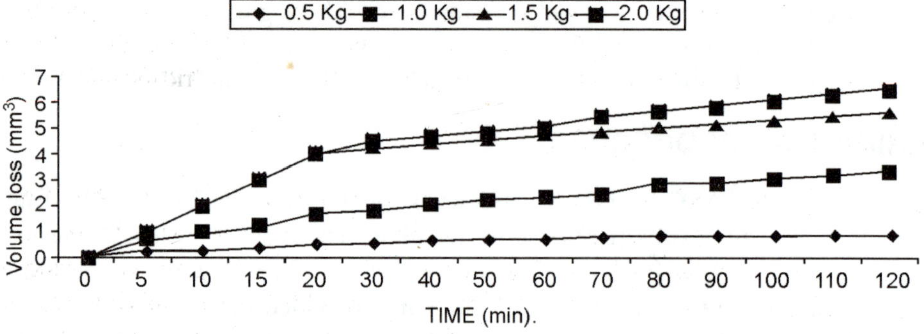

Fig. 2. Variation of Wear Volume with Sliding Time at Different Loads for Specimens Compacted at 589 Mpa Pressure.

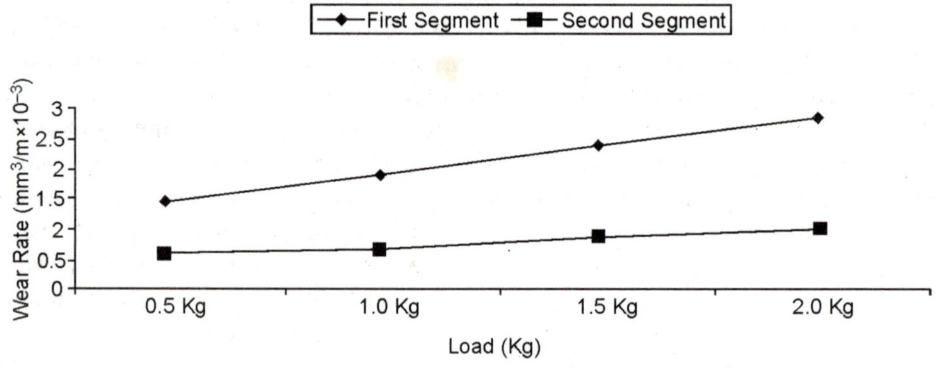

Fig. 3. Variation of Wear Rate with Load for 589 Mpa Pressure Compacted Specimens Corresponding to First and Second Segment.

The variation of wear volume with sliding time for the higher density sintered samples compacted at 700 MPa pressure is shown in Fig. 4. Two segments are observed here as in the case of the lower density sintered compacts. However, the wear rate is lower due to higher density and consequently higher hardness. The variation of wear rate with increase in load for both the segments is shown in Fig. 5. The second linear segment shows a relatively lower wear rate as compared to the first segment for the same load. Wear rate at higher load in the second segment for higher density compacts shows linear behaviour with increasing load similar to that observed in low density samples. However, the variation is non-linear corresponding to the wear rate observed in the first segment. The difference in wear volume loss with sliding time when load is increased from 0.5 to 1.0 kg is less in the case of higher hardness compacts.

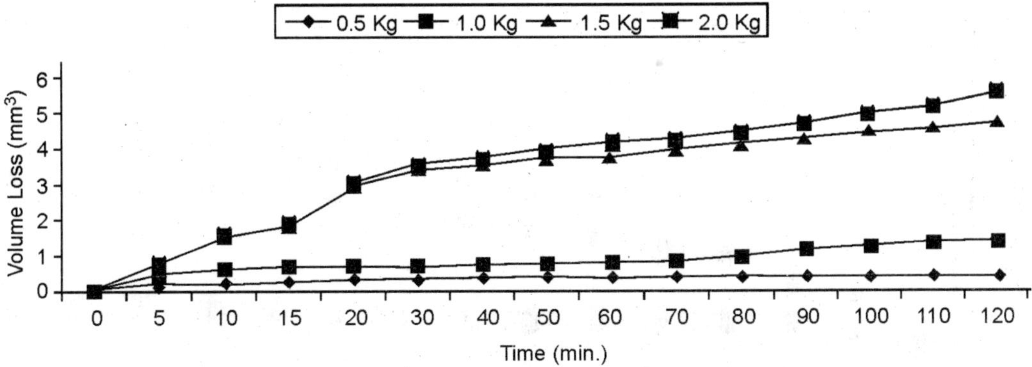

Fig. 4. Variation of Wear Volume with Sliding Time at Different Loads for Specimens Compacted at 700 Mpa Pressure.

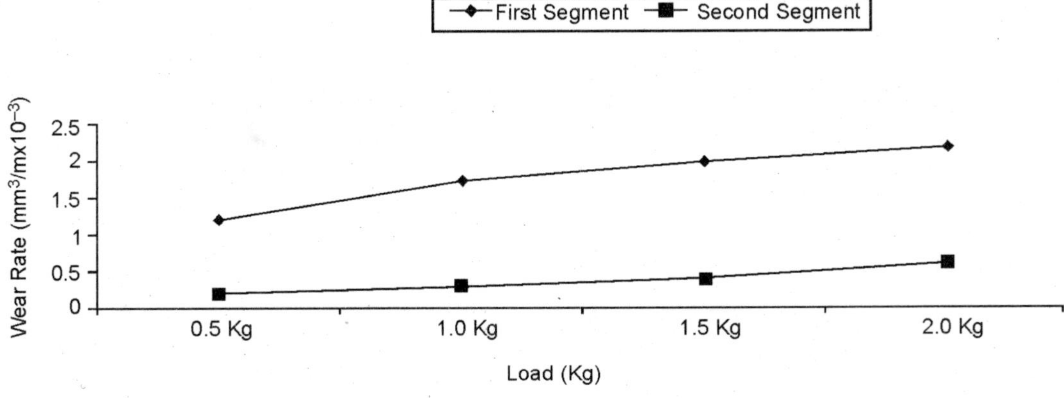

Fig. 5. Variation of Wear Rate with Load for 700 Mpa Pressure Compacted Specimens Corresponding To First and Second Segment.

The variation of wear volume with sliding distance for case-carburized steel (Fig. 6) shows two segments as observed in the case of low and high density sintered compacts. The wear rate at different loads observed for these segments is shown in Fig. 7. The variation of wear rate with load for both the first and second segments is qualitatively similar to that observed in high density compacts except that linearity with increasing load is reduced.

The wear coefficients obtained at different loads for three types of compacts are shown in table 2. The wear coefficient is calculated using Archard's wear law: $V = kWS/3H$, where V is the wear volume loss, W the load applied, S the sliding distance, and H is the (apparent) bulk hardness of the material under test [6]. To obtain the wear coefficients (k), wear rate (V/S) is divided by the contact area (W/H) estimated from applied load and indentation hardness and multiplying the result by the geometrical factor of 3. The first segment is observed to have a higher wear coefficient as comparesd to the second segment for all the compacts at different loads.

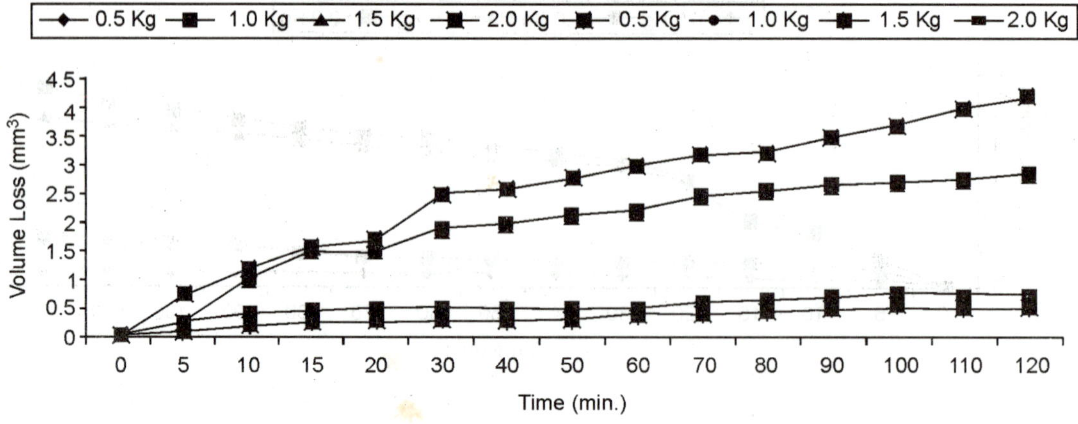

Fig. 6. Variation of Wear Volume with Sliding Time at Different Loads for Case Carburized Specimens Compacted at 700 Mpa Pressure.

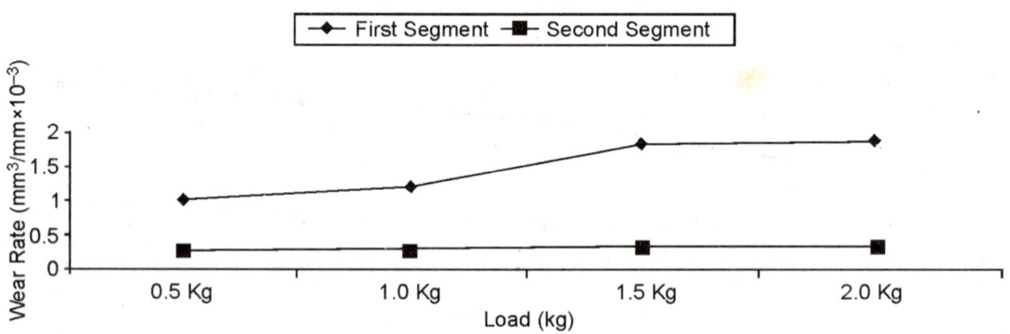

Fig. 7. Variation of Wear Rate with Load for Case Carburised Specimens Compacted at 700 Mpa Pressure Corresponding to First and Second Segment.

Table 2. Wear coefficient for sintered and heat treated samples

Sample	*Linear*	*Wear coefficient ($K \times 10^{-4}$) calculated for different segments at different segment loads*			
		0.5 kg	1.0 kg	1.5kg	2.0 kg
589 MPa	First	–	–	7.0	7.2
	Second	0.5	–	1.4	1.42
700 MPa	First	–	–	6.6	7.2
	Second	0.8	–	2.8	2.9
Carburized	First	–	–	4.6	4.8
	Second	–	–	2.4	2.6

5. DISCUSSION OF RESULTS

Phosphorus containing iron alloys have been in commercial use for the last few decades for soft magnetic components and to a lesser extent for structural components [7] because of enhanced magnetic and mechanical properties. Phosphorus is the only element among commonly added alloying elements in sintered iron and steels which simultaneously improves strength, hardness, ductility, [8-10] machinability [11] and wear and friction properties [2,3]. These properties are acheived using phosphorus additions due to (*i*) formation of liquid phase at usual sintering temperatures (*ii*) its high diffusivity (*iii*) ferrite stabilization (*iv*) solid solution strengthening effect (*v*) formation of rounded and isolated pores in as sintered parts (*vi*) densification due to shrinkage because of activation of sintering due to ferritic and liquid phase sintering. Typical phosphorus additions range from 0.3 to 0.8 m%.

The previous studies on friction and wear properties of P-containing sintered alloys [2,3] containing Ni, Cu, Mo, Mn, Cr showed improved strength hardness, and ductility due to phosphorus additions in the range 0.3 to 0.6 m%. However, the low wear rate of about 10-15 $\times 10^{-6}$ gm^{-1} and low coefficient of friction of 0.2 to 0.4 cannot be attributed due to phosphorus alone. Often, developments are made to improve a particular property for specific needs required by a specific customer. Such studies gave always been phenomenological in nature and often it is difficult to make generalized statements about the results.

In this investigation a limited reverse engineering concept has been used resulting in a combined effect of higher hardness with capability to be work hardened during wear experiments. Although observations and results are as expected, (*i.e.* similar to wrought materials) it is significant to note that by limiting alloying elements to *C* and *P* at a fixed amount, it is possible to generate data for wear rate and wear coefficient with a simple, linear relationship under the conditions of this experiment.

6. CONCLUSIONS

1. This preliminary study has shown that it is possible to generalize wear results if standard, reproducible test conditions are maintained. At least in simpler alloy systems with a sintered density of more than 7.2 Mgm^{-3}.
2. A contact load of more than 1 kg is required to produce appreciable wear– a requirement for wear study.
3. The volume loss in wear has been found to increase with sliding time i.e sliding distance and can be described in terms of two segments:
   (*i*) the wear rate in the first segment is higher and may be linear at higher loads
   (*ii*) the second segment exhibits a lower wear rate and indicates linearity, at least at loads of more than 1 kg in moderately hard and strong alloys.

Work is in progress to complete microstructural observation of the specimen wear scars and wear debris which will provide information regarding the mechanisms of wear. However, it can be concluded that the wear mechanism changes after some time of run-in due to hardening of the surface (probably work- hardening).

REFERENCES

1. B.M. Kataria, Evolution of PM Industry in India, Paper presented at International Conference on Powder Metallurgy for Automotive components (PMO2), India Habitat Centre, New Delhi, January 22-24, 2002.

2. Md. Hamiuddin, K. Gopinath and G.S. Upadhyaya, Wear Behaviour of Sintered Ferrous Alloys Containing Phosphorus, Transactions of Powd. Met. Assoc. of India, Vol. 8, 1981 pp. 63-65.
3. Md. Hamiuddin, Wear Behaviour of Sintered Phosphorus Containing Ternary Iron Alloy Powder Compacts, Powder Metall. Int. Vol. 17, No. 1, 1985, pp. 20-22.
4. Mohd. Asif, M. Arif Siddiqui and Md. Hamiuddin, Wear Behaviour of Sintered Ferrous Alloys Containing Phosphorus Proceed. Of the 4th International Conf. On Mechanical Engg. ICME 2001, Dhaka pp. 201-207.
5. T.S. Eyre, Wear Mechanisms, Powder Metallurgy Vol. 24, No.2, 1981, pp. 57-63.
6. J.F. Archard, Contact and Rubbing of Flat surfaces, J. Appl. Phys. Vol. 24, 1953, pp. 981-88.
7. R.M. German and K.A.D' Angelo, Enhanced Sintering Treatments for Ferrous Powders, International Metals Review. Vol. 29, No. 4, 1984, pp. 249-272.
8. Md. Hamiuddin, Increasing Mechanical Properties of Steels Containing Transition Metal Carbides Master Alloy MCM and MVM, International Journal of Powder Met. and Powder Technology, Vol. 20, No. 4, 1984, pp. 243-250.
9. N.A. Siddiqui and Md. Hamiuddin, Comparative studies of the Effects of Alloying Elements on Sintered Properties of Steels, Science of Sintering, Vol. 17, No. 3, 1985 pp. 175-192.
10. Md. Arif Siddiqui and Md. Hamiuddin, Mechanical Properties of Sintered Steels based on Atomized Iron Alloy Powders, Powder Metallurgy, Vol. 29, No. 3, 1986, pp. 217-229.
11. Md. Hamiuddin and Qasim Murtaza, Machinability of Phosphorus Containing Sintered Steels, Materials Chemistry and Physics, 67, 2001, pp. 79-84.

8

Influence of Mechanically Alloyed Nano Iron Aluminide Intermetallics Particles on the Wear Behaviour of P/M Based Metal Matrix Composites

K. Venkateswaran, M. Kamaraj and K. Prasad Rao*
Department of Metallurgical and Materials Engineering
Indian Institute of Technology Madras, Chennai 600 036, India.

ABSTRACT

The influence of mechanically alloyed iron aluminide nano particles on the wear behavior of P/M based metal matrix composites was studied using a pin-on-disk wear testing machine in dry condition. Sintered iron and a copper-tin alloy made by powder metallurgy were used as matrix. The intermetallics Fe_3A1 nano particles prepared by mechanically alloying were characterized by SEM and TEM and used as particle reinforcement with 5 and 10% volume in the matrix. The processing of the composites includes mixing and cold compacting followed by sintering at 1120°C and 850°C respectively. The presence and distribution of intermetailic Fe_3A1 phase on the surfaces were identified with optical microscopy. The influence of mechanically alloyed Fe_3A1 intermetallics additions on the wear behavior was studied with different sliding conditions. The experimental results indicated that improvement in dry sliding wear properties in MMCS compared to their plain alloys as standard material. Wear tracks were characterized by SEM techniques. Studies revealed that unreinforced sintered alloys showed lower wear resistance, while the composites with mechanically alloyed Fe_3A1 were the most wear resistant and the iron aluminide provided self lubrication properties.

Key words: iron aluininides, mechanical alloying, metal matrix composites, dry sliding wear, self lubrication.

1. INTRODUCTION

Powder metallurgy based metal matrix composites are finding more applications in advanced structural, aerospace, automotive and domestic appliances industries in recent years. These materials can become alternatives to conventional cast materials since they posses specific mechanical properties such as low weight, self-lubrication and excellent wear resistance which are important considerations for the designers of P/M parts. Interest in intermetallic composites has grown significantly over the past several years with the development of metal-matrix composites for tribological applications. Intermetallic compounds are attractive in tribological applications not only for their intrinsic properties such as hardness, stability, wear resistance etc. but also because of their compatibility with metallic matrices which may confer other desirable properties like self-lubrication or thermal diffusivity[1]. The principal advantage of intermetallic composites is the metal-like properties of the matrix. Also, intermetallic-based metal matrix composites reinforced with carbides, nitrides or borides have received a great deal of attention in the recent years. The behavior of these composites is influenced by mechanical and metallurgical properties and volume fraction of the reinforced intermetallic particles. Potential intermetallic composites can be developed and used in automobile as well as domestic appliances industries where low density, low cost of manufacture and better tribological properties are needed.

Among the various intermetallics, iron aluminides can become an alternative for the other reinforcing materials considering their high strength, high stability, good oxidation resistance, low density and good wear resistance. The iron aluminides are particularly suitable from the raw materials cost point-of-view because they consist primarily of abundant and inexpensive elements (iron and aluminium). Hence, Iron aluminides have been considered for a variety of their applications [2]. Also, iron aluminides are highly resistant to oxidizing or sulfidizing environments; they exhibit comparatively low densities; they have high work hardening rates, suggesting high wear resistance[3].

In the past decades, mechanical alloying (MA), in particular, high energy ball milling has been highly developed and widely applied to prepare various materials after the development of oxide dispersion strengthened alloys by Benjamine [4]. Mechanical alloying also produces iron aluminide powders of consistent quality with homogeneous distributed products [5].

Iron and copper based alloys are used extensively in powder metallurgy industries to make various self lubricating bearings and structural parts such as gears, levers etc. These parts possess low wear resistance compared to ingot materials and hence improvement of wear properties is to be explored. Also, the self lubrication property of the oil impregnated P/M materials looses its self lubrication properties in due course of time. The closure of the pores under working conditions and the exposure to environmental conditions such as dust, foreign particles leads to failure of effective wet lubrication in service. The failure of effective wet lubrication leads to high wear in the P/M parts. Hence the development of the P/M based materials with better dry lubrication properties is the need of the hour. This will reduce the unexpected breakdowns and replacement of critical components due to devoid of oil lubrication. Hence the present investigation aims at studying the influence of mechanically alloyed Fe_3A1 nano particle reinforcement on sintered iron and copper based alloy produced by powder metallurgy by conducting dry-sliding wear studies on P/M-based metal matrix composites.

2. EXPERIMENTAL DETAILS

2.1 Preparation of Fe_3Al Powders by Mechanical Alloying

In the mechanical alloying process, a Fritsch PS planetary mill was used at 300 rpm speed. The ball- to-powder weight ratio (BPR) is 10:1. Pure iron and aluminium powders were used as the starting materials. The specification of elementary powders and initial powder mixture is shown in Table 1. Powder mixtures were mechanically alloyed for 42 hours. Mechanically alloyed samples were characterized by X-Ray. Diffraction analysis and a typical intensity pattern is shown in Fig. 1.The SEM and TEM photomicrographs are shown in Fig. 2 (*a*), (*b*) and Fig. 3 (*a*), (*b*) respectively.

Table 1. Specification of Powders.

	Fe powder	***Al powder***
Size (mesh)	–100+325 mesh 82% –325 mesh 14% –80+100 mesh 3.7%	–200 mesh
Composition of mixture (vol.%)	70	30

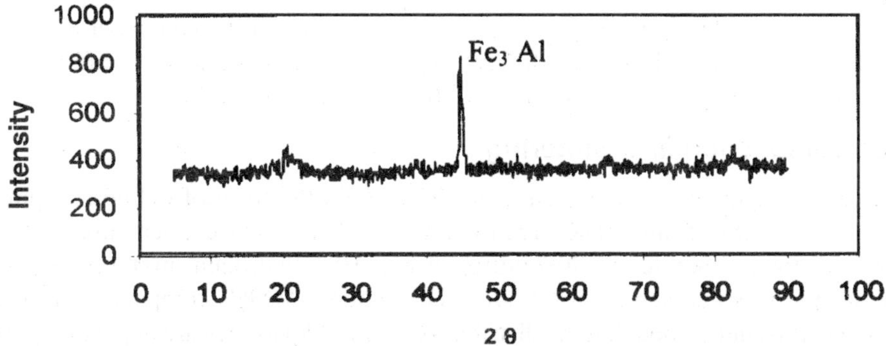

Fig. 1. X-ray diffraction pattern of the mechanically alloyed Fe_3 Al Powders.

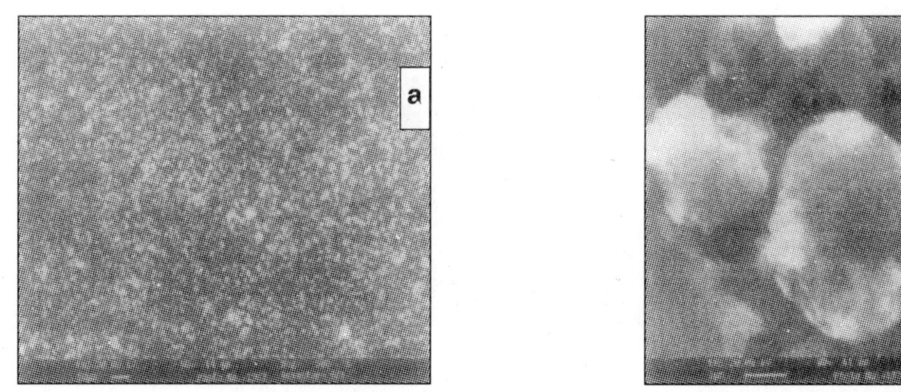

Fig. 2. SEM photomicrographs of iron aluminide powders.

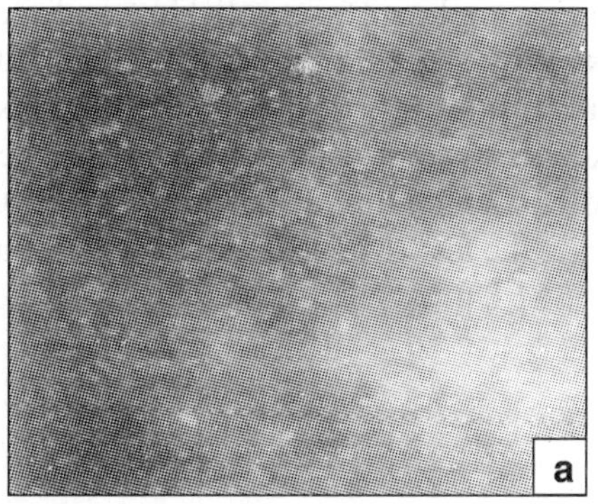

Fig. 3. TEM Photomicrograph of iron aluminide.

2.2. Preparation of Metal Matrix Composites

Metal matrix composites mixture was prepared by using pure iron and copper alloy powder with mechanically alloyed Fe_3Al powders under different volume fractions namely 5% and 10%. The mixing was carried out in the Fritsch P5 planetary mill at 100 rpm for 30 minutes.

2.3. Powder Compaction and Sintering

The mixtures were uniaxially compacted at 400 MPa in the form of cylindrical pin samples of diameter 6 mm and length 15 mm. The green compacts of iron and copper alloy were sintered at 1120°C and 850°C for 30 minutes in 90% nitrogen and 10% hydrogen mixtures respectively. The density of the sample was maintained at 6.6 g/cc. The sintered MMCs compacts were studied under optical microscope, it is understood that the distribution of Fe_3Al particles are depending on the volume fractions considered during processing of MMCs. The distribution is higher in higher volume fractions namely in 10% compared to 5% of Fe_3Al in the present study. The photomicrographs of the matrix with 5 and 10% volume fraction for iron are shown in Fig. 4. (*a*) and (*b*) respectively.

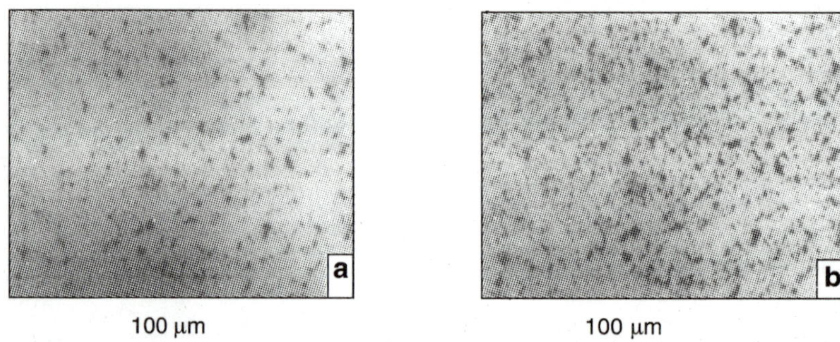

Fig. 4. Photomicrograph of the matrix at 5 and 10% volume fractions Magnification :100x.

2.4. Dry Wear Studies

The wear behavior of all the MMCs and base material was evaluated through a pin-on-disc test as per ASTM Standard G-99. Wear test prerequisites including sample surface polishing and cleaning were carried out. Pin specimens were made to slide against alloy steel disc of hardness 62 HRC at a constant load of 20 N up to the sliding distance of 5400 m. In all the cases three experiments were performed for each volume fraction and the results are taken as average values. Wear resistance was determined by measuring the weight loss of the specimen after each sliding distance 2160, 4320, 5400 m. The friction coefficients were measured continuously during the test. The wear tracks were examined using Scanning Electron Microscope (SEM), JEOL model JSM-840 A.

3. RESULTS AND DISCUSSIONS

3.1. Mechanically Alloying

From Fig. 1, which shows the XRD pattern, it is evident that after 42 hours of mechanical alloying, the iron aluminides has formed. Also, from Fig.2 and 3, it is clearly understood that the particles formed by mechanical alloying are of nano size ranges. The TEM measurements of nanoparticles of the iron aluminides indicates the particle size ranges from 10 to 25 nanometers. The agglomeration of the particles shown in Fig. 3(*b*) supports the formation of nanoparticles of the iron aluminides during mechanical alloying.

3.2. Wear Loss Behaviour

The measured weight loss with sliding distance for copper alloy and iron samples A, B and C and the variation of wear loss with sliding distance is presented in Fig. 5 (*a*) and (*b*).

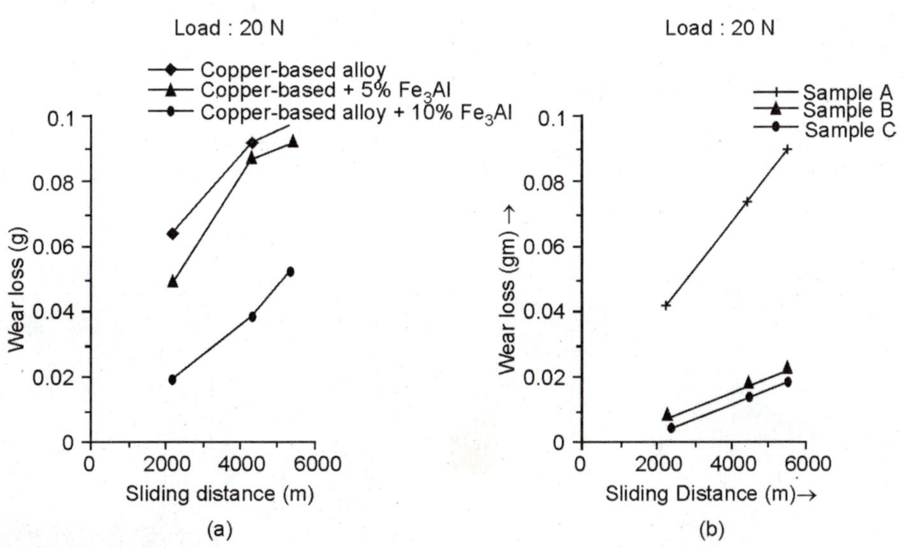

Fig. 5. Variation of wear loss with sliding distance for copper alloy and iron.

From the above results it is clear that as sintered copper alloy and iron shows significant weight loss when compared to 5% and 10% MA Fe_3Al nano particle reinforced MMCs. However, there is a major difference between 5% and 10% Fe_3Al nano particle reinforced MMCs. From the observations, it is clear that the wear loss increases with increase in sliding distance. Also, the reduction in extent of wear loss with increase in sliding distance was observed. The wear behaviour of the two systems exhibits the same wear trends. It is very evident from the wear data that FC_3Al nano particle reinforcement had significant influence on the wear behavior of the material under study. This is attributed to high hardness of Fe_3Al particles uniformly distributed in the MMCs compared to unreinforced sintered copper alloy as. well as iron. Also, the unreinforced matrix has lower hardness values and porosity in the matrix. The pores act as stress concentration zones and reduce the mechanical strength and ductility. These properties attributes more wear in the matrix without particle reinforcements. The reduction in extent of wear loss with increase in sliding distance may be due to the self lubrication properties of the intermetallics and also due to the combined effect of heat generated due to friction as well as work hardening and dynamic reerystallisation effects during wear process [6]. Also, under abrasive and sliding wear conditions, it is reported that increasing volume fraction of particulate in MMC increases wear resistance properties [7], which confirms the observations. From the Fig. 5, it is understood that the wear loss is less in 10% Fe_3Al volume fraction in both the systems. The wear loss decreases with increase in volume fraction of particulate because the matrix is strengthened and thus the hard particles reduce the tendency of the composites to the adhesion of the counter face [7]. Also, being P/M based MMC the porosities in the metallic matrix are closed due to the plastic deformation of Fe_3Al nano particles reinforced in the matrix also, the nano size ranges of the particles can reduce the porosity of the matrix effectively in higher volume fractions by accommodating itself into the pores and reduces the porosity of the matrix considerably. As the sliding distance increases, plastic deformation of Fe_3Al particles starts due to friction and it closes the pores further adjacent to the reinforced particles. This may contribute to the low wear loss observed in the matrices containing higher volume fractions of the Fe_3Al particles. The closure of porosities due to the plastic deformation of the particles was also observed in the SEM studies.

3.3. Morphology of Worn Surface

Scanning electron micrographs of worn surfaces of 10% particle reinforcement in iron and copper shown in Fig.6. The micrograph of iron Fig. 6a and Fig. 6b reveals wear due to the mechanism

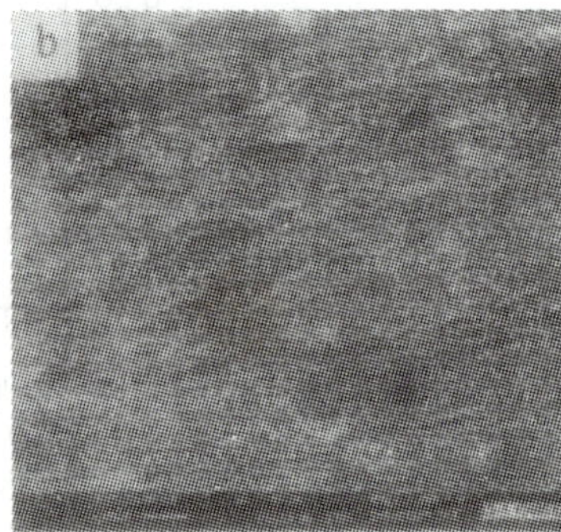

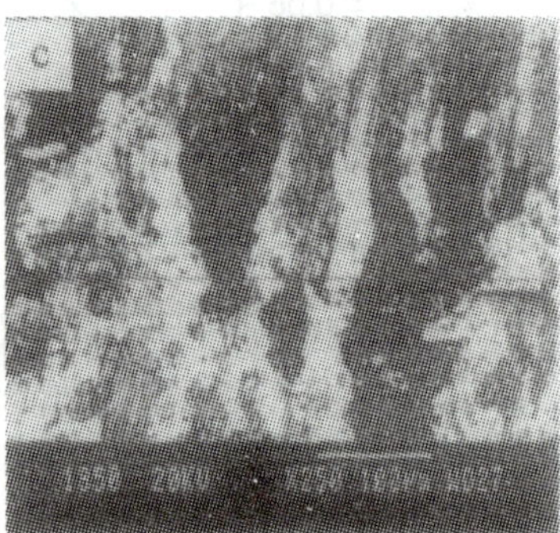

Fig. 6. SEM micrographs of worn surfaces.

of delamination. Deformed layers are noticed in the sliding direction suggesting to severe plastic deformation during wear. This is due to accumulated strain in sliding surfaces by repeated contact surfaces which subsequently forms delamination of layers with increased dislocation density. Also, localized plastic deformation and strain hardening lead to delamination of the surface layers which resulted in large amount of material removal [6]. The examination of wear tracks of iron (Fig. 6b) reveals microploughing was the prevailing and the dominant wear mechanisms. It is reported that microploughing and micro cutting were the dominant wear mechanisms during abrasion wear of iron aluminides [9]. Also, distinguished region of plastic deformation of iron aluminide in copper alloy (Fig. 6.c) is noticed on the wear tracks which may be due to nano size of iron aluminide particles supports microploughing is the wear mechanism. Though the iron aluminides are brittle in nature, the plastic deformation occurred during sliding indicates microploughing and the formation of surface platelets. The plastically deformed layer of iron aluminides formed on the wear surface also acts as lubricant layer during the dry sliding conditions of the matrix. During the entire duration of the wear tests, the friction coefficient was observed in the range of 5.5 to 6.5 for the particle reinforced matrixes and 4.0 to 4.5 for the unreinforced matrixes. This was due to the higher hardness of the iron aluminide particles which was reinforced in the matrix.

4. CONCLUSIONS

1. Mechanical alloying produces nano iron aluminides particles.
2. Dry sliding wear property improves considerably with the addition of mechanically alloyed nano Fe_3Al compared to sintered copper alloy and iron without reinforcement. In addition, there is significant increase in wear resistance with the increase in volume fraction of the particles in the matrix.
3. Microploughing and plastic deformation of iron aluminides were the mechanism for the removal of material on the matrix.
4. Nano size range and self lubrication properties of the Fe_3Al particles can reduce wear under dry lubrication conditions in P/M based iron and copper composites.

REFERENCES

1. J.H. Westbrooke: Intermetallic Compounds, Volume 2.
2. D.E. Alman, J.A. Hawk, J.H. Tylczak, C.P. Dogan, R.D. Wilson, Wear 251, 875-883 (2001).
3. K. Venkateswaran, M. Kamaraj, K. Prasad Rao, International Conference on Advances in Surface Treatment: Research & applications, Nov. 3-6, 2003, Hyderabad, India.
4. J.S. Benjamine, Metall. Trans.1,2943 (1970).
5. K. Wolski, F. Thevenot, J. Le Coze, Intermetallics 40, 299-307 (1996).
6. B. J. Johnson, F. E. Kennedy, I. Baker, Wear 192, 241-247 (1996).
7. Garish Sharma, M. Sundararaman, N. Prabhu, G. L Goswami Bull. Mater. Sci., Vol. 26, No. 3, 311-314, 4, 2003.
8. Yong-Suk Kim, Yong-Hwan Kim, Materials Science and Engineering A 258, 319- 324(1998)
9. J.P. Tu, L. Meng, M.S. Liu, Wear 220, 72-79 (1998).

Unit–IV

Powder Metallurgy for Electrical/Electronic Industry

1

Evolution of Powder Processing Techniques for Production of Silver-Metal Oxide Electrical Contact Materials

P. B. Joshi
Asstt. Professor Department of Metallurgical Engineering, M. S. University of Baroda, Baroda, India

P. Ramakrishnan
Professor Emeritus, Department of Metallurgical Engineering & Materials Science, Indian Institute of Technology Bombay, Mumbai-400 076, India

ABSTRACT

Silver- metal oxide (Ag-MeO) type composites such as Ag-CdO, Ag-ZnO, Ag-SnO_2, Ag-SnO_2-In_2O_3, Ag-SnO_2-Bi_2O_3, Ag-NiO, etc. are used as electrical contacts in a variety of switchgear devices as low to medium power arcing contacts. As uniformly dispersed discrete particles, the oxide phase improves the properties of silver matrix by increasing its hardness and reducing the tendency to contact welding.

Such Ag-MeO contact materials are generally produced by Powder metallurgy (PM) processes involving powder production, consolidation, and sintering followed by repressing /hot pressing to bulk solid. Several processes have been already developed and some more are under developmentt to produce silver-metal oxide composite powders, with high degree- of homogeneity of dispersion between silver and metal oxide powder particles. A few frequently used processes include Mechanical Mixing or Blending, Ball Milling, Atomization Co-precipitation Method, Reaction in-Spray roasting process, Freeze Drying, Electroless Coating, I.O.A.P. process, Internal Oxidation processes, High-energy milling or Mechanical alloying (MA) and Reactive milling (RM).

An attempt has been made here to briefly review some of these processes for synthesis of Ag-CdO, Ag-SnO_2, Ag-ZnO composite powders for electrical contact application.

1. INTRODUCTION

As a member of the switchgear device and an integral part of the electrical circuit, electrical contacts are employed in low as well as high-voltage switchgear devices such as circuit breakers, a.c. and d.c. contactors, relays, different types of switches, etc.

Metal-metal oxide type composites, is the largest group of contact materials for applications in switchgear devices operating in the range of a few amperes to about 5000A. Amongst metal-metal oxides group silver-metal oxide (Ag-MeO) are the most widely used contact materials. They are used as low to medium power arcing contacts for more than seven decades. The development in the silver-metal oxide group of materials as electrical contacts dates back to as early as 1930's. The first patent for these materials was filed in United States by F.R. Hensel in 1937 and was granted in 1939 [1]. A historic review of developments in Ag-MeO contacts has been presented by Shen *et al.* [2].

Of all the Ag-MeO type contact materials, Ag-CdO has proved to be a pioneer and a most promising material as contacts in switchgear devices because of its tendency to offer low arc-erosion tendency, low contact resistance and an excellent anti-welding characteristics as a contact material. However, the environmental agencies such as Occupational Safety and Health Administration (OSHA) and Environmental Protection Agency (EPA) in countries like Japan, North America & some parts of Europe have put restrictions on materials incorporating cadmium in view of its toxicity [3]. Eventually, attempts are made to replace CdO in Ag-CdO completely or partially by non-toxic environment-friendly substitutes [4-6]. A significant amount of work is going on in this direction since late 70's. As an outcome of all this, despite its few limitations, silver-tin oxide system has emerged as a main substitute to silver-cadmium oxide [7,8].

Though silver-tin oxide has emerged as an appropriate replacement material to conventional silver-cadmium oxide, contact systems incorporating oxides other than CdO and SnO_2 are also developed. They include materials like Ag-ZnO, Ag-NiO, Ag-CuO, Ag-$1n_2O_3$, Ag-PbO, AG-FCO, Ag-Al_2O_3, Ag-SiO_2, etc. [9-11] By and large, these alternative materials cater to only a specific nature of switchgear application.

Powder metallurgy techniques are widely used in the production of electrical contact materials because of their edge over the conventional casting and other forming processes such as rolling, forging, etc. The powder metallurgy process permits the constituent materials to retain their properties in such composites and the properties of composite by and large follow the rule of mixture.

2. POWDER PROCESSING TECHNIQUES

Various techniques used/under development for production of AG-MeO powders for electrical contact applications are briefly summarized below.

2.1. Atomization

Among several physical methods of powder synthesis, atomization has a special importance as a commercially viable, fast, reproducible and relatively less expensive technique on large scale. It can be used for production of metal and alloy powders for contact application and can also he used to produce even composite powders under certain specific conditions. For example, Ag-CdO or Ag-ZnO powders can be produced by oxidation of Ag/CdO or Ag/Zn alloy powder produced by atomization route.

2.2. Mechanical Mixing or Blending Process

Blending or mixing of constituent powders is a starting step in preparation of any PM part and in particular electrical contact materials., e.g., blending of powders in systems like silver and metal oxide powders or copper with metal oxide powder or refractory metal/metal carbide. The main objective of mixing or blending operation is to ensure uniform distribution of different constituent materials having difference in size, shape, specific gravity and other particulate characteristics. However, uniform mixing of powders having variedly different densities or particle size is rather difficult and hence optimization of process variables such as speed of rotation of the mixer/blender or the mixing time is a must for specific system.

2.3. Ball Milling

It is one of the oldest methods of powder synthesis for brittle, friable, tough and hard materials and a pulverization process for malleable and ductile materials. There are several drawbacks of milling as a powder synthesis route, particularly for synthesis of Ag-MeO powders for electrical contact application. They are: (*i*) excessive work hardening of powder particles leading to difficulties in their compaction (*ii*) segregation of powders particularly during milling while the constituents have large difference in their specific gravities (*iii*) excessive tendency to leafing or flaking of powders of malleable and ductile metals like silver (*iv*) pyrophoricity of powders when they are milled to ultrafine size and (*v*) natural tendency of powders to agglomeration with increasing fineness.

2.4. High-energy Milling

High-energy milling or mechanical alloying (MA) was originally developed by J. S. Benjamin in late 1960's at INCO Alloys International and has traditionally been used for production of oxide-dispersion strengthened superalloys for high temperature applications. e.g. ODS Fe, Ni Al, Ti, etc. [12].

Mechanical alloying is a dry, high-energy ball milling process for producing composite powders with controlled microstructure. The equipment used for MA consists one of several type of high-energy mills such as a horizontal ball mill, planetary ball mill, stirred ball mill or an attrior, shaker mill, vibratory mill, etc. During MA the metal powder particles are repeatedly flattened, fractured and rewelded resulting in an extremely fine and uniform dispersion of second phase particles in the matrix.

The application of mechanical alloying to synthesize composite powders for the manufacture of electrical contact materials is relatively recent. MA is an upcoming technology for production of Ag-MeO powders such as Ag-CdO, Ag-ZnO, Ag-SnO_2, Ag-Cr_2O_3 etc. on a commercial scale [13-16]. The major limitation of the process is contamination of the powder from process control agent (PCA) and the iron pick- up from the attritor vessel and the grinding media.

2.5. Reactive Milling

The scope of MA technique has been widened in recent years to the extent that it can even be used to initiate oxidation/reduction type of chemical reactions under normal ambient temperature conditions, thus turning an attritor into an active chemical reactor [17,18]. This application of mechanical alloying is popularly termed as mechano-chemical synthesis or reactive milling.

H. Zoz and coworkers have applied the. concept of reactive milling to synthesis of Ag-SnO_2 composite powders for electrical contact application [19]. The starting materials for this synthesis are silver-tin alloy (Ag_3Sn) and the silver oxide (Ag_2O). The stoichiometric amount of the powder mix is subjected to reactive milling in a horizontal rotary ball mill called 'Simoloyer'. Very fine dispersion of tin oxide in silver matrix in a record time of only a few minutes to a couple of hours with minimum contamination from the attritor mill is attainable by this process.

The authors have also worked in the area of reactive synthesis of Ag-MeO powders for electrical contact application. The study involves reactive milling of Ag_2O and Cd, Sn or Zn powders, respectively in stoichiometric amounts in an attritor followed by their X-ray diffraction for phase identification. It is observed that during reactive milling in a short span of time the silver oxide reduces to silver and the liberated oxygen oxidizes the base metal particles to their respective oxide. A fine and uniform dispersion of metal oxide in silver matrix can be achieved in relatively much shorter times as compared to that achieved through conventional high-energy milling of silver and metal oxide powders under similar processing conditions [20]. The major advantages of reactive milling process for synthesis of composite powders are: (*i*) almost total freedom from contamination (*ii*) reduced plastic deformation and hence less work hardening of powder particles offering improved powder compressibility and (*iii*) more efficient process on time scale.

2.6. Coprecipitation Method

This is one of the most extensively used methods to produce silver-metal oxide composite powders for electrical contact application. Douglas *et. al.* have patented the process for making Ag-MeO powders by co-precipitation [21]. The composite powders of Ag/CdO, Ag/ZnO, Ag/MgO, Ag/SnO_2,Ag/Li_2O, Ag/CoO, Ag/CuO, Ag/MnO, Ag/Fe_2O_3, Ag/CaO, etc. can be produced by the procedures described in the patent. For production of Ag-CdO powders the process comprises dissolution of silver and cadmium in nitric acid. The homogeneous nitrate solution is then converted into precipitates of composite powders of silver and cadmium carbonates or hydroxides by reaction with precipitant solution such as carbonates or hydroxides, respectively. The precipitates are washed dried and calcined to produce Ag-CdO composite powders. Instead of simple mixing, the nitrate solution may be sprayed in the form of fine atomized droplets into the precipitant solution so as to the compositional variation within the powder so produced. The process is then named as 'spray-co-precipitation'. A conventional glass capillary spray nozzle is used on laboratory scale to ensure geometric mean diameter of sprayed droplets of ~ 20 micron size. The major advantages of the spray-co-precipitation method are: (*i*) fine and uniform dispersion of oxide particles in silver matrix. (*ii*) a wide range of composition with respect to both quantity and type of oxide (*iii*) case of doping with beneficial ternary additions without influencing the microstructure of the composite contact.

Production of Ag-CdO, Ag-ZnO, Ag-SnO_2, Ag-NiO, Ag-La_2O_3 powders by co-precipitation method has been reported in the literature [22, 23].

2.7. Spray Roasting or Reaction Spraying Process

Spray- roasting is the most suitable method for production of doped silver-metal oxide composite powders [24]. The starting step in the production of composite powder by this method is to prepare a homogeneous aqueous solution of the constituent elements in stoichiometric proportions. The, typical starting solution could be a mixed nitrate solution. The mixed nitrate solution is then atomized into small droplets and the droplets are thermally decomposed by passing through a gas flame or hot air wherein a reaction zone is maintained at a temperature above the decomposition temperature of the constituent nitrates. (for example the decomposition temperature in the case of silver and cadmium nitrate is 550°C. Due to spraying of solution into a hot reaction vessel, instantaneous evaporation of water from the solution droplets leads to formation of powder particles. These particles, which first consist of salts or hydroxides, react with water vapor in reaction vessel and become oxides, mixed oxides or oxide mixtures. The oxides of silver being thermally unstable; the silver metal oxide composite powder is produced. The important advantages of reaction spray process are: (*i*) ability to produce compound powders in wide variety of composition (*ii*) properties of the powders produced can be varied by controlling the process parameters like solution chemistry, reaction temperature, size of atomized droplets, cooling rates, etc. (*iii*) effective doping of desired dopant in powder form (*iv*) high degree of homogeneity between the various constituents within individual powder particles.

2.8. Freeze Drying

Freeze drying or cryochemical method for preparation of ceramic oxides was introduced by Schnettler *et al.* [25]. The technique involves the separation of a solution into small droplets by spraying it through hydraulic nozzle at a pressure of few psi into a bath of immiscible liquid, such as hexane chilled by Dry Ice- acetone or spraying directly into liquid nitrogen. The rapid freezing is essential to prevent ice-salt segregation. The frozen mass is then subjected to drying, by sublimation of water from the frozen solution without melting. This is done by introducing the frozen sample into a vacuum chamber that is evacuated to a pressure of about 1 torr or less to maintain the partial pressure of water below triple point. In order to keep the partial pressure of water in the system as low as possible, a refrigerated condensing coil is introduced to collect the sublimed water.

D.J. Anderton and F.R. Sale [26] used the freeze-drying technique to make silver- cadmium oxide powders from aqueous nitrate solutions to produce contacts having improved microstructural homogeneity. However, the maximum benefits of freeze-drying could not be achieved in such a system because of the melting of silver nitrate prior to its decomposition, which ultimately led to loss of freeze-dried morphologies.

2.9. Electroless Coating

Electroless deposition of metals has received full recognition as a valuable coating technique over the last decade. Electroless coating technology can be used to produce composite powders of silver-metal oxide type also. It is based on the principle of deposition of a metal or an alloy by controlled chemical reduction of metal ions being deposited. For example, it is possible to carry out electroless coating of silver using proprietary reducing baths like Brashear (sugar), Rochelle salt or Formaldehyde. In order to develop silver coating on the substrates like powder particles such as those of metal oxide, excess charge is first developed on the oxide particle surfaces. The excess positive or negative charge on the oxide particles promotes interaction with positive or negative ions from the solution. The ions from the solution are attracted and absorbed over the oppositely charged surfaces of oxide particles leading to formation of composite powders.

H. Chang and co-workers [27] have developed a novel method of coating tin oxide particles with silver by an electroless coating process to produce silver-tin oxide electrical contact materials. The method consists of *in-situ* reduction of silver from silver nitrate solution over the negatively charged surfaces of highly dispersed particles of tin oxide in aqueous medium in the form of a sol. The coating over the substrate particles (in the form finely dispersed sol) is carried out under controlled conditions of temperature, metal ion concentration, metal ion feed rate and pH of the solution containing the substrate particles.

P. B. Joshi *et al.* [28] have developed a method for producing composite powders of Ag-CdO and Ag-ZnO using sodium formate as the reducing agent for *in-situ* reduction of silver nitrate to silver ions. The reduced silver ions in turn get attracted towards the negatively charged surfaces of cadmium oxide or zinc oxide particles in the suspension leading to silver coating. The electroless coating offers improved dispersion between silver and metal oxide phases and therefore superior contact properties in general.

2.10. I.O.A. P. Process

Internal Oxidation of Alloy Powders (I.O.A.P.) process was developed by Pedder, Douglas and McCarthy [29] in order to obtain a very fine and uniform dispersion of metal oxide in silver matrix at relatively low temperature and substantially short period of internal oxidation treatment. The process consists of preparation of silver-base alloy powder of fine particle size by low temperature reduction and alloying of a silver and metal oxide powder mixture. The metal oxide particles get reduced to metal and diffuse into silver to form silver-base alloy. The reduced powder is usually in the form of loosely sintered spongy mass, which can be converted to powder by light pulverizing action. Internal oxidation of the pulverized powder can then be carried out at relatively low temperatures and in times typically 10^3 times shorter than are required in conventional internal oxidation.

Later on Yuan- Shou Shen and Lawrence Gould [30] also used this technique to produce electrical contacts grade silver-base composites by oxidation of Ag-Cd and Ag-Sn alloy powders obtained by water atomization.

Recently, J. Troxell *et al.* have developed dispersion- strengthened silver by controlled oxidation of silver/aluminium alloy powder in an oxidizing atmosphere [31]. The resultant Ag-Al_2O_3 nano-composite powders are in turn consolidated by press-sinter-repress, hot- isostatic pressing (HIP), or hot- extrusion route to high-density. Such dispersion- strengthened composites produced by I.O.A.P. route offer high strength, high conductivity and superior resistance to softening.

P. B. Joshi and P. Ramakrishnan have carried out a detailed investigation on the scope and limitations of various processing routes for production of Ag-MeO powders discussed above [32]. The influence of powder characteristics for some of the processing routes on the physical, microstructural and electrical properties of final contacts as well as their behaviour as a member of switchgears has also been discussed.

REFERENCES

1. F.R. Hensel, U.S. Patent, No. 2,145,690
2. Yuan-Shou Shen, William D. Gote and Lawrence J. Gould, A Historic Review of Ag-MeO Materials, Proc. of Holm Conf on Electrical Contacts, (1986), p.71.
3. Ed.L. Bertherick Hazards in the Chemical Laboratory, The Royal Society of Chemistry, London, 4th Edition, September (1986), p. 222.
4. M. Poniatowski, E. Schulz, and A. Wirths, The Replacement of Silver-Cadmium Oxide by Silver-Tin Oxide in Low Voltage Switching Devices, Proc. of Holm Conf. on Electrical Contacts, (1976), p. 359.
5. K.H. Schroder, Silver-Metal Oxides as Contact Materials, /EEE Trans. on Comp., Hybrides, and Manuf Technol, Vol. 10, No. 1, (1987), p. 127.
6. P.C. Wingert and C.H. Leung, The Development of Silver-Based Cadmium-Free Contact Materials, Proc. of Holm Conf. on Electrical Contacts, Vol. 12, No. 1, March (1988), p. 31.
7. N. Behrens, J. Bohm, P. Braumann, and G. Kle00, Experiences with Contact Material Ag/SnO_2, Proc. of Holm Conf on Electrical Contacts, (1984), p. 185.
8. C. Lambert, D. Weber, S. Coupez, and J.P. Gurlet, A New Doped Silver Tin Oxide Material with Improved Electrical Behavior, Proc. of IEEE-Holm Conf. on Electrical Contacts, (1989), p. 69.
9. P.B. Joshi, P.S. Krishnan, R.H. Patel, N.S.S. Murti, V.L. Gadgeel and P. Ramakrishnan, Improved P/M Silver-Zinc Oxide Electrical Contacts, Int. J Powder Met., Vol. 34, No. 4, May 1 June (1998), p. 63.
10. M. Ohta, S. Matsuda, K. Ogawa, and A. Shibata, New Contact Materials of Silver- Metal Oxide System, Proc. of 16th Int. Conf. on Electrical Contact Phenomena, (1992), p. 421.
11. Kishio Arita, Koichi Fujiwara, and Shigeyuki Tsurumi, Eutectic Silver Silicon Internally Oxidized Contact, Proc. of Holm Conf. on Electrical Contacts, (1980), p. 299.
12. J.S. Benjamin, Dispersion Strengthened Superalloys by Mechanical Alloying, Met. Trans., Vol. 1, October (1970), p. 2943.
13. G.H. Upadhyay, G.R. Marathe, P.B. Joshi, V.L. Gadgeel, P.M. Raole, and P. Ramakrishnan, Nanocrystalline Ag-CdO Powders Produced by Mechanical Alloying, Proc. of Int. Conf. on Trends

in Mechanical Alloying. Science Technology & Applications, Jaipur, February (2001), p. 149.

14. P.B. Joshi, R.H. Patel, V.L. Gadgeel, V.K. Kaushik and P. Ramakrishnan, Effect of Powder Processing Technique on Structure and Properties of Silver-Tin Oxide. Electrical Contact Material. Proc. of Powder Metall. World Congress, Spain Vol. (1998), p. 533.

15. U. Grundmann, M. Heilmaier and L. Schultz, High Temperature Deformation Behaviour of Mechanically Alloyed Microcrystalline ODS Silver, Scripta Materialia, Vol.37, No. 1, (1997), p. 103.

16. U. Grundmann, M. Heilmaier and L. Schultz, Manufacturing of Oxide Dispersion Strengthened Silver by Mechanical Alloying, Proc. of Powder Metall World Congress, Spain (1998)

17. G.B. Schaffer and P.G. McCormick, Reduction of Metal Oxides by Mechanical Alloying, Appl. Phys. Lett., Vol. 55(1), July (1989), p. 45.

18. G.B. Schaffer and P.G. McCormick, Displacement Reactions During Mechanical Alloying, Met. Trans., Vol.21 A, October (1990), p.2789.

19. H. Zoz, H. Ren and N. Spath, Improved Ag-SnO_2 Electrical Contact Material Produced by Mechanical Alloying, Metall., Vol. 53, No. 7-8, (1999), p. 423.

20. P.B. Joshi, G.R. Marathe, V.L. Gadgeel, G.H. Upadhyay, V.K. Kaushik, and P. Ramakrishnan, Reactive Milling-An Improved Process for Synthesis of Silver-Tin Oxide Composite Powders for Automotive Contacts, Proc. of Int. Conf on Powder Metall. for Automotive Applications, New Delhi, January (2002), p. 113.

21. P. Douglas and T.I. Stewart, Method of Producing Powdered Materials, U.S. Patent No. 4, 023, 961, May (1977).

22. V. Saravanan, S. Vyanatheya and Parwati Ramaswamy, Electrical Contact Materials by Coprecipitation Route, CPRI News, No. 56, October (1995), p. 9.

23. P.B. Joshi, V.L. Gadgeel and P. Ramakrishnan, Electrical Contacts Grade Silver- Zinc Oxide Powders by Spray Coprecipitation Technique-Processing and Characterization, Trans. Powder Metall. Assoc. of India, Vol. 21, (1994), p. 27.

24. V. Behrens, T. Honig, A. Kraus, R. Michal, K. Saeger, R. Schmidberger, and T. Stannef, An Advanced Silver/Tin Oxide Contact Material, IEEE Trans. on Comp. Packag., and Manuf Technol., Part A, Vol. 17, No. 1, March (1994), p. 24.

25. F.J. Schnettler, F.R. Monforte, W.H. Rhodes: A Cryochemical Method for Processing Ceramic Materials, Science of Ceramics, Vol. 4, and Edited by G.H. Stewart, The British Ceramic Society, Stoke-on-Trent, England (1968), p. 79.

26. D.J. Anderton and F.R. Sale, Application of Freeze-Drying Process to the Production of Silver-Cadmium Oxide Contact Material, Proc. of Holm Conf. on Electrical Contacts, (1978), p. 183.

27. H. Chang, C.H. Pitt and G.B. Alexander, Novel Method for Preparation of Silver-Tin Oxide Electrical Contacts, Jr. of Mat. Engg. and Performance, Vol. 1 (2), April (1992), p. 255.

28. P.B. Joshi, R.H. Patel, Prabha S. Krishnan, V.L. Gadgeel, V.K. Kaushik, and P. Ramakrishnan, Powder Metallurgical Silver-Metal Oxide Electrical Contacts by an Electroless Coating Process, Adv. Powder Technol., Vol. 7, No. 2, (1996) p. 121.

29. D.J. Pedder, P. Douglas and J.P. McCarthy, I.O.A.P. – A Novel Process for the Manufacture of Long-Life Silver-Cadmium Oxide AC Contact Material, Proc. of Holm Conf. on Electrical Contacts, (1976), p. 109.

30. Yuan Shou Shen and Lawrence Gould, A Study of Manufacturing Ag-MeO.

Contacts from Oxidized Alloy Powders, Proc. of Holm Conf. on Electrical Contacts, (1983), p. 185.

31. J. Troxell and Abrams, Dispersion Strengthened Silver, *Adv. Mat. & Processes,* January (2001), p. 75.

32. P. B. Joshi and P. Ramakrishnan, Materials for Electrical and Electronic Contacts-Processing, Properties and Applications, Science Publishers Inc., USA, (2004), p. 48.

2

Comparative Routes of Solid-Solution-Formation by MM of Ag-70Cu (at%)

H. Zoz,
Zoz GmbH, D-57482 Wenden, Germany.
ESIQIE, National Polytechnic Institute, Mexico City, DF 07300, Mexico.

I. Vernet,
ESIQIE, National Polytechnic Institute, Mexico City, DF 07300, Mexico.

D. Jaramillo V.,
ESIQIE, National Polytechnic Institute, Mexico City, DF 07300, Mexico.

ABSTRACT

The solid-solution-formation (SSF) of Ag-70Cu (at%) *has been chosen as* the goal to be "eyed by Mechanical Milling (MM) in a SPEX8000-Mixer/mill and alternatively by a Simolover CMO I-21. In case of the SPEX-Mixer/mill, this is reached after about 43.2 ks, in case of the Simoloyer after 18 ks. The observation, that processing at given parameters in the Simoloyer leads to a flaky geometry already after 0.060 ks versus the SPEX-Mixer/mill after 10.8 ks or not at all is concluded to be caused by a higher collision rate. In result this matches to the long known attempt, that the collision itself is the main event for energy transfer in MA. Additionally in case of the Simoloyer, Cycle Operation has been applied which results here in rapid flaking with quasi-disappearing ductility followed by shear-stress to break the flaky structure repeatedly.

Since a higher collision rate versus shear and friction effect is expected to cause less abrasive wear on the milling tools, the Fe-contamination of both systems was investigated and showed a lower contamination for the Simoloyer which matches to the expectation, too.

Material's characterization is given by SEM, XRD and TEM.

1. INTRODUCTION

The aim of the present paper is to compare alternative routes of Mechanical Milling (MM) to achieve solid solution formation (SSF) in the example of Ag-70Cu (at%) by using different devices (Simoloyer and SPEX-Mixer/mill) with different kinetic energy impact levels.

In case of the Simoloyer, we did not use HKP (High Kinetic Processing) which would refer to a relative velocity of grinding media in the order of 9 m/s and higher but medium kinetic processing in the order of 4 m/s since the kinetic in the SPEX-Mixer/mill was expected to be around or below that level.

The Ag-Cu-system is attractive to investigate SSF by alternative MM-routes since it was one of the first systems studied by Mechanical Alloying (MA) and is perfectly defined by low energy milling and has perfectly been characterized [1–3]. It is a system very noble and easy to work with when a matter of comparison wants to be made. SSF is favoured because Cu and Ag belong to the same group in the periodic table. On the other hand, Ag and Cu do not form a SS in conventional melting route because of their significantly different atomic radii. Finally we can expect high deformation rates at relatively low sheer stress due to the fcc-structure of both of the materials which is expected to favour the SSF by quasi-thin-layer formation on micro-structural base.

The comparison of Simoloyer and SPEX-Mixer/mill is interesting because their have up to now not been any direct comparative investigation been reported. This has been done for Simoloyer and Planetary Ballmill [4] but for the theme here, all known data is confidential since this is always related to the scale-up res. to the commercialization of some advanced materials in industry. And of course the major issue is here, that the SPEX-mixer/mill cannot he scaled up at all due to strict technological barriers and in so far is limited to 5-10 g with powder yields usually at less than 50 %. That means any kind of direct application of results achieved by the SPEX-Mixer/mill is impossible. On the other hand, a number of scientists are and/or have been using this device for fundamental research and in so far it is attractive to be able to determine potential relations of scalability by the Simoloyer for materials systems that fundamentally already exist.

In the following. we give one of these "blue" examples where we have been confirmed in 2002, to report some limited information about the initial phase of a scale up of a metal-matrix-battery-material by a US-company where we neither can give the company-name, nor the composition of the MMC nor the pattern-scales of the given XRD-results. Figure 1a gives 2 XRD-patterns of the MMC with the received statement that the material referred to the upper pattern has been synthesized in a SPEX8000D-Mixer/mill in a processing time of 10 h at a batch of 10 g and the lower pattern refers to the same material but produced in a CMOI-Simoloyer in a processing time of 2 h at a batch of 100 g where the XRD-investigation is reported to exhibit nearly identical performance. This information came along with 2 pictures of the two defferent devices being used where figure lb shows the SPFX-Mixer/mill and figure 1c the laboratory-scale Simoloyer at this company.

Fig. 1. XRD-patterns of MMC (a), SPEX8000D-Mixer/mill (b), laboratory-scale Simoloyer CM01-2I (c)

What makes the Simoloyer interesting here, is the high kinetic process (HKP) based not on shear and friction (low kinetic processing) but mainly on the collision of grinding media at relative velocities up to 14 m/s [5-6] which first promises a high level of energy impact and second a lower level of impurities res. of contamination by the milling tolls since collision is naturally less abrasive than shear and friction effects. The systems are available in large scale of several hundred liters volume and are economically and ecologically favorable in particular in case they can be operated semi-automatically if they are combined with a continuous or semi-continuous (auto-batch) powder separation system [7-8]. Atmosphere and cooling seems non-problematic since these mills can be operated, loaded and unloaded under vacuum or inter *gas* and are equipped with efficient cooling or cooling and heating systems. Figure 2a shows an industrial-scale CM100-Simoloyer that is used for the processing of nanocrystalline metal-hydride and zinc-oxide, Figure 2b the same size but in semi-continuously mode used for the processing of ductile metal flakes (Cu and Ag) [9-10].

(a) (b)

Fig. 2. Simoloyer CM100 at IREQ- Hydro Quebec/HERA Hydrogen (a) and at Fukuda Co. (b).

2. EXPERIMENTAL

2.1. Ag- and Cu-powder

Comparatively, the MM-process was applied in both, in a SPEX-Mixer/mill and in a Simoloyer (see chapter 1) to the Ag-Cu powder-mixture where the detailed composition is given in Table 1. Both of the starting materials were based on the chemical production route and therefore of high purity.

Table 1. Ag-Cu-starting Powder Mixture

starting powder mixture for the MM-process				
Element	***purity [%]***	***PS [μm]***	***comp. [at%]***	***comp. [wt%]***
Ag	99.9	1-10	bal.	bal.
Cu	99.5	1-5	70	57.88

In case of the SPEX-Mixer/mill, Methanol was used as a PCA in order to avoid sticking and agglomeration of the powder. In case of the Simoloyer, a PCA was not expected to he necessary since Cycle Operation [10-14] was applied. In both devices, MM was performed under argon-atmosphere which means in case of the SPEX-Mixer/mill, the vial was loaded in a glove-box and in case of the Simoloyer, a standard air-lock was used to transfer the starting powders from an argon-container into the evacuated grinding chamber which was after that flooded with argon, to. Figure 3 shows a picture of the air-lock at the grinding chamber. Detailed parameters and procedure are given in Section 2.2 Res. 2.3.

2.2. MM-process in the SPEX-Mixer/mill

The principle of the SPEX-Mixer/mill is, that a small milling vial is manually loaded with some grinding media (1-10 pieces) and powder (5-10 g) and this container will shake on an angle-section for and back. In this kinetic system, it is very difficult to determine the kinetic and no model was found The interaction of grinding media can be imagined as the grinding media is driven by the moving vessel and always at that moment, when the vessel is changing the direction, the milling balls that follow the container with time delay are hit by the vessel and accelerated into the vice-versa direction. If a number of balls is used, collision will take place in the hit front of the imagined ball-packet and shear and friction effects are initiated in between the grinding media. However, if we take into account, that the angle of the shake-like motion describes only a distance of less than 20 mm, in other words, the milling vial is only moving for 2 cm, then the SPEX-Mixer/mill turns out to be simply a kind of a vibration mill.

In the literature, a number for a velocity was found at 18 m/s [15] which must be regarded sceptical since it would not match at all to the numbers given at figure 1. since XRD-investigation here is related to a kind of lattice-destruction, which is in the MM related directly to the kinetic of this process, it could not be explained, that a system at a maximum relative velocity of grinding media of 14 m/s would after 2 h lead to similar result than a system at a velocity of 18 m/s after 10 h.

Table 2. Process Parameters of Ag-Cu-powder Mixture in the SPEX-Mixer/mill

Process parameters of Ag-Cu-powder Mixture in the SPEX-Mixer	
Mixing/milling device	SPEX8000D-Mixer/mill
Mixing/milling vessel	Stainless steel, 80 cc approx.
Grinding media	100cr6, 6 mm
Grinding media load	6 balls (30 g)
PCA (lubricant)	5 drops of Methanol
Starting powder	Ag-70Cu (mixture by at%)
Starting powder load	6 g
Powder/ball weight ratio	1 : 5
Atmosphere	Argon during loading
Frequency	Set by mixer-manufacturer, 5 hz approx.
MM time I-IV	10.8, 21.6, 43.2, 54.0, 75.6 ks (3-21 h)
Milling temperature	<50°C (vessel outside, estimated)
Average powder yield	Approx. 3-4g >> 50-65%

If we follow the attempt to better compare the SPEX-mill with a vibration-mill, then the found number for the velocity must be regarded even more sceptical.

Figure 1b shows a picture of the SPEX8000D-Mixer/mill which is the same model that has been used here. In this device, MM was carried out for 10.8, 21.6, 43.2, 54 and 75.6 ks which refers to 3, 6, 12, 15 and 21 h respectively where the detailed process-parameters are given in Table 2. For processing, the vessel of the mixer/mill is removed and carried into a glove-box in order to load and maintain the Ar-atmosphere in the mixing/milling-vessel. Then it was returned to the mixer/mill for the MM. Problematic is the control of the milling temperature since there is no cooling system of the vial available. The only way is to cool the environment (*e.g.* to place the mill into a fridge or use a van).

After each of the MM-sequence, the vessel was transferred into a glove-box and powder and balls were unloaded onto a sieve in order to receive the powder only. Here it is of course very difficult to remove all of the powder, in particular because some of the material is sticking at the inner vessel-wall and at the milling balls. This is why the methanol was used as a lubricant and by this method we could receive at least a powder yield of about 3-4 g which refers to 50-65% approx.

2.3. MM-Process in the Simoloyer

For the explanation of the Simoloyer as a horizontal high energy ball mill we can refer to the introduction of this paper, where the same model (CMOI-21m) that has been used here is given in figure 1c already. The Simoloyer devices are known from academic as well as industrial applications in MA [16-17], high energy milling (HEM) [7-10] and reactive milling (RM) [14, 18-19] and supply the highest relative velocity of grinding media, which leads to a high level of kinetic energy transfer, an intensive grinding effect and short processing times. Since the grinding media is accelerated by a horizontally arranged rotor inside the grinding vessel, these devices do not have to move unnecessarily any large masses like *e.g.* the entire chamber/mill in case of vibration-, shaker- or planetary ballmills.

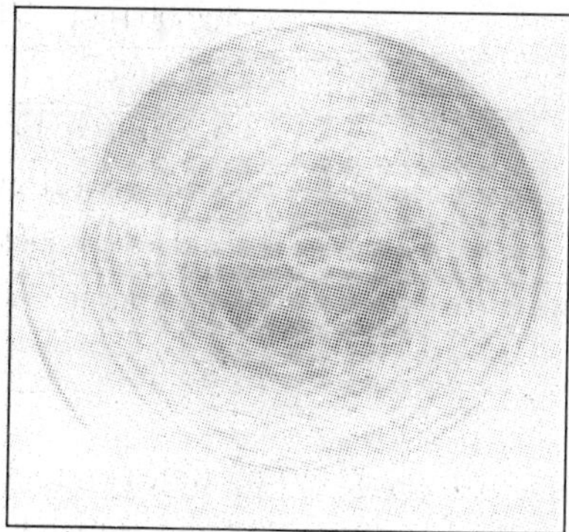

Fig. 3. Collision effects in HKP (Simoloyer).

Figure 3 shows a real image of a horizontal high energy ball mill in operation where the effect of collision in MM is visualized and the working principle can be imagined where the rotor is the tool to transfer the kinetic energy into the grinding media and the grinding media transfers into the powder material. The Simoloyer CMO 1 - 21 can be operated at rotation frequencies up to 1800 rpm, is water-cooled and computer-operated. This laboratory-scale size has been used for the MM at medium kinetic and was carried out for 3.6. 7.2, 10.8, 14.4, 18.0 and 25.2 ks which refers to 1, 2, 3, 4, 5 and 7 hours which is quite long for the Simoloyer but was set since the kinetic was decided to be used at a quite low level in the range of 600 rpm which refers to 3.6 m/s only. In maximum this was still 3 times shorter than in case of the SPEX-mixer mill since still a higher efficiency was expected. Since the definition of high kinetic processing (HKP) requires a relative velocity of grinding media higher than 9-10 m/s, here we might want to talk about medium kinetic processing (MKP). The detailed process-parameters are given in table 3.

In order to not to change the load relations inside the chamber, we did not use the sampling-unit which means for each MM-sequence, a separate MM-process was operated. For loading the starting powder, first the mill that must be in charging/operation position is evacuated to around 10^{-4} hPa and then flooded with Argon. The powder, that was loaded inside a glove-box into a powder-container, was transferred by the air-lock into the vessel at the same time when the vacuum was flooded since then the gas flow can even be used for the powder transportation.

Table 3. Process Parameters of Ag-Cu-powder Mixture in the Simoloyer

Process parameters of Ag-Cu-powder mixture in the Simoloyer	
Milling device Operating software	Simoloyer CM01, 2.7 kW, Maltoz 3.1
Grinding unit	W01-21 (2 liter, water-cooled)
Grinding media	Chromium steel, 100Cr6, 5 mm, 1.5 kg
PCA (lubricant)	No > Cycle Operation
Starting powder	Ag-70Cu (mixture by at%)
Starting powder load	30 g
Powder/ball weight ratio	1:50
Atmosphere	Argon, preceding evacuation
Operation Cycle	600/150 rpm - 14/1 min (refers to 3.6 m/s)
Discharging Cycle	150/600 rpm - 4/1 min
Average discharging time	10-20 min
MM time I-VI	3.6, 7.2, 10.8, 14.4, 18.0, 25.2 ks (1-7 h)
Milling temperature	<25°C (vessel inside by Maltoz)
Feeding system	Standard air-lock DN40-KF
Discharging/separation	Draingrating Ask-01
Average powder yield	Approx. 27-31 g>> 90-103%

After the adjusted MM-time elapsed, the grinding unit is tuned into discharging position and the same air-lock (after cleaning) was used for unloading the powder where this time first the powder container is evacuated via the air-lock. In that way, the gas-flow supported the unloading (a little) and the powder was discharged completely (grinding Media remains in vessel).

Unloading is computerized and automatically operated. In order to increase the powder yield and to decrease the discharging time, Cycle Operation (figure 4) has been applied which means during discharging, the rotational speed of the rotor is changed in a special frequency in the range from here 150-600 rpm which has been proven to tremendously increase the powder yield of materials that tend to stick and agglomerate under MM to the milling tools and to each other[10-14].

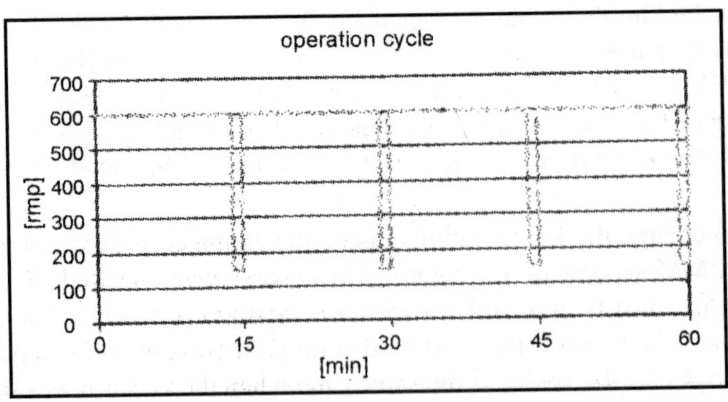

Fig. 4. Operation Cycle for MM.

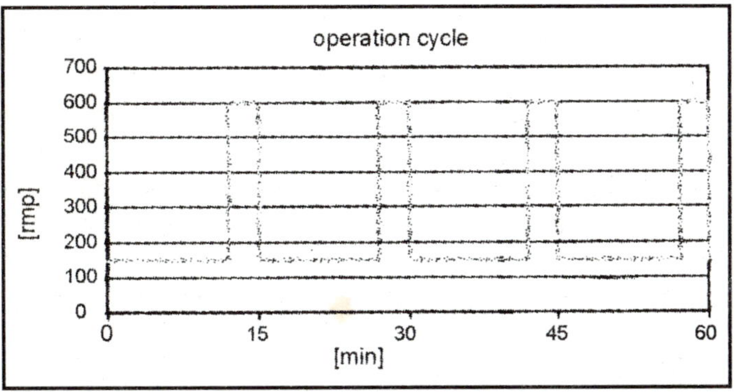

Fig. 5. Discharging Cycle after MM.

By this method in average complete powder yield could be obtained. Since between the each of the different MM-operations, the system has never been opened and since a little amount of powder always remains in the system e.g. sticking to grinding balls and vessel, the powder yield varied between 90 and 103 % after Discharging-Cycle of 10-12 min at 150/600 rpm (figure 5). Both of the cooling systems (vessel and pre-seal-unit) were constantly operated and the computerized measurement of the inner-vessel surface indicated temperatures lower than 25°C.

2.4. Powder Characterization

The initial and the as milled powders were characterized by SEM, XRD and TEM.

For Scanning Electron Microscopy, we applied a Jeol JSM-6300 and prepared the powder samples on a graphite tape in order to observe morphology and size. The iron-contamination of the powder samples has been investigated by Microanalysis using the same device. For X-ray diffraction we applied a Siemens Diffractometer D 5000 using monochromatic Cu-Kα radiation and determined/ estimated the grain size by the Scherrer method and later confirmed by TEM. For Transmission Electron Microscopy we applied a Jeol JSM 2000 FXII and used an ultrasonic bath to de-agglomerate the powder samples and then mounted on copper grids.

3. RESULTS

3.1. SEM and Microanalysis

(*a*)

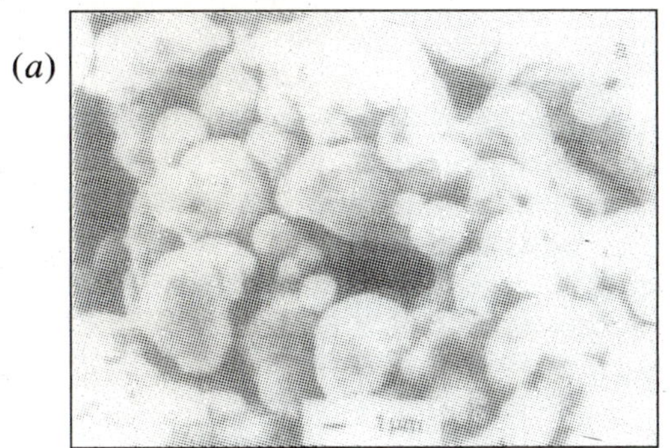

(*b*)

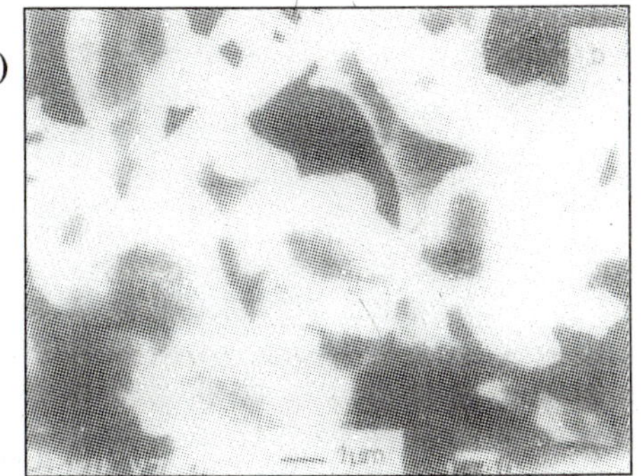

Fig. 6. SEM-micrographs of the starting powders (a) Cu and (b) Ag.

Figures 6a and 6b show the SEM-micrographs of the starting powders (a) Cu and (b) Ag.

Copper presents rounded shape with some porosity and an average particle size of around 5 μm. Silver shows an elongated shape with an average particle size between 1 -10 μm.

Figures 7a-e gives the SEM-micrographs of processed powder by the SPEX-Mixer/mill after 10.8, 21.6, 43.2, 54 and 75.6 ks respectively.

After 10.8 ks a flattened shape of the particles can be observed where the flattening seems to increase until that point when the image after 43.1 ks was obtained. After 21.6 ks, the primary particles appear as a kind of multi-layer agglomerates which looks alike until 54 ks. As of 75.6 ks, the primary particles look more like fissured bulks of agglomerates with irregular shape and not flake-like.

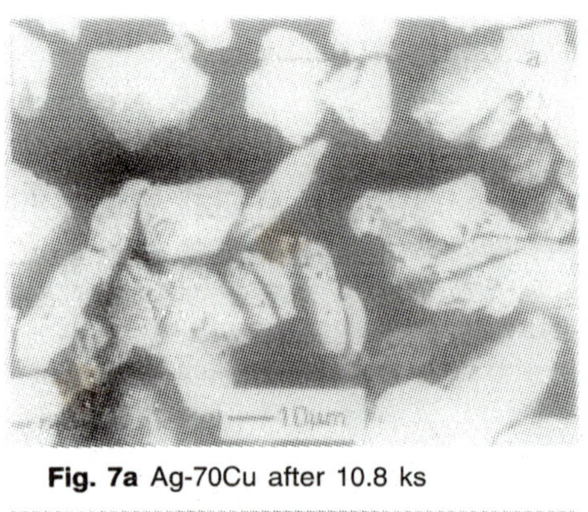

Fig. 7a Ag-70Cu after 10.8 ks

Fig. 7b Ag-70Cu after 21.6 ks

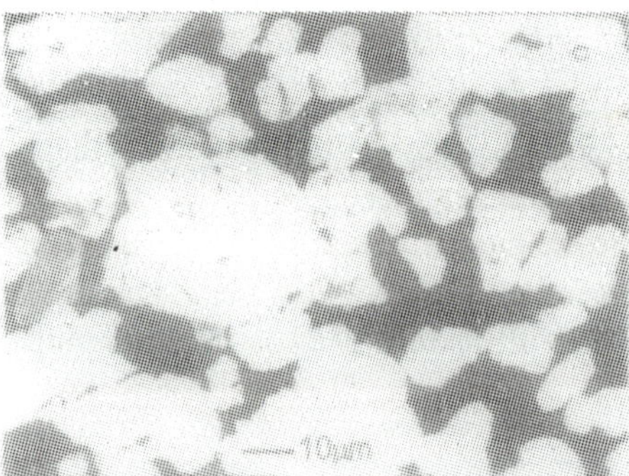

Fig. 7c Ag-70Cu after 43.2 ks

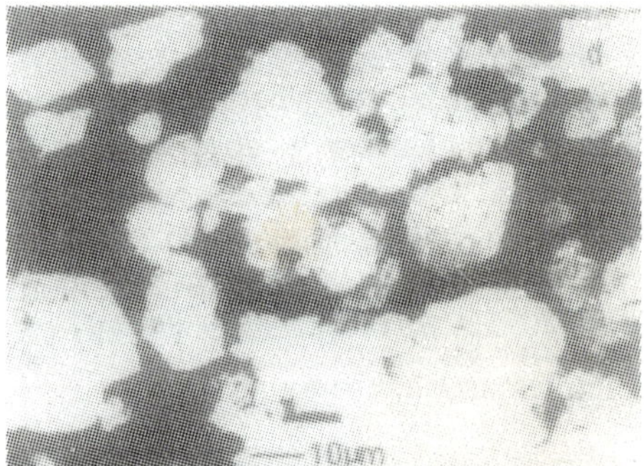

Fig. 7d Ag-70Cu after 54.0 ks

Fig. 7e Ag-70Cu after 75.6 ks

Fig. 7. SEM-micrographs of Ag- 70Cu powders after MM in the SPEX- Mixer/mill at various times.

The most close to flaky-structure seems to appear on the image obtained after 43.2 ks and the thickness is carefully estimated to an average of maybe 5-10 μm at an average particle size of around 10-40 μm.

In table 4 we try to summarize the above where we estimate the morphology evolution, the average particle size and in case of flaky structure the thickness of the particles:

Table 4. Summary of SEM-results Related to Morphology Evolution after MM in the SPEX-Mixer/mill

SEM-results related to morphology evolution after MM in the SPEX-Mixer/mill, estimates				
MM-time		*Particle shape*	*Particle size*	*Particle thickness*
[ks]	[h]	[μm]	[μm]	[μm]
0	0	Spherical-elongated	1–10	
10.8	3	Flattened bulks	30	5–10
21.6	6	Thick flakes	30	5–10
43.2	12	Flattened bulks	10–40	5–10
54.0	15	Flattened bulks	10–30	
75.6	21	Agglomerated bulks	20–100	

Figs. 8a-f gives the SEM-micrographs of processed powder by the Simoloyer after 3.6, 7.2, 10.8. 14.4, 18.0 and 25.2 ks respectively.

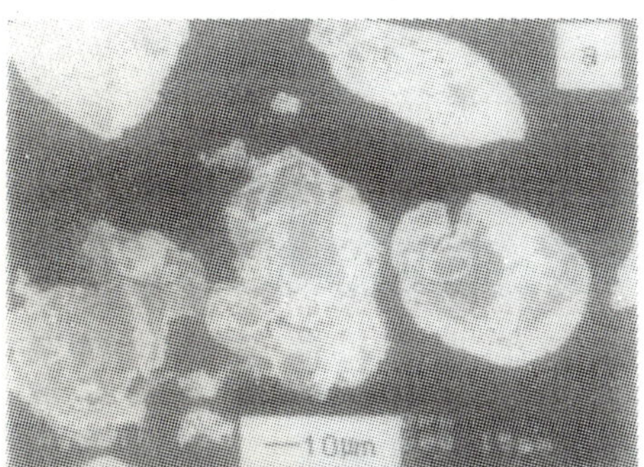

Fig. 8a Ag-70Cu after 3.6 ks

Fig. 8b Ag-70Cu after 7.2 ks

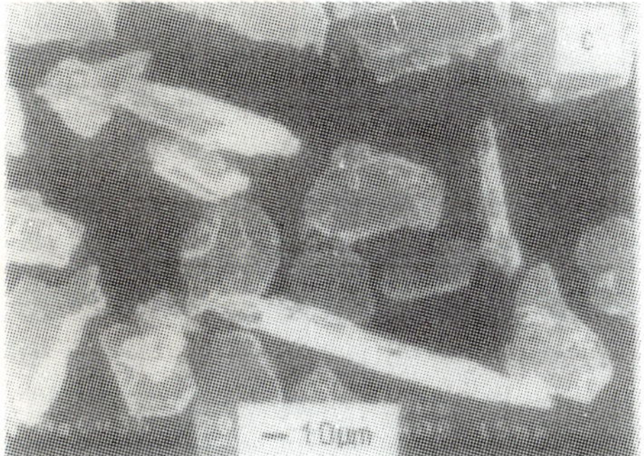

Fig. 8c Ag-70Cu after 10.8 ks

Fig. 8d Ag-70Cu after 14.4 ks

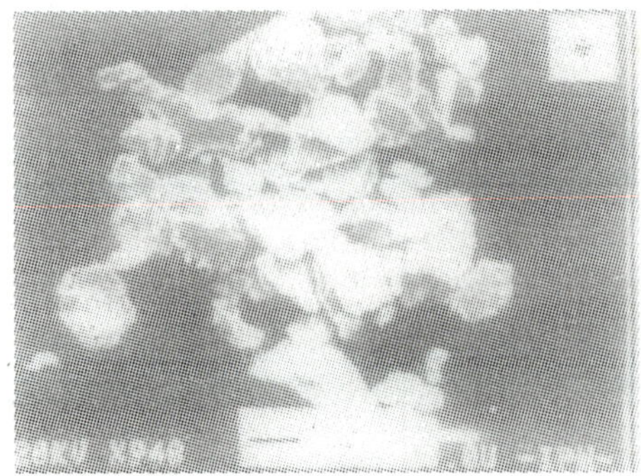

Fig. 8e Ag-70Cu after 10.8 ks

Fig. 8f Ag-70Cu after 21.6 ks

Fig. 8. SEM-micrographs of Ag-70Cu powders after MM in the Simoloyer at various times.

After 3.6 ks a very much flattened and in so far flaky shape appears and the primary particles look like a multi-layered agglomerate but much thinner than ever reached in the SPEX-Mixer/mill. The size can be estimated to an average of 60 μm. At 7.2 ks, the particles look much more condensed and the size got much larger up to more than 250 μm where the thickness is estimated to less than 5 μm. At 10.8 ks, the surface of the flaky bulks is smoother but the edges are still fissured and irregular where the thickness is estimated to about 5 μm and the size got smaller to an average of maybe 50-140 μm. At 18 ks, the morphology looks similar but the size is significantly smaller at an average of 20 μm. Also at 18 ks, the appearance remained similar and size further reduced to an average of maybe 10 μm. At 21.6 ks, the size seemed to slightly increase again and the surface appears much smoother. The edges always remain fissured.

In Table 5 we try to summarize the above where we estimate the morphology evolution, the average particle size and in case of flaky structure the thickness of the particles:

Table 5. Summary of SEM-results Related to Morphology Evolution after MM in the Simoloyer

SEM-results related to morphology evolution after MM in the Simoloyer, estimates				
MM-time		*Particle shape*	*Particle size*	*Particle thickness*
[ks]	[h]		[μm]	[μm]
0	0	Spherical-elongated	1-10	
3.6	1	Thick flakes	60	5
7.2	2	Thinner flakes	300	1-5
10.8	3	Flaky fractures	50-140	5
14.4	4	Flaky fractures	20	3-5
18.0	5	Flaky fractures	10	3-5
25.2	7	Flaky fractures	10-30	3-5

The graph in figure 9 gives the results from the Fe-mapping which we achieved by microanalysis investigating at least 10 areas per spot. Remarkable is the fact, that even at the same processing times the Fe-contamination in case of MM in the SPEX-Mixer/mill is always 1.5-4 times higher.

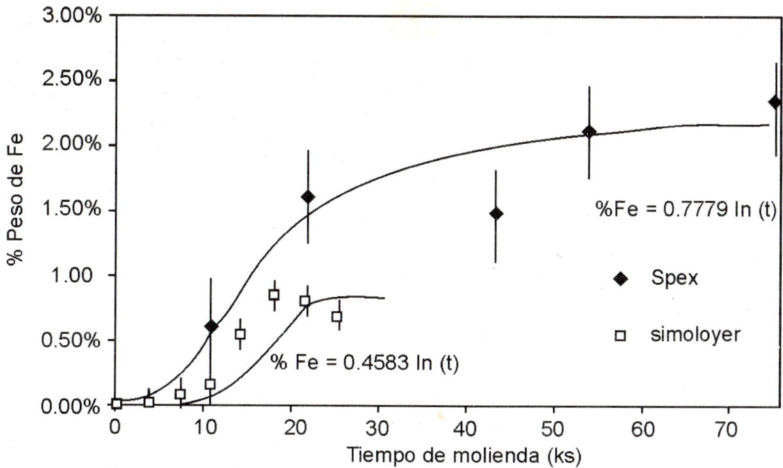

Fig. 9. Fe-contamination by microanalysis.

If we take a fix-point at that time when the SSF was observed, *in case of the* SPEX-Mixer/mill, the Fe-contamination is close to 2% and in case of the Simoloyer 1.2%. It seems. that in case of the Simoloyer, there is a time frame. where flaky structure is given already but Fe-contamination cannot be detected yet.

3.2. XRD-INVESTIGATION

Figures 10a and 10b show the XPD-patterns of the original and the processed powders achieved (*a*) from the SPEX-Mixer/mill and (*b*) from the Simoloyer.

We determine the solid solution formation (SSF) by the detection of the (111), (200) and (220) planes of the SS. In case of the MM in the SPEX-Mixer/mill, this could be expected after 21.6 ks and including the later repeated TEM-investigation, we can determine it to be obtained at nanometric grain size after 43.2 ks.

In case of the MM in the Simoloyer, this could be expected after 10.8/14.2 ks according to the XRD and including the TEM after 18 ks which is about 2.4 times faster. From angle measurement we calculated the lattice parameter in both cases to 3.769 Å.

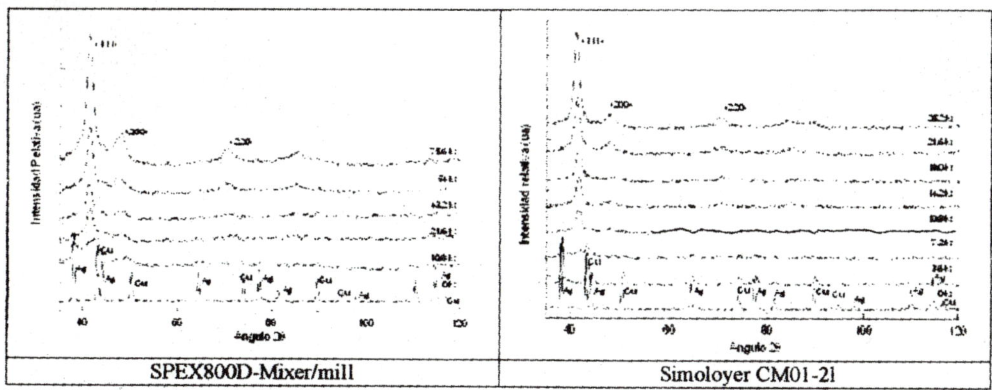

Fig. 10. XRD-patterns of original and processed powder achieved (*a*) from the SPEX-Mixer/mill and (*b*) from the Simoloyer

The grain size determined by the Scherrer-method is given at < 10 nm for both cases, the grain size determined by TEM at 5-50 nm for the SPEX-Mixer/mill-process and at 5-30 nm for the Simoloyerr-process. The results are listed in Table 6:

Table 6. Results of XRD (and TEM) in survey

Results of XRD (and TEM) in survey			
Processing device	***SSF after***	***Grain size by TEM***	***Grain size by Scherrer***
SPEX-Mixer/mill	43.2 ks	5-50 nm	< 10 nm
Simoloyer CM01	18 ks	5-30 nm	< 10 nm

3.3. TEM-Investigation

Since there are natural doubts on the reliability of XRD for the completion of SSF, which is the goal in this work, because we might expect some local re-crystallisation by high energy impact (local heat) and XRD would not detect a weight fraction below 5-10 percent depending on the element, it was necessary to check SSF by TEM.

The TEM-investigation of the MM powders in the SPEX-Mixer/mill after 43.2 ks is given in figure 11 a-c. The dark- and bright field micrographs (a, b) show a grain size at around 5-50 nm where we obtained a value estimate of < 10 nm by the Scherrer-method.

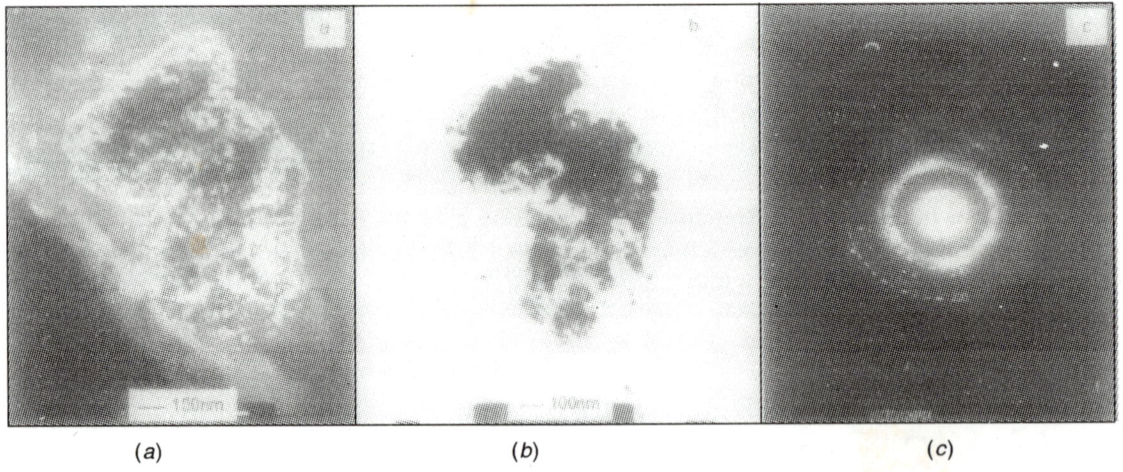

(*a*) (*b*) (*c*)

Fig. 11. TEM-investigation of the MM powders in the SPEX-mill: (*a*) darkfield micrograph, (*h*) brightfield micrograph, (*c*) diffraction pattern after 43.2 ks.

The diffraction pattern (*c*) shows the diffraction rings which are Characteristic for the nanometric grains of the new solid solution formed. However, it seems, that the SSF has not yet been developed completely because according to our TEM-observation, there were still diffraction spots from Cu and Ag identified which can also be seen on figure 11 *c* as spots out of the radii of the 111, 200 and 220 planes respectively.

The TEM-investigation of the MM powders in the Simoloyer after 18.o ks is given in figure 12 *a-c*. The dark- and brightfield micrographs (*a, b*) show a grain size at around 5-30 nm where we obtained a value of 5 nm by the Scherrer-method.

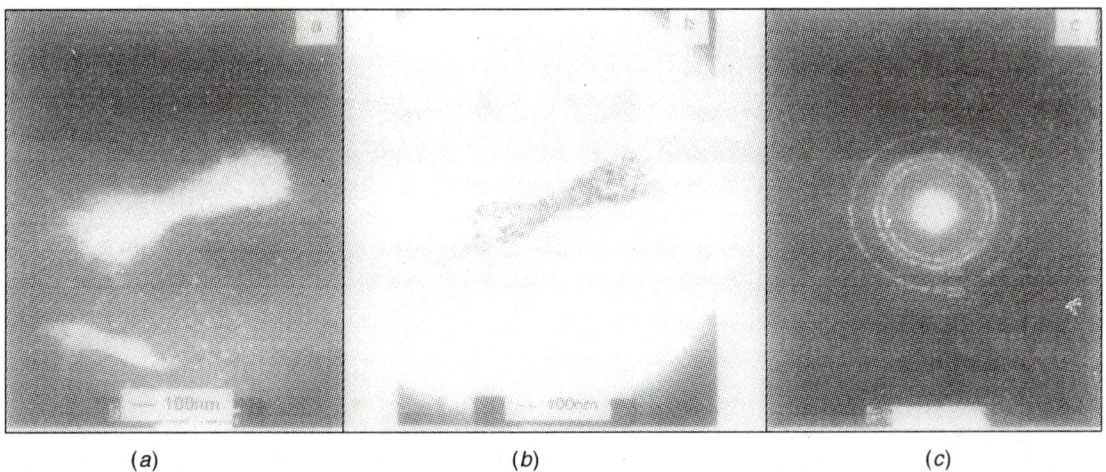

(*a*) (*b*) (*c*)

Fig. 12. TEM-investigation of the MM powders in the Simoloyer: (*a*) darkfield micrograph, (*b*) brightfield micrograph, (*c*) diffraction pattern after 18. 0 ks.

The diffraction pattern (c) shows diffraction rings which are characteristic for the nanometric grains of the new solid solution formed.

Figure 13 a-c shows the TEM-investigation of the MM powders in the Simoloyer after 14.4 ks. The dark and brightfield micrographs (a, b) confirm a grain size at around 100-500 nm.

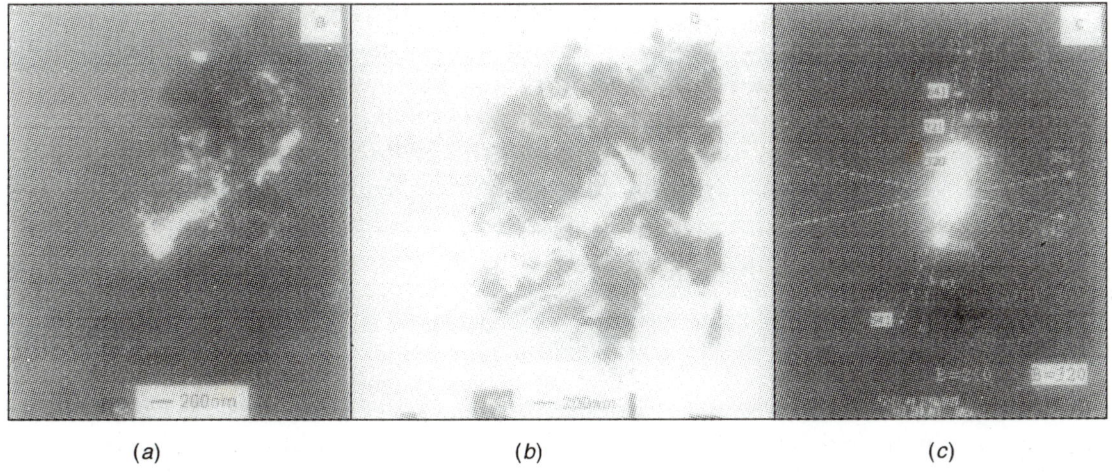

(*a*) (*b*) (*c*)

Fig. 13. TEM-investigation of the MM powders in the Simoloyer: (*a*) darkfield micrograph, (*b*) brightfield micrograph, (*c*) diffraction pattern after 14.4 ks.

The diffraction pattern (*c*) exhibits a dot-pattern that confirms microcrystalline grain size and was identified to belong to the SS of Cu and Ag. In result, at 14.4 ks, the SSF has already been obtained but a nanometric grain size was not achieved yet.

Figure 14 *a-c* shows the TEM-investigation of the MM powders in the Simoloyer after 25.2 ks. The dark- and brightfield micorgraphs (*a*, *b*) confirm a grain size at around 100-500 nm.

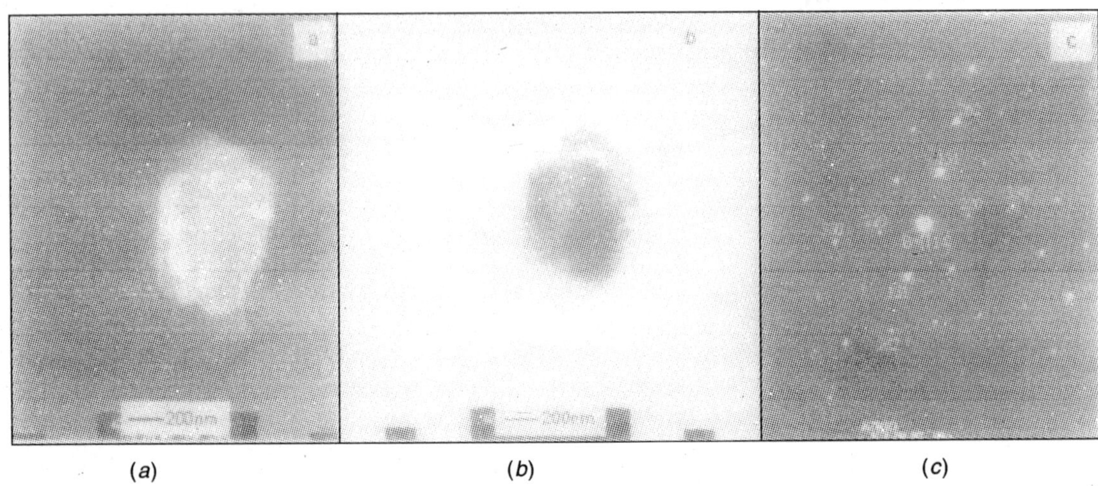

(*a*) (*b*) (*c*)

Fig. 14. TEM-investigation of the MM powders in the Simoloyer: (*a*) darkfield micrograph, (*b*) brightfield micrograph, (*c*) diffraction pattern after 25.2 ks.

The diffraction pattern (*c*) exhibits a spot-pattern that confirms grain growth of the SS which may be explained by to much energy impact into the material which is transformed into (micro) heat.

4. SSF, MORPHOLOGY EVOLUTION, Fe-CONTAMINATION AND MM-TIME

The comparison of the received results by microanalysis, XRD and TEM is given in Table 7:

Table 7. Results of SEM-micrographs and microanalysis XRD and TEM in survey

SEM-, XRD- and TEM-results (including microanalysis) in survey				
MM-device	***SSF starts after***	***SSF completed after (b)***	***Fe-contamination after time b***	***Comment***
SPEX-Mixer/mill	21.6 ks	43.2 ks	1.9 wt%	3.8 x more Fe-contamination
Simoloyer	14.4 ks	18 ks	0.5 wt%	SSF achieved 2.4 × faster

In result, the SSF is achieved in the Simoloyer 2.4 times faster at 3.8 times less Fe-contamination. The comparison of the estimated morphology evolution based on the SEM-results (tables 4 and 5) is given in Fig. 15 where we additionally marked the SSF:

In Fig. 15 we can see impressively a major difference in the evolution. The tremendous increase of particle size in case of the Simoloyer identifies the formation of large flakes that later break- into fractures. This does not happen in case of the SPEX-Mixer/mill where we assume that the kinetic here is to low and matches in so far to the hint that the SPEX-system can be better understood as a vibration mill rather than a system based on collision of grinding media (see chapter 2.2).

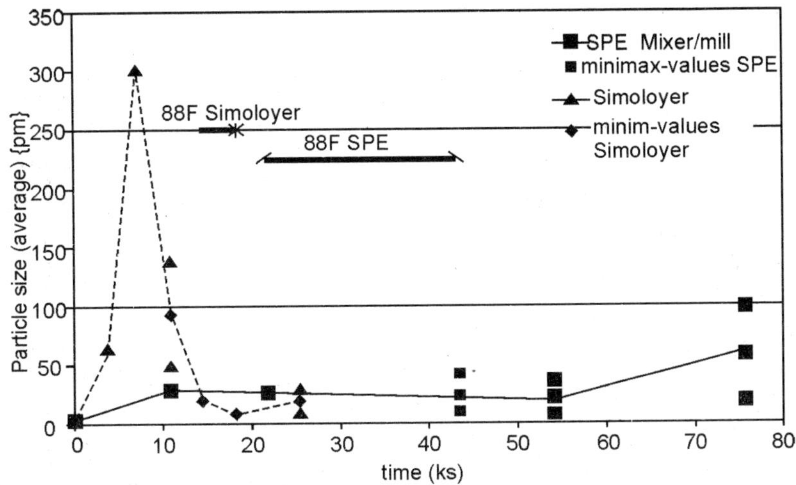

Fig. 15. Morphology evolution based on the SEM-results in survey.

In order to further comment the morphology evolution in case of MM of Ag-70Cu as a mixture of ductile metals we refer for the initial morphology change to a parallel work [20]. Then we can predict the development as follows (table 8):

Table 8. Estimated morphology evolution in Spex-Mixer/mill and Simoloyer

	Spex-Mixer/mill (low kinetic)		
		Simoloyer at medium kinetic (3.6 m/s)	
01	✓	✓	Agglomeration of the starting powders to lose bulks
02	✓	✓	Densification of still small loose bulks
03	✓	✓	Flattening of the bulks to flaky shape, development of some large bulks
04	✓	✓	Agglomeration of remains of stage 01 on the surface of flakes/large bulks
05		✓	Further flattening and break of the flaky bulks/flakes
06		✓	Repeat of stage 05 including cold-welding of flakes to bulks

The evolution stages 01-06 in table 8 give an attempt for the understanding of the formation of flakes when we process ductile metal powder by MM. Stages 05-06 seem not to apply for the process in the Spex-Mixer/mill since the kinetic is to low to create thin enough flakes with a work hardening effect.

We further assume that the initial flake formation, in particular if the flakes are built up by a multi-layer structure (of thinner flakes), does favor the SSF because for both transformations. the kinetic energy impact is determining [20-21].

Since in case of SSF, here, Ag-atoms shall be located in the Cu-lattice intersticially or by substitution, this is provided mainly by kinetic impact which means it is of major importance, how this kinetic impact can effect the lattice structure.

For example, if some grinding media would be thrown into a volume of Ag-Cu-powder, there will be plastic deformation of the volume and there may be plastic deformation of some of the direct hit particles in the front-layer but it is very much unlikely, that this can cause atomic dislocations in the lattice of the materials. This means the dumping effect of the powder plays the same important role for SSF as well as for flake formation.

It must be pointed out again that we used the Simoloyer only at medium kinetic (3.6 m/s) and not at high kinetic (> 9 m/s).

Since the SSF seems to and the flake-formation does depend on the kinetic energy impact, in the future, the work reported here shall be repeated with the Simoloyer not under medium kinetic at around 3.5 m/s but at high kinetic > 10 m/s. It has already been proved in earlier work 19-10, 221, that the formation of Ag- and Cu-flakes by HKP can he obtained already after some minutes processing time and at a flake-thickness less than 1 μm (Figs. 16 and 17 show SEM- and optical micrographs of Cu- and Ag-flakes that are produced at Zoz by HKP).

If we then take into account, that these flakes can be built up by a number of approximately 50 theoretical inter-layers according to the starting particle size [23], then we would end up with a theoretical thickness in 10 mm range.

If we apply this to a powder mixture and use the principle of MA and produce a kind of sandwich-structure, then we probably end up with very short diffusion ways inside thin structure and this could explain why multi-layer flake formation favors SSF and it will be very interesting to find dependencies for this in the ongoing work where one of the next steps will be the application of HKP at 3-4 times higher relative velocities of the grinding media.

(*a*) (*b*)

Fig. 16. Zoz-Cu-flake Z-CUF17, (*a,b*) SEM- and (*b*) optical micrographs.

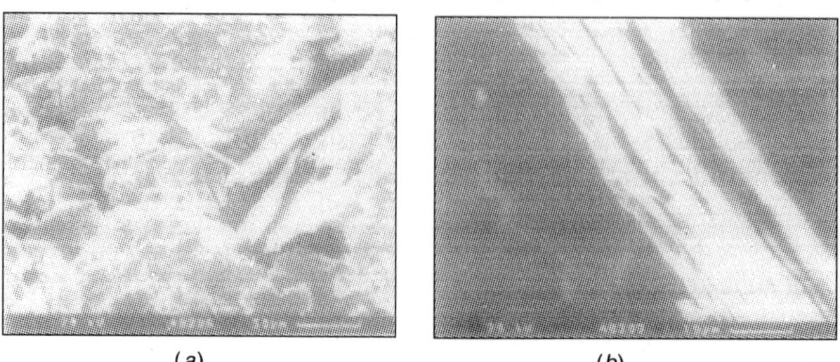

(*a*) (*b*)

Fig. 17. Zoz-Ag-flake Z-AGF34, (*a,b*) SEM- and (*b*) optical micrographs.

The given Cu- and Ag-flakes are manufactured under high kinetic in the Simoloyer at processing times below 30 minutes. The thickness is in average 3-5 times thinner that the thickness of the here processed Ag-70 Cu-flakes at medium kinetic and at 5-40 times longer processing time also in the Simoloyer. Taking this into account, we can presume that the SSF at HKP will be obtained at much shorter processing time and will exhibit a much more flaky morphology.

CONCLUSIONS

The solid solution formation by MM of Ag-70Cu in the SPEX8000D-Mixer/mill after 43.2 ks and in the Simoloyer CMOI -21 after 18 ks could be obtained at nanometric grain size.

For this, the Simoloyer was not used at high but at medium kinetic only (< 4 m/s).

After completion of the SSF, the Fe-contamination by the milling tools in the SPEX8000-Mixer/ mill was about 1.9 wt % and in the Simoloyer about 0.5 %.

It seems that the solid solution formation is related to fake like morphology, which is achieved at shorter times in the Simoloyer.

It seems, that in case of the Simoloyer, there is a time frame, where flaky structure is given but Fe-contamination cannot be detected yet.

In result, the Simoloyer-process leads to the SSF faster at lower contamination than the SPEX-Mixer/mill which should be understood as a vibrating mill rather than a high energy mill.

Most interesting will be the future work, when the Simoloyer will be used at high kinetic <9 m/s.

REFERENCES

1. K. Uenishi, K.F. Kobayashi, K.N. Ishihara, P.H. Shingu: Formation of a supersatured solid solution in Ag-Cu systems by mechanical alloying, Mat. Sci. & Eng. A 134 (1991) pp. 1342-1345.
2. R. Esquivel-Gonzalez, Ph.D. Tesis, Compactacion dinamica de materiales nanocristalinos obtenidos por aleacion mecanica, ESIQIE-IPN, Mexico (2001).
3. R. Esquivel G., I. Vernet P., C. Renero Y., D. Jaramillo V. Dynamic compaction of nanostructured alloys obtained by mechanical alloying, Journal of Materials of Materials Processing technology, Thermec 2000 – Processing & Manufacturing of Advanced Materials, TMS, Elsevier Science. (2001).
4. K. Ameyama, S. Umekawa, H. Ren D. Jaramillo V., *H. Zoz.* Alternative Mechanical Milling routes for grain-refinement of conventional High-Speed Steel powder for later consolidation by SPS, proceedings of PM2TEC 2003, MPIF (2003).
5. H. Zoz: HEM/MA/RM - Devices in use, 3rd Intern. Symp. of The School of Chemical Engineering, IPN, Mexico City, May 27-29, 1998.
6. A. Bose, K. Ameyama, S. Diaz de la Torre, D. Jaramillo V., D. Madang, H. Zoz, Zoz GmbH (Germany & USA), Materials Processing Inc. (USA). Ritsumeikan University, (Japan), CIMAV S.C., ESQIE, (Mexico), F.W. Winter Inc. & Comp. (USA): Review of Applications and Materials processed by Rotating Horizontal High Energy Milling: 2002 World Congress on Powder Metallurgy & Particulate Materials, proceedings (2002).

7. H. Zoz, H.U. Benz, G. Schafer, M. Dannehl, J. Krull, F. Kaup, H. Ren, and D. Jaramillo V.: High Kinetic Processing of Enamel, p. 1-a, INTERCERAM International Ceramic Review, Vol. 50 (2001) [5] pp. 388-395.
8. H. Zoz, H.U. Benz, G. Schafer. M. Dannehl, J. Krull, F. Kaup, H. Ren and D. Jaramillo V.: High Kinetic Processing of Enamel. p. I-b, INTERCERAM International Ceramic Review. Vol. 50 (2001) [6] pp. 470.477.
9. H. Zoz, D. Ernst, T. Mizutani, H. Okouchi, Simoloyer CM100s, semi-continuously Mechanical Alloying in a production scale using Cycle Operation-Part I, Metall. Vol. 51, 10197. pp. 568-572. 1997.
10. H. Zoz, D. Ernst: Simoloyer CM100s, semi-continuously Mechanical Alloying in a production scale using Cycle Operation-Put 11, Metall Vol. 51, 09/98, pp. 521-527, 1998
11. H. Zoz D. Ernst, H. Weiss. M. Magini, C. Powell, C. Suryanarayana, F.H. Froes, Mechanical Alloying of Ti-24AI-11Nb (at%) using the Simoloyer Metall Vol. 50, 09/96, pp. 575-579, 1996.
12. J.Y. Chung, J. Kim and Y.D. Kim: Formation of Nanocrystalline Fe-Co Powders Produced by Mechanical Alloying, Dept. of Metallurgy and Materials Science, Hanyang University, Ansan, Korea, 1999.
13. H. Zoz D. *Ernst,*I S, Ahn. W. H. Kwon:. Mechanical Alloying of Ti-Ni-based Materials using the Simoloyer, TMS Annual Meeting 1997, eds. C.M. Ward-Close, F.H. Froes, S.S. Cho, D.J. Chellman: Synthesis/Processing of lightweight Metallic Materials, proceedings (1997).
14. H. Zoz, H. Ren, N. Spath: Improved Ag-SnO_2 Electrical Contact Material Produced by Mechanical Alloying, Metall Vol. 53. 07-08199. pp. 423-428. 1999
15. R.M. Davis, B. McDermott, C.C. Koch, Mechanical Alloying of Brittle Materials, Metall. Trans. Vol. 19a, 2867 (1988).
16. H. Zoz, H.U.Benz, K. Huttebraucker,, L. Furken, H. Ren: Stellite bearings for liquid ZN-/AI- Systems with advanced chemical and physical properties by Mechanical Alloying and Standard-PM-Route, Part 1, Metall Vol. 54, 11/2000, pp. 650-659, 2000.
17. B. Wielage. J. Wilden, T. Schnick. A.Wank, J. Beczkowiak, R. Schulein, H. Ren, H. Zoz: Mechanical alloyed SiC composite Powders for HYOF applications, ITSC 2002, Conference Proceedings, p, 1047-1052, ASM/DVS March 2002, Essen, Germany.
18. G. Kaupp, 3. Schmeyers, M. R. Naimi-Jamal- H, Ren, H. Zoz : Reactive milling with the Simoloyer: environmentally benign quantitative reactions without solvents and wastes. Elsevier, Chemical Engineering Science 57. 2000 763-765.
19. G. Kaupp, M. R. Naimi-Jamal, H. Ren and H. Zoz : Environmentally Protecting Reactive Milling, CHEMIE TECHNIK- 31. Vol. 6 (2002) p 58-60.
20. H. Zoz, 1. Vernet and D. Jaramillo V. .Solid-solution-formation by MM of the Ag-70ae/OCti alloy, proceedings of PM2TEC 2003, MPIF (2003).
21. BOTCHAROVA, E.; HEILMAIER, M.; FREUDENBERGER, J.; DREW, G.; KUDASHOW, D.; MARTIN. U.; SCHULTZ, L.; Supersaturated solid solution of niobium in copper by mechanical alloying, Journal Alloys and Compounds, 351 (2003), pp. 119-125.
22. H. Zoz, H. Ren, H. U. Benz: Ductile Metal Flakes based on [Au], [Ag], [Cu], [Ti], [AI], [Ni] and [Fe] by High Energy Milling-part 1, PM2TEC 1999, MPIF (1999) proceedings.
23. H.U. Benz, H. Zoz: particle deformation (V = const)-investigation and production of CMB- metal-flakes Au, Ag, Cu, Ti, Ni. Al, Internal Report Zoz GmbH, 1998.

3

Silver-Graphite Nanocomposite Materials for Electrical Contact Applications

Bharati R. Rehani and *P.B. Joshi*
Dept. of Metallurgical Engineering, M. S. University of Baroda, Baroda-390 001, India

ABSTRACT

Silver-base composites are developed as electrical contact materials for switchgear applications in devices such as contactors, relays, circuit breakers, switches, etc. owing to their low are erosion improved antiwelding properties and lower contact resistance compared to contacts made from pure metals and alloys. Silver-graphite composites with graphite percentage in the range of 0.25 wt. percent to about 50 wt. percent find applications as *make and break* type switching contacts and slip ring/sliding contacts in switchgear devices. Ag-C contacts are currently manufactured by conventional powder metallurgy route involving blending or milling of elemental silver and graphite powder particles followed by their consolidation, sintering and extrusion to bulk solids. However, large difference in the bulk density of these two constituent, materials leads to segregation during milling and eventually heterogeneous microstructure responsible for inferior electrical contact properties.

The present work deals with synthesis of silver-graphite nanocomposite powders (with 1 to 5 wt. percent graphite) by mechanical alloying in a high-energy attritor followed by their pressing-sintering-repressing into bulk solids. The compacts so produced are found to exhibit density and electrical conductivity values comparable to those of conventionally produced materials, whereas they offer improved microstructure in terms of dispersion of graphite in silver matrix. 'The silver-graphite contacts so produced from mechanically alloyed Ag-C nanocomposite powders are expected to offer superior electrical contact properties viz. greater resistance to arc erosion and lower tendency to contact welding as well as low contact resistance.

INTRODUCTION

The silver-graphite composites are widely used as switchgear contacts for commercial as well as industrial applications. The low- graphite grades (containing less than 10-15 wt.% graphite) are used as *make and break type* switching contacts whereas high-graphite grades (containing more than 15 wt.% graphite) find applications as sliding contacts in electrical brushes.

The silver-graphite contacts have low contact resistance, high conductivities, self- lubricating properties and anti-welding properties provided contact pressures are low [1]. Silver-graphite contacts are used for applications requiring rubbing speeds less than 0.03 meters per seconds and light pressures [2]. Low-graphite electrical contacts find applications as rheostats, contact brushes, contactors, molded case circuit breakers and as slip ring contacts, etc. [3]

The silver-graphite contacts are manufactured by using conventional powder metallurgy route of blending/milling of elemental silver and graphite powder particles, followed by their pressing, sintering and extrusion as ingot. The extruded ingot is further processed to form wrought silver-graphite strips or wires, which can subsequently be used to produce contact pieces. The Ag-C contacts are also produced by pressing, sintering and repressing of silver- graphite composite powders [3]. However, large difference in bulk density/specific gravity of these two constituent materials give rise to the problem of segregation during blending/milling operation and eventually heterogeneous microstructure responsible for inferior physical, metallurgical, and electrical contact properties.

The present investigation has been concerned with a study involving processing and property evaluation of Ag-1 wt.%C, Ag-3wt.%C and Ag-5wt.%C contact materials produced by mechanical alloying (MA) route.

Mechanical alloying was developed by J.S.Benjamin at the laboratories of INCO Alloys International in the late 1960's for production of oxide-dispersion strengthened super alloys [4]. Later on MA process has been modified and adopted to produce a wide range of materials such as amorphous/nanocrystalline materials, intermetallic compounds, metal-matrix composites, solid-solution alloys, ete [5 – 8] During MA of ductile-ductile system like Ag-C, the powder particles develop a composite layered structure in the early stages of milling. On further milling, the structure is gradually refined and homogenized through the mechanical kneading action that the powder charge is subjected to and also by the diffusion across composite layers. Finally, the composition of individual powder particles reaches the composition of the starting powder blend along with the high degree of dispersion between the constituent phases [5].

EXPERIMENTAL PROCEDURE

Elemental silver powder of 99.9 % purity and 10 μm average particle size and graphite powder of 99.5 % purity were used as the starting materials. The stoicheometric amounts of silver-graphite (Ag-I%C, Ag-3%C, Ag-5%C) were mixed in cylindrical blender at 130 rpm for 30 min. The blended powders were then subjected to attrition milling in a Szegwari type attritor, having stainless steel vial of 200 cc capacity. AISI 52100 grade hardened steel balls of 6.3 mm dia. were used as grinding media. The milling was carried out at a ball to charge ratio of 20:1 and at 400 rpm speed of attrition. The powder samples were drawn at specific time intervals during the course of milling and were characterized by X-ray diffraction (XRD) and Scanning Electron Microscopy (SEM) to monitor the degree of mixing between silver and graphite and also to study the changes in the powder morphology with progressive milling. Figure 1 shows the variation in peak intensity (a.u.) for silver and graphite phase with respect to milling time (hrs) for Ag-3%C composite powders. A Philips Xpertpro X-ray diffractometer with solid state zirconium detector was used for X-ray diffraction analysis. The diffraction profiles were obtained using a Cu target at a scan speed of 3° per min. within the 2θ range of 0 to 70°. The shape morphology of powder particles was examined using Jeol JSM 5610 LV model of Scanning Electron Microscope. The SEM microphotographs for Ag-3%C as- blended and 4 hrs milled powder samples are given in Figs. 2 (a) and (b).

The synthesized powders of three different compositions (viz. Ag-I%C, Ag-3%C, Ag-5%C) were pressed to get the green compacts of 10 mm dia.x 2mm height at 300 MPa pressure on a 100-ton capacity hydraulic press in single action die compaction mode. The green compacts were sintered at 900°C for 60 min. in vacuum in a resistance-heating furnace at a heating rate of 6°C per min. The sintered compacts were repressed at 700 MPa.

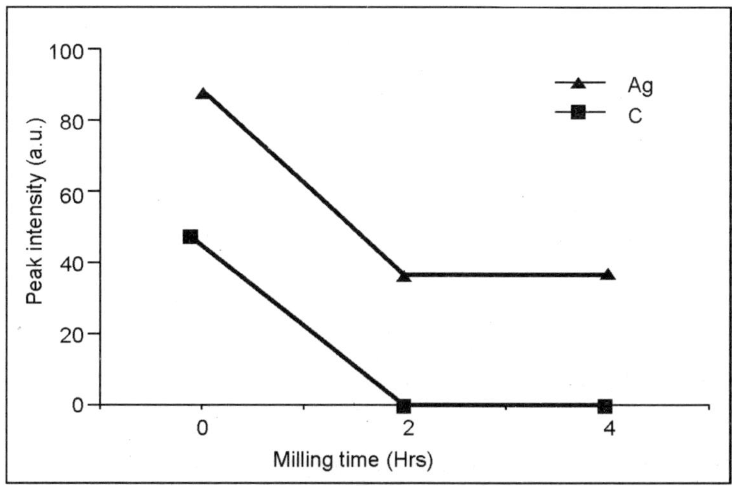

Fig. 1. Variation of peak intensity with respect to milling time for Ag–3% C composite powders produced by MA.

The bulk solid contact pieces were subjected to density, electrical conductivity and microstructural evaluation. For microstructural examination a ground and polished sample in unetched condition was used. The micro section was subjected to examination using SEM in BSE mode without any surface modification. Fig. 2 (*c*) shows the SEM micrograph for Ag-3%C bulk solid sample.

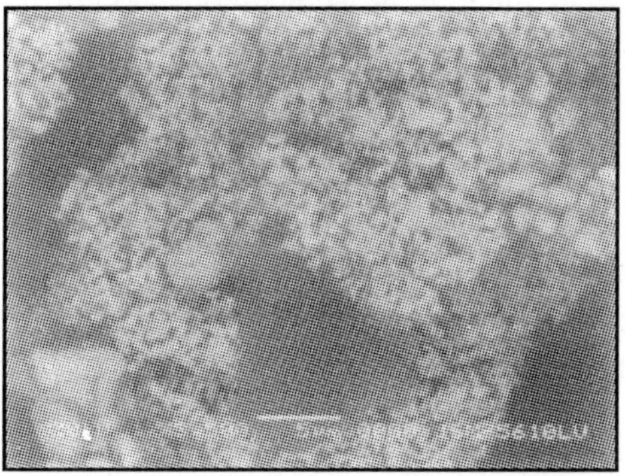

Fig. 2(a). Ag–3%C as-blended powder showing heterogeneously dispersed grey coloured graphite particles in silver matrix.

Fig. 2(b). Ag–3%C attrition milled powder after 4 hrs of milling showing flattened silver particles and intimately oriented grey graphite in silver.

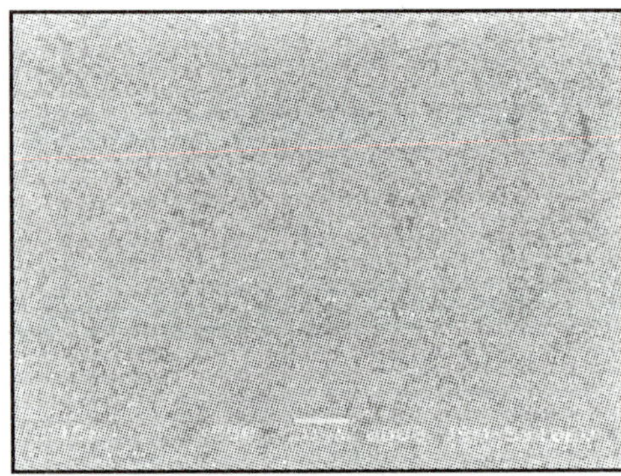

Fig. 2(c). Photomicrograph of Ag-3% C sintered and repressed bulk solid contact material produced from attrition milled powder.

RESULT AND DISCUSSION

The plot for variation of peak intensity v/s milling time for attrition milled powder (Fig. 1) clearly indicates that the peak intensity for silver as well as graphite phase gradually decreases with milling time. The drop in peak intensity is attributed to broadening of X-ray diffraction peaks due to reduction in crystallite size with progressive milling. It is apparent from this plot that the diffraction peaks corresponding to graphite broadened to such an extent that the peak height practically ceases to zero and disappears completely. This is indicative of a very fine level of dispersion of graphite within the silver particles.

The SEM micrographs as shown in Figs. 2(a) and 2(b) for as blended and milled powders, respectively clearly highlight the degree of dispersion of graphite in silver. The microphotographs also indicate the change in the shape morphology of Ag-C composite powder due to milling. The 4 hrs milled powder sample shows extensive flattening of silver particles along with fine and uniform dispersion of graphite within silver.

Table 1 gives a comparative data on properties of silver-graphite composite contacts developed under the investigation and those reported in the literature. The literature values are for silver-graphite contact materials produced by press-sinter-extrusion route using ball-milled powders. A critical comparison of the properties for the compacts produced under this investigation and the values reported in literature clearly indicates that it is possible to attain comparable levels of density and electrical conductivity in case of compacts produced from mechanically alloyed powders and subsequently processed by press-sinter-repress route.

The point to be emphasized here is that the Ag-C contact materials produced from MA powders and processed by press-sinter-repress route give rise to a set of properties which are no way inferior to the properties obtained for Ag-C contact materials produced by press-sinter-extrude route. The process developed under this investigation thus claims to have an edge over the existing manufacturing practice for production of Ag-C contact materials.

Table 1. Comparative Data on Properties of Ag-C Composite Contacts.

Sr. No.	*Composition*	*Property*			
		Density g/cm³		*Electrical conductivity (%IACS)*	
		Measured value PSR	*Literature value PSE*	*Measured value PSR*	*Literature value PSE*
1	Ag-1%C	10.0	10.1	95	95
2	Ag-3%C	9.3	9.5	84	86
3	Ag-5%C	9.7	8.9	80	75

REFERENCES

1. H.H. Hausner, Handbook of Powder Metallurgy, Chemical Publishing Co. Inc., New York, (1973), p. 68.
2. R.H.T. Dixon and A. Clayton, Powder Metallurgy for Engineers, The Machinery Publishing Co. Ltd., London (1971).
3. P.G. Slade, Ed., Electrical Contacts- Principles and Applications, Marcel Dekkar Inc., New York (2002).
4. J.S. Benjamin, Dispersion Strengthened Superalloys by Mechanical Alloying, Met. Trans., Vol. 1, October (1970), p. 2943.
5. C.C. Koch, The Synthesis and Structure of Nanocrystalline Materials Produced by Mechanical Attrition: A Review, Nanostructured Materilas, Vol. 2, (1993), p. 109.
6. B.S. Murty and S. Ranganathan, Novel Materials-Synthesis by Mechanical Alloying/Milling, International Materials Reviews, Vol. 43, No. 3 (1998) p. 101.
7. Li Lu and M.O. Lai, Formation of New Solid Materials in Solid State by Mechanical Alloying, Materials and Design, Vol. 16, No. 1, (1995), p. 33.
8. G.B. Schaffer and P.G. McCormick, Review on Mechanical Alloying, Mat.Forum,16, (1992), p. 91.

4

Solid-Solution-Formation by MM of the Ag-70 at% Cu Alloy

H. Zoz,
Zoz GmbH, D-57482 Wenden, Germany
ESIQIE, National Polytechnic Institute, Mexico City, DF 07300, Mexico

S. Morales
ESIQIE, National Polytechnic Institute, Mexico City, DF 07300, Mexico

D. Jaramillo V.
ESIQIE, National Polytechnic Institute, Mexico City, DF 07300, Mexico

ABSTRACT

The solid-solution-formation of the well known Ag-70Cu (at%) has been chosen to study efficiency, Fe-contamination and particle-morphology evolution in a Simoloyer CMOI-21 high energy mill. Grinding media/powder ratio was varied at constant media load of 1500g. Samples for solid-solution and for morphology-evolution were taken out at different times. The media/powder ratio was compensated by discharging the corresponding load of balls for the extracted sample.

Solid-solution-formation was obtained after 79.2, 43.2 and 18 ks for the 10:1, 20:1 and 50:1 ratio respectively. Fe-contamination was always lower than 1.0%.

Solid-solution-formation mechanism for this ductile material under MM seems to he related to the flake-formation-mechanism occurring during the collision of grinding media. Additionally, Cycle Operation has been applied which resulted in rapid flaking with quasi-disappearing ductility followed by shear-stress to break the flaky structure repeatedly.

Since a higher collision rate versus *shear* and friction effect is expected to cause less abrasive wear on the milling tools, the Fe-.contamination was expected and found in low range.

Material's characterization is given by SEM, XRD and TEM.

1. INTRODUCTION

The aim of the present paper is a better understanding of the mechanism for the solid solution formation (SSF) by MM (Mechanical Milling) where we chose the example of Ag-70Cu (at%). We want to study efficiency in SSF, Fe-contamination and particle-morphology evolution in a Simoloyer CM01-21 high energy mill. In this work we did not use HKP (High Kinetic Processing) which would refer to a relative velocity of grinding media in the order of 9 m/s and higher but medium kinetic processing in the order of 4 m/s since we want to study the flake formation which at HKP in case of ductile metals is performed to fast.

The Ag-Cu-system is attractive to investigate SSF since it was one of the first systems studied by Mechanical Alloying (MA) and is perfectly defined by low energy milling and has perfectly been characterized [1-3]. It is a system very noble and easy to work with when a matter of comparison wants to be made. SSF is favoured because Cu and Ag belong to the same group in the periodic table. On the other hand, Ag and Cu do not form a SS in conventional melting route because of their significantly different atomic radii. Finally we can expect high deformation rates at relatively low sheer stress due to the fcc-structure of both of the materials which is expected to favour the SSF by quasi-thin-layer formation on micro-structural base.

What makes the Simoloyer interesting here, is the high (or medium) kinetic process based not on shear and friction (low kinetic processing) but mainly on the collision of grinding media at relative velocities up to 14 m/s [4-5] which first promises a high level of energy impact and second a lower level of impurities res. of contamination by the milling tools since collision is naturally less abrasive than shear and friction effects. Since it is well known that the mode of impact by the Simoloyer can lead to very thin flakes when processing ductile metals [6-8], this is exactly what we expect to need for a fast and complete SSF. This is the expected future application and this is why we did chose this system but at lowered medium kinetic for the present work.

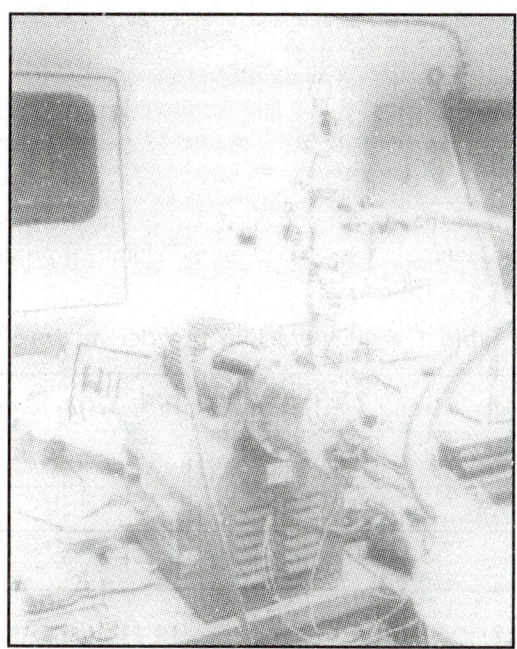

Fig. 1. Horizontal rotary ballmill (Simoloyer CMO1-21) with air-lock for loading, operation and unloading under vacuum and/or inert-gas.

Figure 1 shows a Simoloyer CM01-21 which is a horizontal rotary ballmill with a 2 liter chamber volume that can be placed on a table next to the process controlling computer which is operated with water cooling or heating at rotation frequencies up to 1800 rpm. Figure 2 shows a real image of a Simoloyer CM01 with a transparent chamber in operation where the effect of collision in MM is visualized and the working principle can be imagined where the rotor is the tool to transfer the kinetic energy into the grinding media and the grinding media transfers into the powder material.

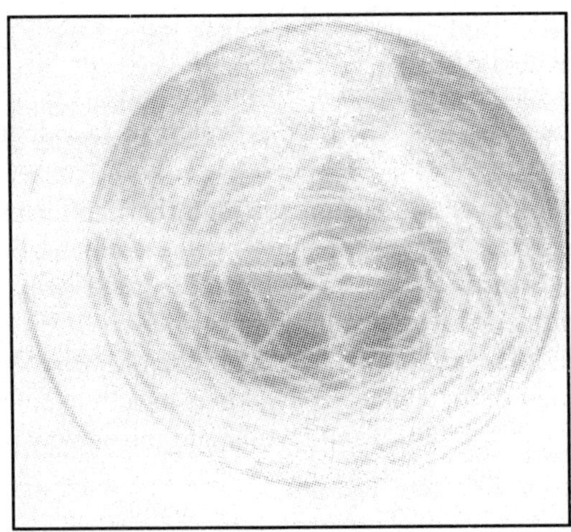

Fig. 2. Collision effects in HKP (Simoloyer)

2. EXPERIMENTAL

2.1. Ag- and Cu-powder

The detailed composition of the Ag-Cu powder-mixture that we used for the processing is given in Table 1. Both of the starting materials were based on the chemical production route and therefore of high purity.

Table 1. Ag-Cu-starting powder mixture

Starting powder mixture for MM to achieve flake formation and SSF				
Element	*Purity [%]*	*PS* [μm]	*Comp [at%]*	*Comp. [wt%]*
Ag	99.9	1-10	bal.	bal.
Cu	99.5	1-5	70	57.88

Since both materials are classified as CMB-materials [9-11] due to their ductility and are known to cause problems in terms of sticking and agglomeration tendency in a mechanical milling (MM) process, we decided to use Cycle Operation [9-14] and since of that we did not use any PCA as a lubricant in the process.

MM was performed under argon-atmosphere which means a standard air-lock was used to transfer the starting powders from an argon-container into the evacuated grinding chamber which was after that flooded with argon. Detailed parameters and procedure are given in chapter 2.2.

2.2. MM-process

The laboratory-scale Simoloyer CM01-21 has been used for the MM but only at medium kinetic and this was carried out for 0.9, 1.8, 3.6, 14.4, 18.0, 36, 54, 72, 75.6, 79.2, 82.8 and 86.4 ks which refers to intervals from 15 min to 24 h. In order to investigate expected early flake formation we took samples after 3, 5, 7, 10, 15 and 30 minutes within the first 3 reported operations (0.9, 1.8, 3.6 ks). In order not to change the powder/ball weight ratio, we extracted next to a small powder sample always the corresponding mass of grinding media in these 3 tests. All together, 18 samples have been taken in the early stage. Here we also varied the powder/ball weight ratio and used 1:10, 1:20 and 1:50.

The detailed process-parameters are given in Table 2 and Tables 3 and 4 give the processing-schedules for the investigation of flake-formation and SSF respectively.

Table 2. HKP parameters of Ag-Cu-powder in the Simoloyer

Process parameters of Ag-Cu-powder mixture in the Simoloyer	
Milling device	Simoloyer CM01, 2.7 kW
Operating software	Maltoz 3.1
Grinding unit	W01-21 (2 liter, water-cooled)
Grinding media	Chromium steel, 100Cr6, 5 mm, 1.4-1.5 kg varied
PCA (lubricant)	No> Cycle Operation
Starting powder	Ag-70Cu (mixture by at%)
Starting powder load	Varied at 30, 75 and 140 g
Powder/ball weight ratio	1:10, 1:20 and 1:50 varied
Atmosphere	Argon, preceding evacuation
Operation Cycle	600/150 rpm-14/1 min (refers to 3.6 m/s)
Discharging Cycle	150/600 rpm-4/1 min
Average discharging time	10-20 min, sampling time 1 min
MM time I-VI	0.9–86.4 ks (15 min-24 h) varied
Milling temperature	<25 °C (vessel inside by Maltoz)
Feeding system	Standard air-lock DN40-KF
Discharging/separation	Draingrating Ask-01
Average powder yield	90-103%

For loading the starting powder for each single test, first the mill that must be in charging/operation position is evacuated to around 10^{-4} hPa and then flooded with Argon. The powder, that has been loaded inside a glove-box into a powder—container, was then transferred by the air-lock into the vessel at the same time when the vacuum was flooded since then the gas flow can even be used for the powder transportation.

After the adjusted MM-time elapsed, the grinding unit is tuned into discharging position and the same air-lock (after cleaning) was used for unloading the powder where this time first the powder container is evacuated via the air-lock. In that way, the gas-flow supported the unloading (a little) and the powder was discharged completely (grinding media remains in vessel).

Table 3. Process schedule to investigate flake formation

Sampling schedule to investigate flake-formation			
Test no.	*Pow./ball ratio (wt)*	*Sample [g]*	*Sample time [min]*
01	1:10	0.5	1, 5, 15, 30, 60
02	1:20	1	1, 5, 7, 10, 13, 15, 30
03	1:50	1	1, 3, 5, 7, 10, 15

Table 4. Process schedule to investigate solid solution formation (SSF)

Sampling schedule to investigate SSF			
Test no.	*Pow./ball ratio (wt)*	*Sample mass [g]*	*Sample time [h]*
10	1:10	1	5, 20, 24
11	1:20	1	13
12	1:50	1	5

Fig. 3. Sampling unit of air-lock DN40-KF.

The samples were taken by following the same procedure but instead of the powder container, a sampling unit was used, which is a small glass-tube with a ball valve and an adapter to the proper KF-dimension (Fig. 3).

Unloading is computerized and automatically operated. In order to increase the powder yield and to decrease the discharging time. Cycle Operation (Fig. 4) has been applied which means during processing and discharging, the rotational speed of the rotor is changed in a special frequency in the range from here 150-600 rpm which has been proven to tremendously increase the powder yield of materials that tend to stick and agglomerate under MM to the milling tools and to each other [9-10, 12].

By this method in average complete powder yield could be obtained. Since between the each of the different MM-operation, the system has never been opened and since a little amount of powder always remains in the system e.g. sticking to grinding balls and vessel, the powder yield varied between 90 and 103% after Discharging-Cycle of 10-20 min at 600/150 rpm (Fig. 5). Both of the cooling systems (vessel and pre-seal-unit) were constantly operated and the computerized measurement of the inner-vessel surface indicated temperatures lower than 25°C.

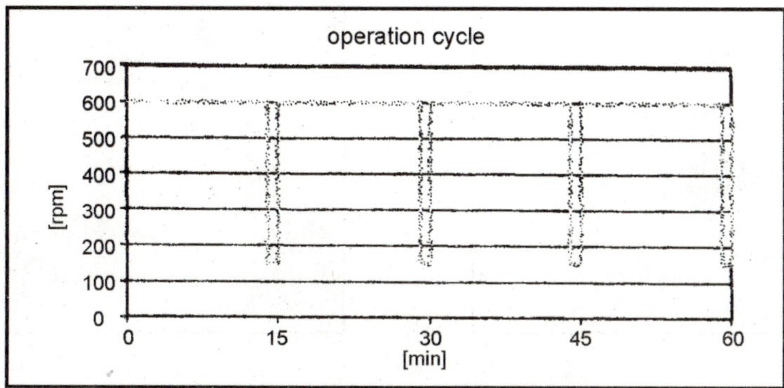

Fig. 4. Operation Cycle for MM.

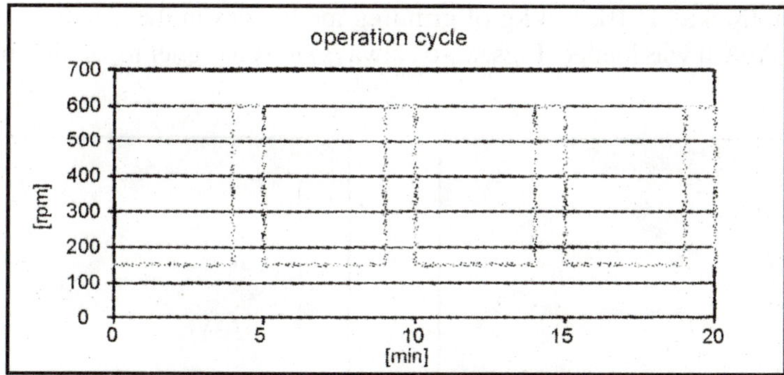

Fig. 5. Discharging Cycle after MM.

2.3 Powder Characterization

The initial and the as milled powders were characterized by SEM, XRD and TEM.

For Scanning Electron Microscopy, we applied a Jeol JSM-6300 and prepared the powder samples on a graphite tape in order to observe morphology and size. The iron-contamination of the powder samples has been investigated by Microanalysis using the same device.For X-ray diffraction we applied a Siemens Diffractometer D 5000 using monochromatic Cu-Kα radiation and determined/estimated the grain size by the Scherrer method and later confirmed by TEM. For Transmission Electron Microscopy we applied a Jeol JSM 2000 FXII and used an ultrasonic bath to de-agglomerate the powder samples and then mounted on copper grids.

3. RESULTS

3.1. SEM and Microanalysis (Flake Formation)

Figs. 6a and 6b show the SEM-micrographs of the starting powders (*a*) Cu and (*b*) Ag. Copper presents rounded shape with some porosity and an average particle size of around 5 μm, Silver shows an elongated shape with an average particle size between 1-10 μm.

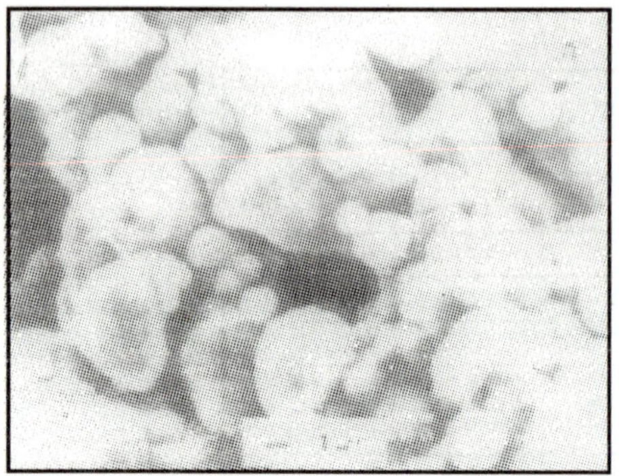

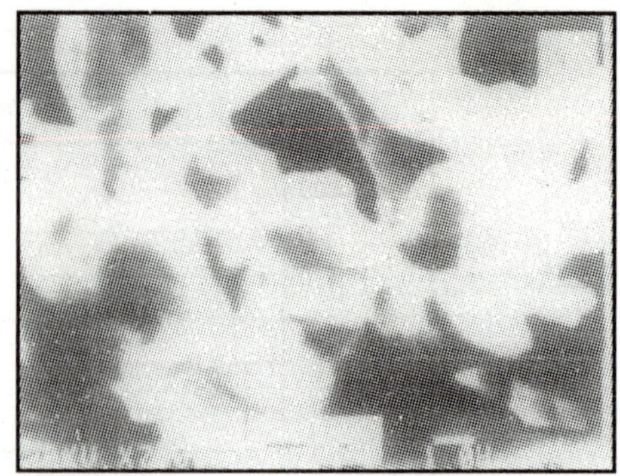

Fig. 6. SEM-micrographs of the starting powders (a) Cu and (b) Ag

Figs. 7al/a2 give the SEM-micrographs of the sample taken after 0.06 ks (1 min), Fig. 7 bl/b2 after 0.3 ks (5 min) and Fig. 7 cl/c2 after 0.9 ks (15 min). This data refers to test 01 which means the powder/ball weight ratio was 1: 10, 1. 4 kg of grinding media was in the grinding unit and in total 140 g of powder mixture Ag/Cu was loaded. Figs. a2-c2 always show a higher resolution of the corresponding a1–cl SEM-image.

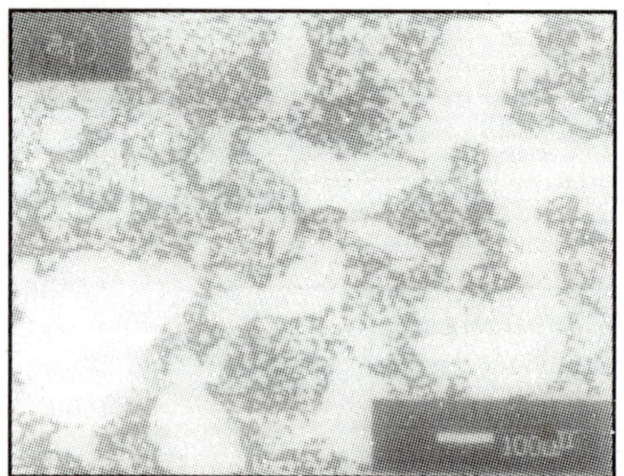

Fig. 7a1 Ratio 1:10, sample after 1 min

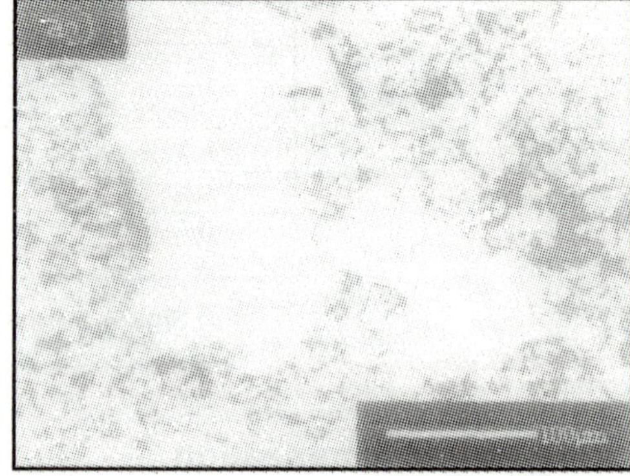

Fig. 7a2 Higher resolution

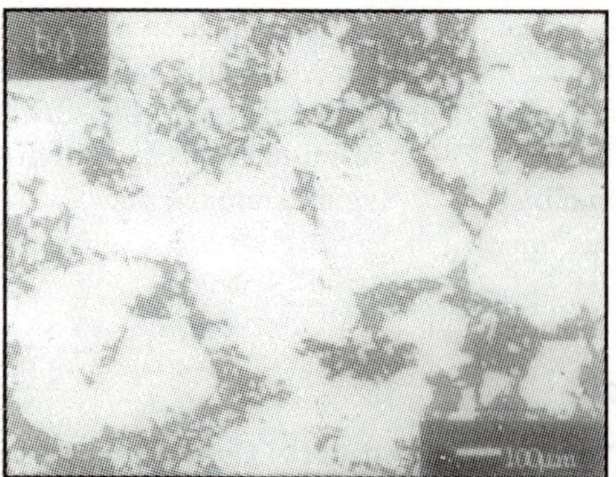

Fig. 7b1 Ratio 1:10, sample after 5 min

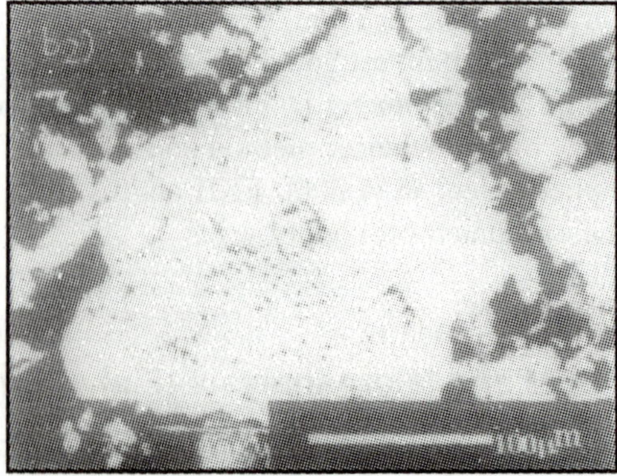

Fig. 7b2 Higher resolution

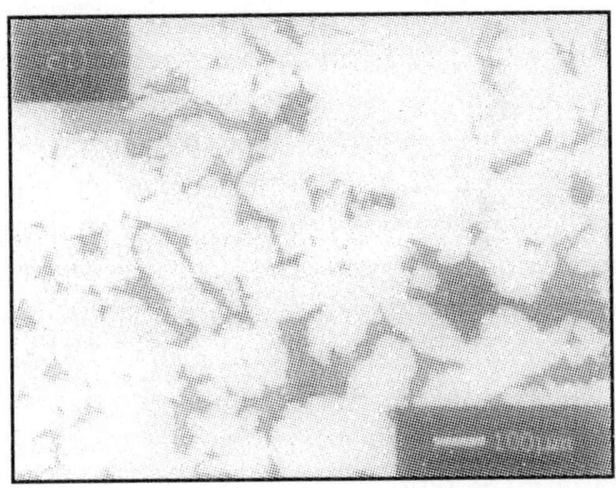

Fig. 7c1 Ratio 1:10, sample after 15 min

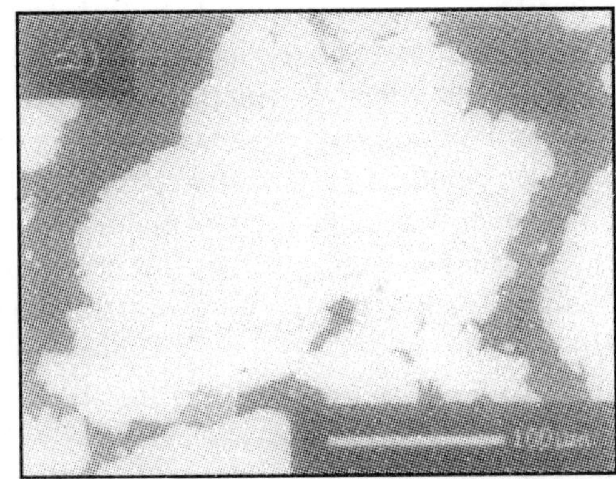

Fig. 7c2 Higher resolution

Fig. 7a-c. SEM-micrographs of Ag-70Cu powder samples after HKP at a load ratio of 1:10.

The given SEM-investigation confirms after just 1 minute of Processing time (Figs. 7al and 7a2) already a complete change of morphology of the Ag/Cu-powder. Following the expected evolution, we would describe first the still most present fine powder fraction which also dramatically changed the appearance which can be seen at Fig. 7a2 at higher resolution. This fraction may be described as loose low dense powder-agglomerates where the primary particle size (PPS) is close to the original size of the starting powders and the secondary particle size (SPS), so the size of the bulks is in the order of 10-50 µm. Then we can *see some large and very irregular bulks in the size of 100-200 µm that must* have been formed from the lower dense agglomerates and finally a few huge flakes as large as about 200-350 µm can be observed. Most probably these flakes, that are already quite thin in the order of only a few µm will have been formed by flattening the huge bulks.

So in principle we expect the fine powder to be agglomerated and maybe already slightly cold welded to huge bulks and then those ones are flattened and get into the. flaky shape. A basic idea here is, that the kinetic impact does not apply at highest efficiency to fine particles because of the dumping effect of the same. So here we only expect agglomeration. Once we have bigger agglomerates in the process, then the colliding grinding media will compact these very low dense agglomerates quite rapidly and since there is not such significant dumping effect any longer, density will increase and cold-welding will start, At this stage we actually would not assume to much interaction of deformation and fracture since fracture can only start once high dense flakes are available.

After 5 minutes MM, as seen in Figs. 7bl and 7b2 respectively, we can clearly see that the procedure we expected before seems to go on. Now we have much less fine fraction, not a significantly higher amount of huge bulks but a very much increased fraction of huge flakes. Additionally it seems, that we see some fractures of former larger flakes and also we see that flakes are agglomerated with other flakes which can be seen very, nicely at figure b2 at higher resolution. And most important, here we can see, that particles of the remaining fine fraction are agglomerated to the surface of already compacted flakes. By this these small particles will be exposed to the kinetic impact of the grinding media and by this very rapidly a kind of multi-layer structure is expected to be built. The size of the flakes is far different in the order of 100-350 µm.

After 15 minutes, as seen in Figs. 7c1 and 7c2, again. we can observe a tremendous change and this time the entire morphology appears different. The originally fine fraction completely disappeared and the large flakes have in comparison to earlier evolution stage a very smooth surface. At higher resolution, in Fig. 7c2, we can actually determine two different appearances since in the center of the image we see some large flakes where earlier particle border can he seen which means these flakes most probably have been built up by multi-layer structure. The flakes to be seen on the left do not show this kind of surface and are either less or further in evolution which means cold welding and deformation went on for the first or the second type. The size of the flakes is not that much different in the order of 80-150 μm.

In Table 5 we try to summarize the above where we estimate the flaky fraction in proportion to the material covered image area and call this ID. Additionally to the size we very carefully estimated the flake thickness and then conclude to the average aspect or cross-ratio:

Table 5. Summary of SEM-results related to flake-formation at the powder/ ball weight ratio of 1:10

SEM-results related to flake-formation at the powder/ball weight ratio of 1:10, estimates				
MM-time [*min*]	*Flaky fraction 1D* [%]	*Flake diameter* [μ*m*]	*Flake thickness* [μ*m*]	*Cross-ratio*
1	10	200-350	5-10	36.6
5	30	100-350	10-20	15
15	95	80-150	10-40	4.6

From the numbers carefully estimated in Table 5. we can of course and very clearly sec, that the fraction of flakes is increasing with the processing time and that the flakes become smaller and thicker which maybe also explained by an embrittlement of the material by surface hardening due to cold welding effects.

Looking at the cross-ratio. we have to notice. that we have no *ductile metal* flakes in common sense here since then we would need to obtain cross-ratios in the order of 200 and above. But at the same time we have to take onto account that here we have flakes that are built up by smaller and thinner flakes where in conventional flake processing, exactly this. is avoided usually by applying a low kinetic process in a drum ball mill and by adding lubricating PCA's in to the pro^ often by wet-processing e.g. in alcohol.

Figures 8al/a2 give the SEM-micrographs of the sample taken after 0.06 ks (1 min), Fig. 8 bl/b2 after 0.3 ks (5 min) and Fig. 8 cl/c2 after 0.6 ks (10 min). This data refers to test 02 which means the powder/ball weight ratio was 1:20, 1.5 kg of grinding media was in the grinding unit and in total 75 g of powder mixture Ag/Cu was loaded. Figs. a2-c2 always show a higher resolution of the corresponding al–c I SEM-image.

Since in case of a higher powder/ball weight ratio. we can expect a higher effect of the kinetic impact on our material since the dumping effect of the same is significantly reduce and we expected flake formation in shorter time and therefore extracted and characterized powder samples after shorter time as 1.5 and 10 minutes.

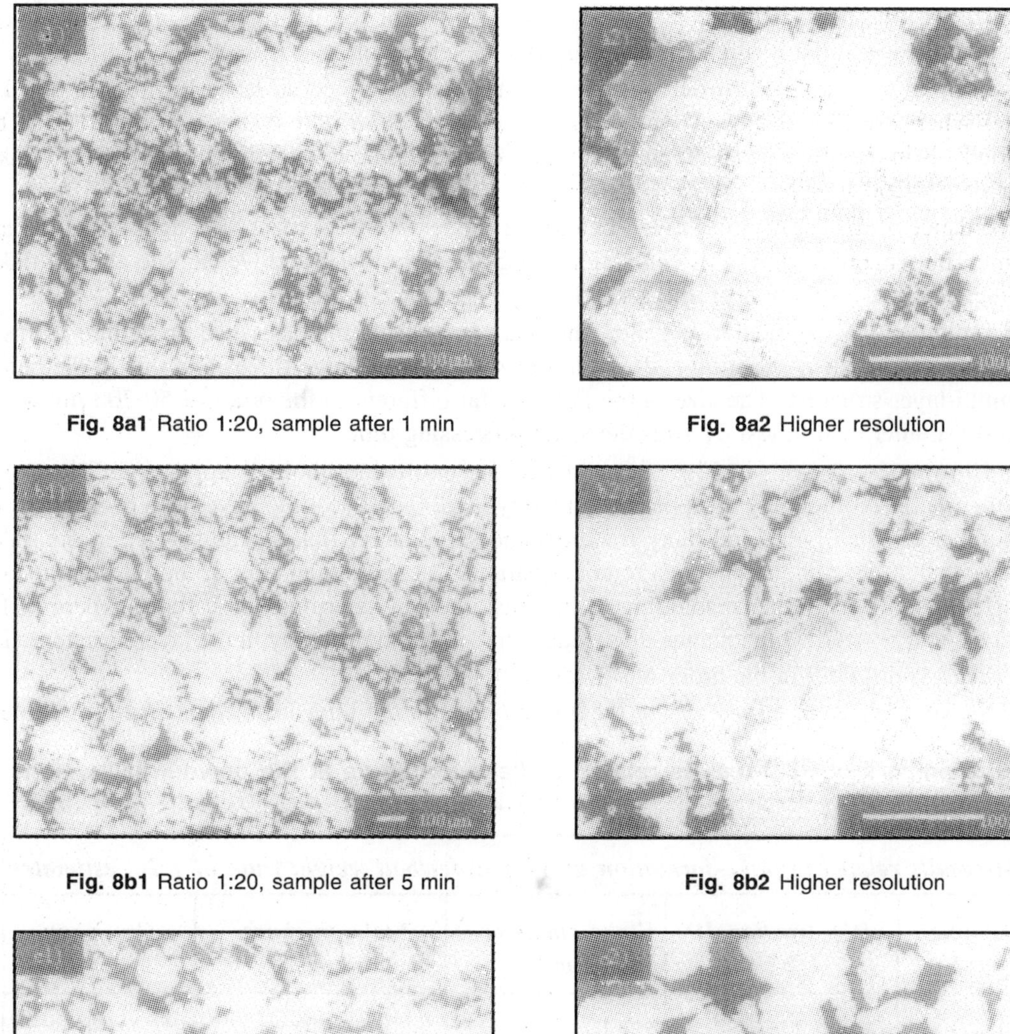

Fig. 8a1 Ratio 1:20, sample after 1 min

Fig. 8a2 Higher resolution

Fig. 8b1 Ratio 1:20, sample after 5 min

Fig. 8b2 Higher resolution

Fig. 8c1 Ratio 1:20, sample after 15 min

Fig. 8c2 Higher resolution

Fig. 8a-c. SEM-micrographs of Ag-70Cu powder samples after HKP at a load ratio of 1:20.

The given SEM-Investigation at the higher powder bail/weight ratio of 1:20 confirms after 1 minute of processing time (Figs. 8al and 8a2) also a complete change of morphology of the Ag/Cu-powder. The evolution is expected similar to what has been described before for the ratio 1:10, however, the appearance we find here, is different from before. We do not see the reported large bulks and it seems. that here the evolution is already further in time development.

The still most present fine powder fraction may still be described as loose low, dense powder-agglomerates where the PPS is close to the original size of the starting powders and the SPS. so the size of the bulks is in the order of 10-50 µm. Then we can see some few medium to large sized flakes in the order of 100-200 µm.

In accordance to the expectation given before for the test 01 we expect here agglomeration and cold welding. The remaining fine powder fraction is significantly lower and we can see already flakes with agglomerates of this fraction on the surface.

After 5 minutes MM, as seen in Fig. 8b1 and 8b2 respectively the described procedure seems to go on. We find much less fine fraction and an increased fraction of huge flakes and this in the before described multi-layer structure. The size of the flakes is far different in the order of 50-200 µm which is significantly smaller than in test 01 after the same processing time.

After 10 minutes, as seen in Figs. 8el and 8c2 the entire morphology appears different. The originally fine fraction completely disappeared, the very large flakes seem to have been broken and we can see a number of small bulks and flaky fractures and some larger flakes. The surface of the flakes got more smooth but they appear to have a relatively higher thickness than after 15 minutes at the ratio 1:10 (Figs. 7cl and c2). At higher resolution, in Fig. 8c2. we can determine large flakes where earlier particle borders can be seen which means these flakes have been built up by multi-layer structure. The size of the flakes is different in the order of 40-150 µm.

Table 6 gives the *estimate summary of the above based on the same criteria as given* for Table 5.

Table 6. Summary of SEM-results related to flake-formation at the powder/ball weight ratio of 1:20

SEM-results related to flake-formation at the powder/ball weight ratio of 1:20, estimates				
MM-time [min]	*Flaky fraction 1D [%]*	*Flake diameter [µm]*	*Flake thickness [µm]*	*Cross-ratio*
1	20	100–200	5–10	20
5	40	50–200	10–20	8.3
10	95	40–150	10–20	6.3

From the numbers carefully estimated in Table 6, we can again see, that the fraction of flakes is increasing with the processing time, that the flakes become smaller and thicker and that the estimated cross-ratio is again decreasing.

Figs. 9al/a2 give the SEM-micrographs of the sample taken after 0.06 ks (1 min). Fig. 9 bl/b2 after 0. 18 ks (3 min) and Fig. 9 cl/c2 after 0.3 ks (5 min). This dam refers to test 03 which means the powder/ball weight ratio was 1:50, 1.5 kg of grinding media was in the grinding unit and in total 30 g of powder mixture Ag/Cu was loaded. Figs. a2-c2 always show a higher resolution of the corresponding al-cl SEM-image.

The given SEM-investigation at the higher powder ball/weight ratio of 1:50 confirms after 1 minute of processing time (Figs. 9al and 9a2) a complete change of morphology of the Ag/Cu-powder.

The evolution is expected similar to what has been described before for the ratio 1.10 and at ratio 1:20, however, the appearance we find here, is different from before. We cannot see the reported large bulks in case of ratio 1:10 and in *comparison* to the ratio 1:20 the fine powder fraction described as loose low dense powder-agglomerates is almost not present where the expectation is, that it already disappeared. This would mean, that here the evolution is already much further in time development.

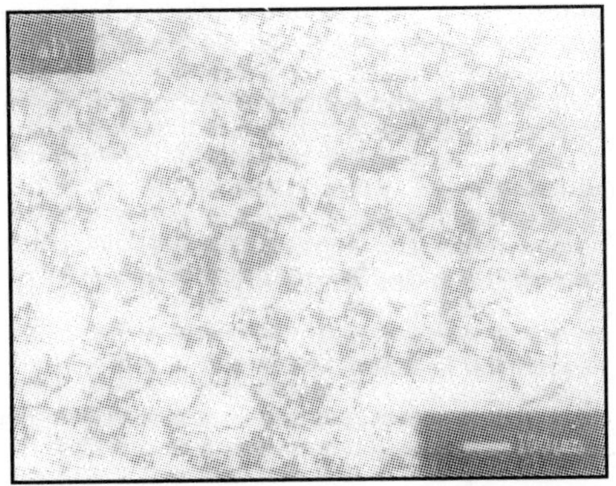

Fig. 9a1 Ratio 1:50, sample after 1 min

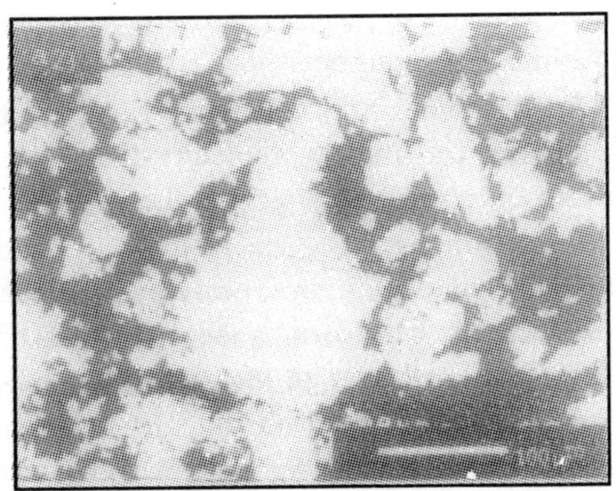

Fig. 9a2 Higher resolution

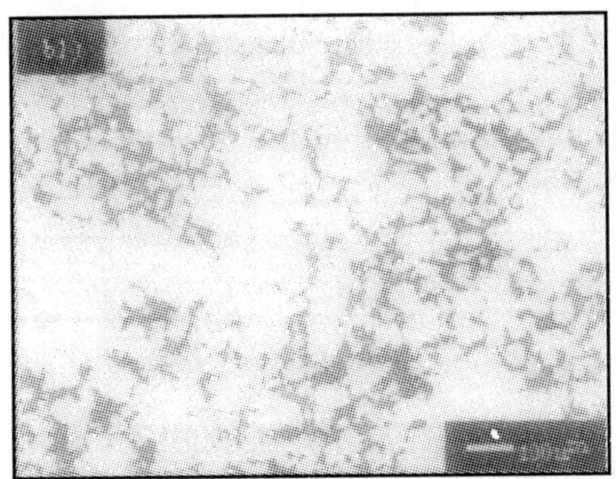

Fig. 9b1 Ratio 1:50, sample after 5 min

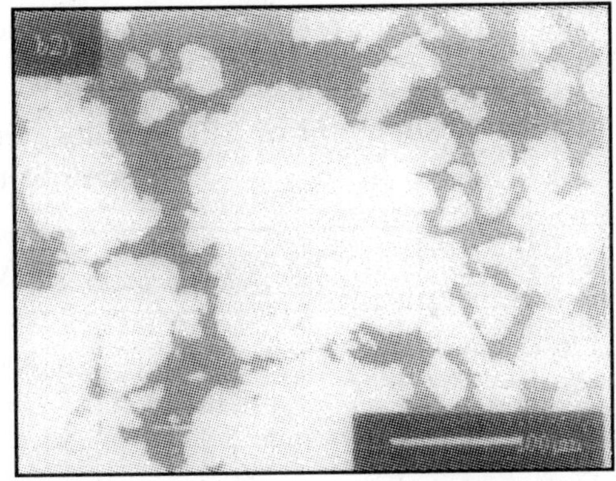

Fig. 9b2 Higher resolution

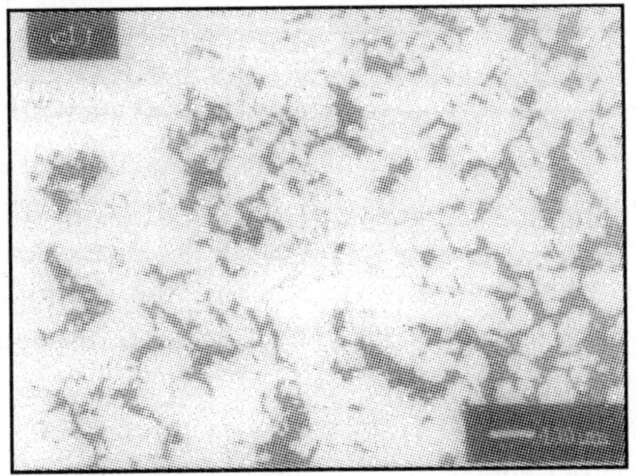

Fig. 9c1 Ratio 1:50, sample after 15 min

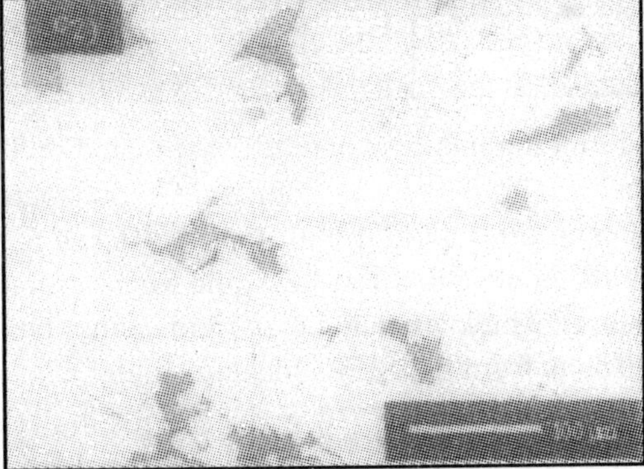

Fig. 9c2 Higher resolution

Fig. 9a-c. SEM-micrographs of Ag-70Cu powder samples after HKP at a load ratio of 1:50

For the expected remains of the powder-agglomerates, the PPS is expected to be close to the original size of the starting powders and the SPS, so the size of the bulks is in the order of 10-50 μm. The most present fractions are flakes in the order of 30-150 μm. In accordance to the expectation given before for the test 01 and test 02 we expect here agglomeration and cold welding. The remaining fine powder fraction is almost disappeared and we can see flakes in different stadiums flattened agglomerates and cold welded multi-layer structured flakes with agglomerates of the remains of the fine fraction on the surface.

After 3 minutes MM, as seen in Figs. 9bl and 9b2 respectively, the described procedure seems to go on. The fine fraction is completely disappeared and we can only find flakes in both stadiums where the surface becomes smoother and the size increased to the order of 30-250 μm.

After 5 minutes, as seen in Figs. 9cl and 9c2 respectively, the flakes described as flattened agglomerates completely disappeared and we can only see cold welded multi-layer structured flakes. The size seems to be grown to the order of 50-350 μm.

At higher resolution, in Fig. 9c2, we can see only a very few number of residues of the originally fine fraction agglomerated on the surface of some of the flakes where in all cases earlier particle borders can he seen which proves the multi-layer structure.

The evolution of flake development seems further that in test 01 after 15 and in test 02 after 10 minutes.

Table 7. Gives the estimate summary of the above based on the same criteria as given for Tables 5 and 6:

Table 7. Summary of SEM-results related to flake-formation at the powder/ball weight ratio of 1:50.

SEM-results related to flake-formation at the powder/ball weight ratio of 1:50, estimates				
MM-time [min]	*Flaky fraction 1D [%]*	*Flake diameter [μm]*	*Flake thickness [μm]*	*Cross-ratio*
1	50	30–150	no	no
3	95	30–200	10–20	9.3
5	95	50–250	5–15	20

From the numbers carefully estimated in Table 7, we can again see, that the fraction of flakes is increasing with the processing time. On the contrary to the earlier tests 01 and 02 we cannot determine at the chosen sampling-times, that the flakes become smaller which does probably not mean that we find a different evolution here and will be commented when we directly compare the results (see Chapter 4). Thickness and cross-ratio cannot be commented here since we cannot draw corresponding conclusions from the SEM-images 9al and a2.

The graph in Fig. 10 gives the results related to tests 01-03 from the Fe-mapping which we achieved by microanalysis investigating at least 10 areas per spot.

It must be noticed, that the given graph in Fig. 10 covers the complete measurement not only for the shorter terms up to 15 min (0.90 ks) for the investigation of flak formation but also for longer terms for the investigation of solid solution formation (SSF).

By experience, it is known, that in case of the Simoloyer, almost no contamination is caused by the vessel of the grinding chamber since this is not exposed to high kinetic impact and that only a small fraction (< 5 %) is caused by the rotor which is exposed to high kinetic impact but has a small surface only. The main contaminant is the grinding media (> 95 %) which is exposed to high kinetic impact and exhibits large surface. Since in the tests 01-03, the grinding media load was relatively constant (1400-1500 g), the contamination variation is expected to be caused by different grade of kinetic impact. On the other hand, the powder/ball weight ratio has a direct influence on the kinetic in the system since the powder itself provides a dumping effect against the kinetic impact (acts like liquid).

Next to this, contamination of the grinding media in case of the Ag-Cu-system will then be expected higher, if not all collisions of the grinding media will take place at in-between powder since the hardness of the grinding media is significantly higher than the one of both starting powder materials. In other words, a direct collision of grinding media without powder will cause highest contamination of the same and this means a) with respect to Fe-contamination and b) with respect to kinetic impact that the load-ratio shall not be to high The results in Fig. 10 match with the explanation before since we can see that the Fe-contamination increases with the powder/ball weight ratio. The crossing of the curve 1:50 with the curve 1:20 at approximately 3 h (10.8 ks) and the crossing of the curve 1:20 with the curve 1: 10 at approximately 50 min (3 k) are explained with the general measurement range of the mapping by microanalysis and may also depend on the disappearing of the fine powder fraction in early stage up to 10 minutes and on the expected work-hardening on the surface of the formed flakes. For the here observed flake-formation, the Fe-contamination is always detected below 0.05%.

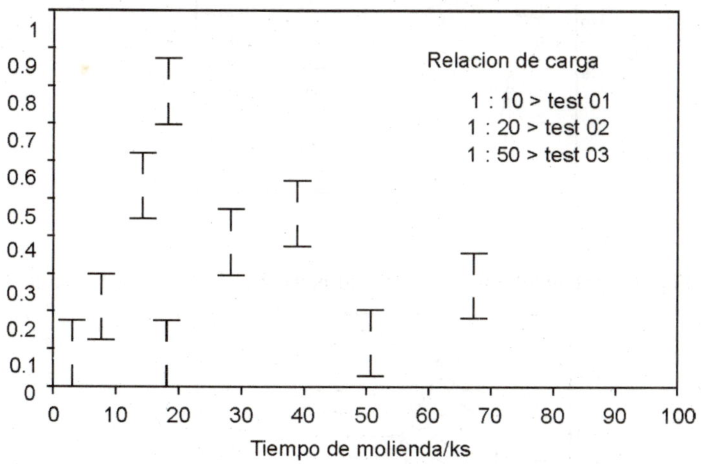

Fig. 10. Fe-contamination by microanalysis (min. 10 areas each).

In order to conclude a comparison of flaks formation (and Fe-contamination) at different powder/ball weight ratios, we determine the flake formation to that point when the fine powder fraction is completely disappeared. This attempt is shown in Table 8.

Table 8. Results of SEM-micrographs and microanalysis in survey.

SEM-results including microanalysis in survey			
Powder/ball weight ratio	*Flake formation after*		*Fe-contamination*
	[*min*]	[*ks*]	[*wt %*]
1:10	15	0.90	<0.05
1:20	10	0.60	<0.05
1:50	5	0.30	<0.05

3.2. XRD (SSF)

The powder characterization by XRD is related to the tests 10-12. These separate tens at same conditions as the previous tests 01-03 were chosen because much *longer* processing times for the formation of a solid solution in the Ag-Cu-system are expected. Fig. 11-13 show the XRD-patterns of the processed powders achieved after MM at various times 3.6 to 86.4 ks at the different applied powder/ball weight ratio of 1:10, 1:20 and 1:50 respectively.

We determine the solid solution formation (SSF) by the detection of the (111), (200) and (220) planes of the SS. This gives us good data for the beginning of SSF at least.

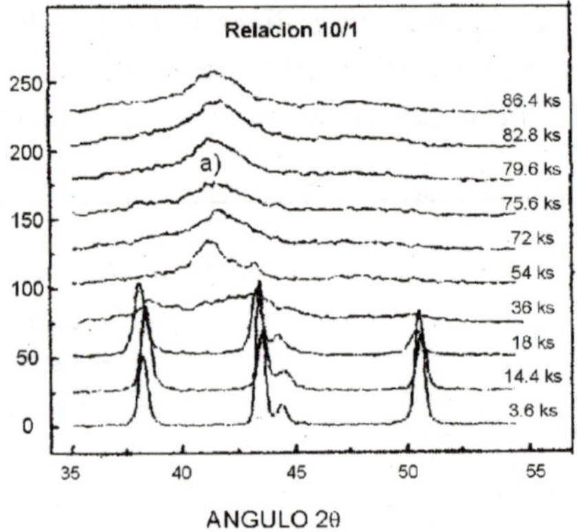

Fig. 11. XRD-patterns of processed powders at p/b-wt ratio of 1:10

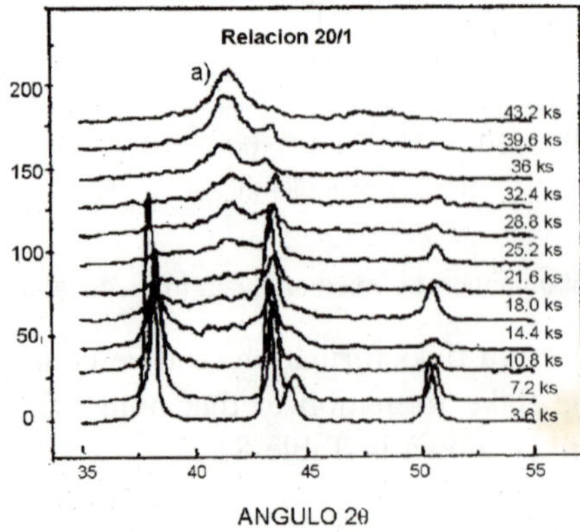

Fig.12. XRD-patterns of processed powders at p/b-wt ratio of 1:20

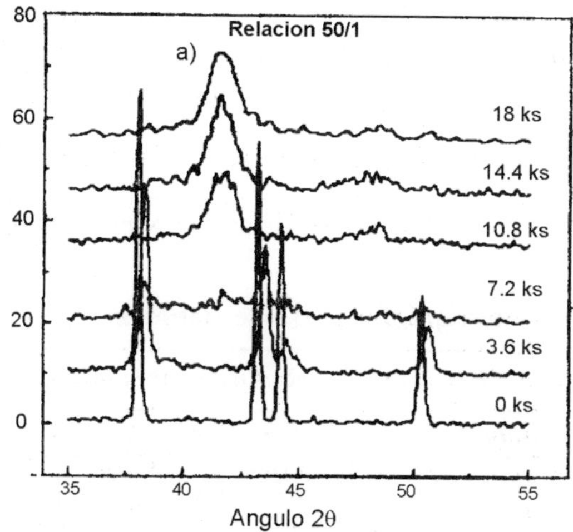

Fig. 13. XRD-patterns of processed powders at p/b-wt ratio of 1:50

In Fig.11 we see the XRD-patterns which are related to the powder/ball weight ratio of 1:10 and the SSF is detected at 54 ks (ignition) and seems to be completed at 79.2 ks. In Fig. 12 we see the XRD-patterns which are related to the powder/ball weight ratio of 1:20 and the SSF is detected at 36 ks (ignition) and seems to be completed at 43.2 ks. In Fig. 13 we see the XRD-patterns which are related to the powder/ball weight ratio of 1:20 and the SSF is detected at 14.4 ks (ignition) and seems to be completed at 18 ks.

3.3. TEM (SSF)

We applied TEM-investigation to check if the SSF detected by XRD is completed. This could be expected *e.g*, in case of test 12 after 14.4/18 ks according to the XRD-data. The TEM-results after 18 ks are given in Fig. 14a-c, all other investigation can he found parallel work [15].

From angle measurement we calculated the lattice parameter to 3.769 A. The grain size determined by the Scherrer-method is given at < 10 nm, the grain size determined by TEM at 5–30 nm.

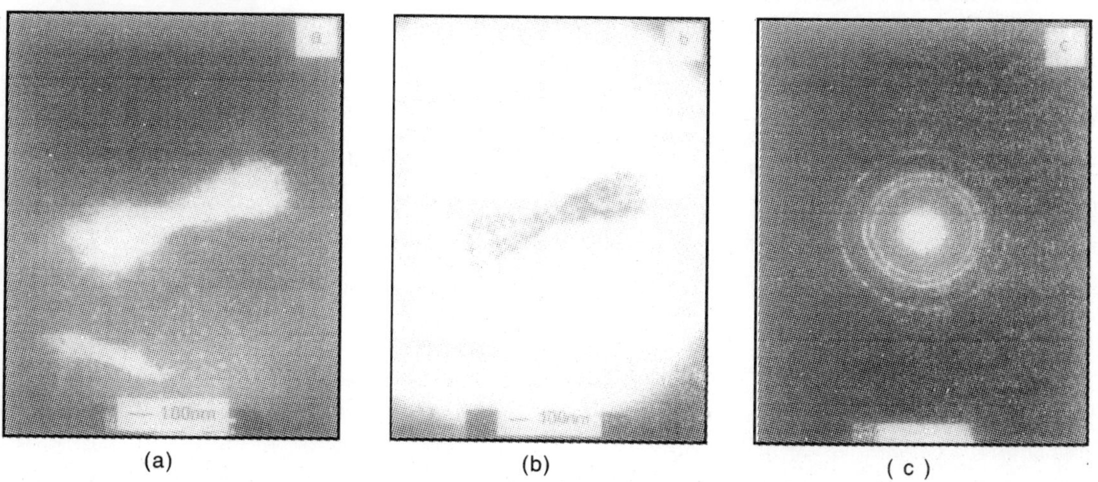

Fig. 14. TEM-investigation of the MM powder in the Simoloyer: (a) darkfield micrograph, (b) brightfield micrograph, (c) diffraction pattern after 18.0 ks.

The dark- and brightfield micrographs (a, b) show a grain size at around 5-30 nm where we obtained a value of 5 nm by the Scherrer-method. The diffraction pattern (c) shows diffraction rings which are characteristic for the nanometric grains of the new solid solution formed.

4. RELATION OF FLAKE-FORMATION, SSF AND Fe-CONTAMINATION

The comparison of the received results by SEM-micrographs and microanalysis, XRD and TEM is given in Table 9:

Table 9. Results of SEM-micrographs and microanalysis, XRD and TEM in survey.

SEM-XRD-and TEM-results (including microanalysis) in survey						
Powder/ ball weight ratio	*Flake formation*		*Fe-contamination after time a*	*SSF*		*Fe-contamination after time b*
	Starts after	*Complete after [a]*		*Starts after*	*Completed after [b]*	
	[*ks*]	[*ks*]	[*wt%*]	[*ks*]	[*ks*]	[*wt%*]
1:10	0.06	0.90	<0.05	54	79.2	0.25
1:20	<0.06	0.60	<0.05	36	43.2	0.45
1:50	<0.06	0.30	<0.05	14.4	18	0.75

The graph in Fig. 15 gives the dependence of flake-formation and SSF at different powder/ball weight ratios. The measured data at each of the different ratios was concluded to a curve which shows that the slope of the 2 curves for flake formation SSF are almost equal which was expected since the driving force for both formations was earlier defined as being dependent on the kinetic impact and the kinetic impact was described to be determined by the powder/ball weight ratio in these experiments.

Fig. 16 gives die dependency of SSF and Fe-contamination at different powder/ball weight ratios 1:10, 1:20 and 1:50.

What we see here is, that in case of higher powder/ball weight ratio, the solid solution is completed earlier, but the Fe-contamination is higher even at the shorter time needed for SSF-evolution. For lower ration, time is longer and contamination remains lower.

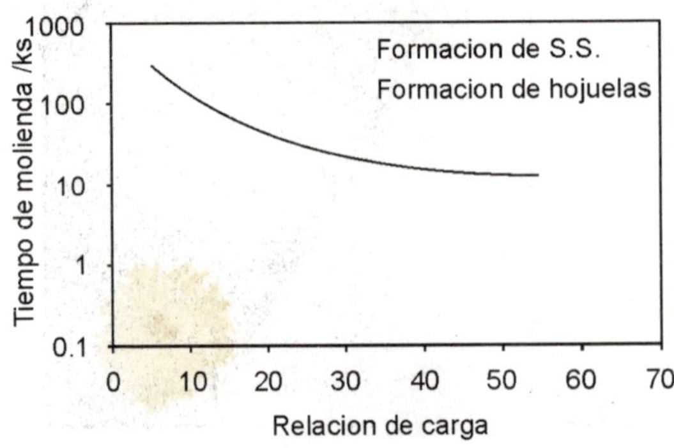

Fig. 15. flake-formation and SSF at different powder/ball weight ratios.

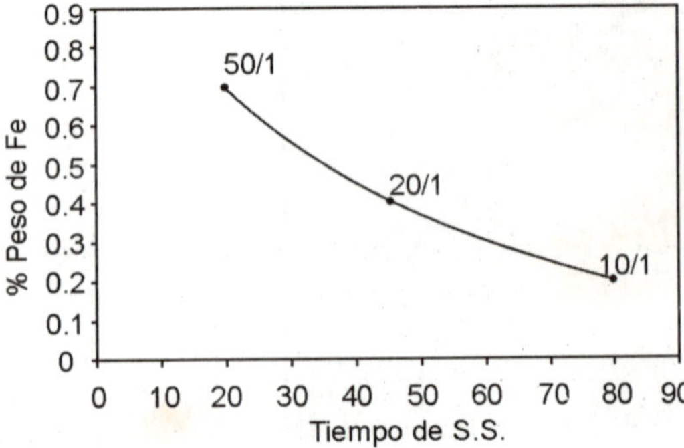

Fig. 16. SSF and Fe-contamination at different powder/ball weight ratios.

CONCLUSIONS

The Ag-70Cu (at%) has been MM in a Simoloyer CMO1-21 high energy ball mill at different powder/ball weight ratio and not under high but under medium kinetic.

The flake formation started after a processing time of 0.06 ks and below and was observed to be completed after 0.3, 0.6 and 0.9 ks at load ratios 1:50, 1:20 and 1:10 respectively. The Fe-contamination by the milling tools which means here mostly by the grinding media was detected to be very low at < 0. 05 wt% m all cases:

Table 10. Completion of flake-formation and comments

Completion of flake formation						
Powder/ball weight ratio		*Flake formation completed*		*Fe-contamination*		*Comment*
[wt%]	*Factor*	*[ks]*	*Factor*	*[wt%]*	*Factor*	
1:10	1	0.90	3	<0.05	(1)	contamination always < 0.05 wt%
1:20	2	0.60	2	<0.05	(1)	2× higher ratio, 2× shorter time
1:50	5	0.30	1	<0.05	(1)	5× higher ratio, 3× shorter time

From these results we may conclude that the ratio of 1:50 is unnecessarily high for the formation of flakes and may need to he only 1:30 since we can only achieve a 3 times shorter time by the ratio 1:50. This shall be checked by further work and from today's point of view could be explained since a to high ratio will lead to a number of quasi useless collisions where no flake is formed but contamination expected to be increased.

The solid solution formation by MM of Ag-70Cu in the Simoloyer CMO1-21 at medium kinetic impact could be obtained at nanometric grain size after 18, 43.2 and 79.2 ks at load ratios 1:50, 1:20 and 1: 10 respectively. The Fe-contamination did increase with the MM-time from 0.25 to 0.45 to 0.75 wt% and did not reach 1 wt% at any time:

Table 11. Completion of SSF and comments

Completion of SSF						
Powder/ball weight ratio		*SSF completed*		*Fe-contamination*		*Comment*
[wt%]	*Factor*	*[ks]*	*Factor*	*[wt%]*	*Factor*	
1:10	1	79.2	4.4	0.25	1	contamination always <1.0 wt%
1:20	2	43.2	2.4	0.45	1.8	2× higher ratio, 2.4× shorter time 1.8× higher contamination.
1:50	5	18	1	0.75	3	5× higher ratio, 4.4× shorter time, 3× higher contamination.

From these results we may conclude that the ratio of 1:50 is slightly to high for the SSF and may need to be only 1:44 since we can only achieve a 4.4 times shorter time by the ratio 1:50. This shall be checked by further work and from today's point of view could be explained since a to high ratio will lead to a number of quasi useless collisions where no kinetic impact on Ag-Cu-particles is performed but contamination increases.

What must be pointed out here is, that the contamination is higher for the higher load ratio even at the shorter times. This means if in absolute numbers the contamination shall be reduced then a lower ratio shall be chosen which is not valid owe looking for flake formation only.

In result of this work, we can see that flake formation and SSF do depend both on the powder/ ball weight ratio which is a determination of the kinetic in the process. Shifted in processing time, the evolution of both seems to be relatively linear.

The first explanation is of course that for both transformations, kinetic energy impact is determining. Next to this, the formation of flakes by thin multi-layers seems to favor SSF.

Since in case of SSF, here, Ag-atoms shall be located in the Cu-lattice inersticially or by substitution', this is provided mainly by kinetic impact which means it is of major importance, how this kinetic impact can effect the lattice structure.

For example, if some grinding media would be thrown into a *volume of* Ag-Cu-powder, there will be plastic deformation of the volume and there may be plastic deformation of some of the direct hit particles in the front-layer but it is very much unlikely, that this can cause atomic dislocations in the lattice of the materials. This means the dumping effect of the powder plays the same important role for SSF as well as for flake formation. If we now take into account- that ductile metal flakes based on Ag and Cu and others can easily be produced by HKP in short times at flake thickness of 0.5 μm and below [6-8], and that these flakes can be built up by a number of approximately 50 theoretical inter-layers according to the starting particle size [4], then we would end up with a theoretical thickness in 10 nm range.

If we apply this to a powder mixture and use the principle of MA and produce a kind of sandwich. then we probably end up with very short diffusion ways inside this structure and this could explain why multi-layer flake formation favors SSF and it will he very interesting to find dependencies for this in the ongoing work where one of the next steps will be the application of HKP at 3-4 times higher relative velocities of the grinding media.

REFERENCES

1. K. Uenishi, K.F. Kobayashi, K.N. Ishihara, P.H. Shingu: Formation of a supersatured solid solution in Ag-Cu systems by mechanical alloying, Mat. Sci. & Eng. A 134 (1991) pp. 1342-1345.
2. R. Esquivel-Gonzalez, Ph D. Tesis, Compactacion dinamica de materiales nanocristalinos obtenidos por aleacion mecanica, ESIQIE-IPN. Mexico (2001).
3. R. Esquivel G., I. Vernet P., C. Reneroy D. Jaramillo V.— Dynamic compaction of nanostructured alloys obtained by mechanical alloying, Journal of Materials of Materials Processing Technology, Thermec 2000-Processing & Manufacturing of Advanced Materials, TMS, Elsevier Science, (2001).
4. H. Zoz: HEM/MA/RM— Devices in use, 3rd Intern. Symp. of The School of Chemical Engineering, IPN, Mexico City, May 27-29, 1998
5. A. Bose, K. Ameyama, S. Diaz de la Torre, D. Jaramillo V., D. Madang, H. Zoz, Zoz GmbH (Germany & USA), Materials Processing Inc. (USA), Ritsumeikan University (Japan). CIMAV S.C., ESQIE,

(Mexico), F.W. Winter Inc. & Comp. (USA): Review of Applications and Materials processed by Rotating Horizontal High Energy Milling: 2002 World Congress on Powder Metallurgy & Particulate Materials, proceedings (2002).

6. H. Zoz, D. Ernst, T. Mizutani, H. Okouchi, Simoloyer CM100s, semi-continuously Mechanical Alloying in a production scale using Cycle Operation-Part 1, Metall Vol. 51, 10/97, pp. 568-572, 1997.
7. H. Zoz, D. Ernst: Simoloyer CM100s, semi-continuously Mechanical Alloying in a production scale using Cycle Operation-Part II, Metall Vol. 51. 09198, pp. 521-527, 1998.
8. H. Zoz, H. Ren, H. U. Benz: Ductile Metal Flakes based on [Au], [Ag], [Cu], [Ti], [Al], [Ni] and [Fe] by High Energy Milling - part 1, PM2Tech 1999 MPIF, Vancouver, (1999) proceedings.
9. H. Zoz, D. Ernst, H. Weiss, M. Magini, C. Powell, C. Suryanarayana, F.H. Froes, Mechanical Alloying of Ti-24Al-llNb (at%) using the Simoloyer Metall Vol. 50, 09/96, pp. 575-579, 1996.
10. H. Zoz, D. Ernst: I. S. Ahn, W. H. Kwon: Mechanical Alloying of Ti-Ni-based Materials using the Simoloyer, TMS Annual Meeting 1997. eds. C.M. Ward-Close, F.H. Froes, S.S. Cho, D.J. Chellman: Synthesis/Processing of lightweight Metallic Materials, proceedings (1997).
11. J.Y. Chung, J. Kim and Y.D. Kim: Formation of Nanocrystalline Fe-Co Powders Produced by Mechanical Alloying, Dept. of Metallurgy and Materials Science Hanyang University Ansan, Korea, 1999.
12. H. Zoz, H. Ren'. N. Spath: Improved Ag-SnO2 Electrical Contact Material Produced by_Mechanical Alloying, Metall Vol. 53, 07-08/99, pp. 423-428, 1999.
13. G. Kaupp, M. R. Naimi-Jamal, H. Ren and H. Zoz : Environmentally Protecting Reactive Milling, CHEMIE TECHNIK, 31, Vol. 6 (2002) p. 58-60.
14. H. Zoz, H.U. Benz, K. Huttebraucker, L. Furken, H. Ren: Stellite bearings for liquid ZN-/AI- Systems with advanced chemical and physical properties by Mechanical Alloying and Standard-PM-Route. Part I. Metall Vol. 54, 11/2000, pp. 650-659, 2000.
15 H. Zoz, I. Vernet and D. Jaramillo V: Comparative of solid-solution formation by Mechanical Milling of Ag-70Cu (at%), proceedings of PM2TEC'2003, NWIF (2003).

5

Synthesis and Characterization of PM Processed Al-SiC Composites for Electronic Package Applications

V.V. Bhanu Prasad
Defence Metallurgical Research Laboratory, Kanchanbagh, Hyderabad-58

Sweety Kumari
Defence Metallurgical Research Laboratory, Kanchanbagh, Hyderabad-58

T. Madhavi Latha
National Institute of Technology, Warangal

Narkale R. Asish
National Institute of Technology, Warangal

K.Somaraju
Defence Metallurgical Research Laboratory, Kanchanbagh, Hyderabad-58

K.S.Prasad
Defence Metallurgical Research Laboratory, Kanchanbagh, Hyderabad-58

ABSTRACT

Metal Matrix Composites (MMC) have a unique set of properties that are ideally suited for electronic packaging applications requiring thermal management solutions. The matching coefficient of thermal expansion (CTE) value of Al / SiC with that of the devices make them ideal for this application. Conventionally these packages are made by melt infiltration technique. In the present study powder metallurgical route using Vacuum Hot Pressing (VHP) is employed to synthesize high volume fraction composites (2124 Al / 70 %SiCp.). Size of the reinforcement particles were varied (36#, 220# and 600#) and their proportion also have been varied to make different MMCs like mono-modal, bi-modal and tri-modal composites. These composites were studied for their mechanical and physical properties like density, flexural strength etc. Structure property correlation has been drawn among these composites. Necessity of higher processing temperatures for finer reinforced composites, achievement of higher densification in multi modal composites over mono-modal composites and lowered and matching CTE are some of the salient observations of the present study.

1. INTRODUCTION

Metal Matrix Composites (MMCs) reinforced by discontinuous ceramic particulate/whiskers offer attractive properties such as high specific strength, stiffness, excellent wear resistance and tailorable coefficient of thermal expansion (CTE) (1-2). The reasons for their success over the continuous fibre reinforced composites are their remarkable isotropic properties, amenability to conventional metal working processes and comparatively low cost. (3). MMCs have been widely considered for both structural applications and functional applications. While higher specific strength and modulus are the prime requirements of the structural MMCs, lower α/k ratio (where to is -CTE and k is thermal conductivity) is the requirement of functional MMCs (4). The size and volume flaction of the reinforcing phase control these properties of the composites. In general, MMCs containing lower contents (<20%)

of finer SiC (<20 μm) are suitable for structural applications. MMCs containing 30-40% SiC (<3 μm) are suitable for dimensionally stable applications such as laser mirror substrates, where the CTE of the substrate matches with that of the nickel coating. In case of MMCs containing 60-70% of SiC, the CTE matches with that of the Si and GaAs chips and hence most considered to replace the conventional packaging materials like Cu, Al, Kovar, Cu/W (5).

The recent developments in the semiconductor technology is centred around the high powered GaAs and Si chips where thermal dissipation is a problem. It is often stated that the growth of the semiconductor technology is hindered by non-availability of suitable packaging materials (6). Though Cu and Al packages are excellent heat dissipaters, they are non-reliable due to high CTE mismatch between the packages and the semiconductors. Kovar is widely used for this application due to its matching CTE with the GaAs and Si chips. However, its low thermal conductivity results in poor thermal dissipation. More over high density of Cu, Kovar, Cu/W makes them not suitable for air born applications. Al base MMCs containing 60-70% SiC can offer an integrated solution to all these problems due to its low density, low and tailored CTE and high thermal conductivity. Table 1 shows the thermo-physical properties of the various materials used for electronic devices and their packaging.

Table 1. Thermophysical properties of materials used for electronic chips and their packaging

Material	*Composition*	*Density (g/cc)*	*CTE ppm/°C*	*Thermal conductivity (W/m-k)*
Aluminium	Al	2.7	23	171
Copper	Cu	8.9	17.6	391
Molybdenum	Mo	10.2	5.1	140
Tungsten	W	19.3	4.5	167
Cu-Mo	Mo/20 Cu	10	8.0	170
Cu-W	W/20 Cu	15	8.3	200
Silicon	Si	2.3	4.3	151
Galium arsinide	GaAs	5.3	5.3–6.5	54
Alumina	Al_2O_3	3.6	6.7	17
Berilia	BeO	2.3	7.6	250
Aluminium Nitride	AlN	3.3	4.5	160–200
Kovar	Fe-Ni-Co	8.2	5.8	17
Silvar	Fe-Ni-Ag	8.8	6.5	153
MMC	Al/50-60 SiC	3.0	6.5	160

Most of the fabrication techniques used to make Al/SiC packages adopt infiltration of molten Al-Si alloy into the near net shaped SiC performs. Common ceramic processing techniques such as die pressing, injection moulding, slip casting followed by partial sintering are used to fabricate the SiC performs. Alcoa infiltrates Al alloy into a SiC performs, made by injection moulding and sintering, using a high pressure die casting technique (7). Lanxide Corporation follows pressureless infiltration in N_2 atmosphere (8). Attempts have been made to infiltrate the preforms by encapsulating them in Al-Si alloy block followed by melting (9). In all these infiltration techniques, Si is added to the matrix Al to suppress the formation of Al_4C_3 (10). More over Si provides fluidity and wettability to the molten metal to flow into the capillaries between the reinforcements. There is no information in open literature on making these MMCs by PM route. This route is not matrix composition specific unlike liquid route. More over wide range of reinforcement particles can be used. Due to the combination of stronger alloys and fine particles it is expected to develop stronger MMCs for electronic packaging applications.

2. EXPERIMENTAL

SiC particulates (36#, 220# and 600#) having the average particle size of 480 μm, 53 μm and 12 μm, respectively were used to make different reinforcement mixtures as per the compositions shown in table 2. These optimum compositions were arrived based on the tap density studies carried in a different work using a wide range of mixtures using three grades of SiC (10). Composite powders were made by blending the above reinforcement mixtures with 2124 Al powder (200#) having the nominal composition of 4.1% Cu, 1.5% Mg, 0.2% Si, 0.3% Fe and having the average particle size of 40 μm. The composite powders were vacuum hot pressed (VHP) into 125 mm diameter discs of 10 mm thickness using an standard two stage hot pressing technique (11). During 1st stage, powders are densified at 500°C as a function of compaction pressure up to 115 MPa. In 2nd stage, compact is further densified by increasing the temperature, at constant pressure of 115 MPa, till the compact is fully densified. In both the stages, % theoretical density (%TD) pattern was recorded. These composites were studied for their bend strength using 5 × 5 × 70 mm bar samples. The microstructures of the polished samples and the fractographs of the fractured samples were studied using JEOL SEM. NETZSCH make dilatometer was used for the measurement of CTE in N_2 atmosphere while the sample is heated with 5 K/min.heating rate.

Table 2. Composition of different reinforcement mixtures used in making MMCs by VHP.

Sample	*Composite*	*36#*	*220#*	*600#*
Mono-modal (M)	M1	100	–	–
	M2	–	100	–
	M3	–	–	100
Bi-modal (B)	B1	70	–	30
	B2	60	40	–
	B3	–	80	20
Tri-modal (T)	T	50	40	10

3. RESULTS AND DISCUSSION

The shape and topography of 2124 Al powder and 36#, 220#, 600# SiC powders are shown in Fig. 1. The atomised Al alloy powders are irregular in shape with dendratic structure. All the SiC powders are angular in nature with flat surfaces and sharp edges. The size distribution of all the powders are of Gaussian distribution.

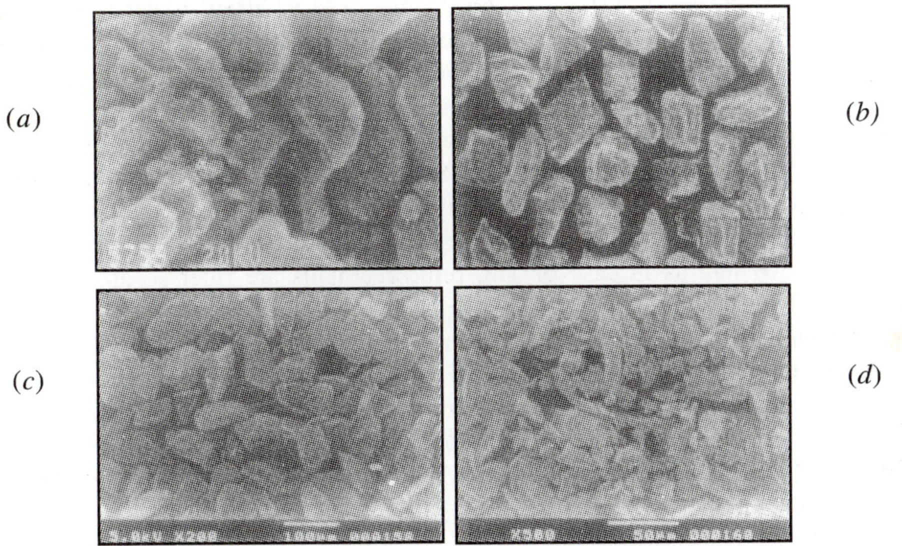

Fig. 1. Shape and topography of raw materials . (a) Al b) 36# SiC (c) 220# and (d) 600#.

Fig. 2 a and b show the 1st and 2nd stage compaction curves drawn for the matrix 2124 Al and composites M2 and M3. From this figure it can be observed that 2124 Al matrix sample could be compacted to its full density at 500°C, which is lower than the solidus temperature of the alloy. Whereas, the M2 and M3 composites could be compacted to the extent of 92% and 86% TD, respectively, during the 1st stage. Lower % TD in case of M3 indicates the lower compactability of the MMC powder reinforced with finer SiC powder (600#). During 2nd stage both the composites M2 and M3 were heated up to 590°C. While M2 got compacted to its full density, M3 could be compacted to only 97%. M3 could be fully densified when the temperature was raised to 620°C, indicating the need for more liquid phase in the matrix and the same could be obtained by raising the temperature further. It can be concluded that finer the reinforcement particles or larger their volume fraction more is the second stage processing temperature to get full density. This observation is consistent with the hot pressing mechanism published elsewhere (11).

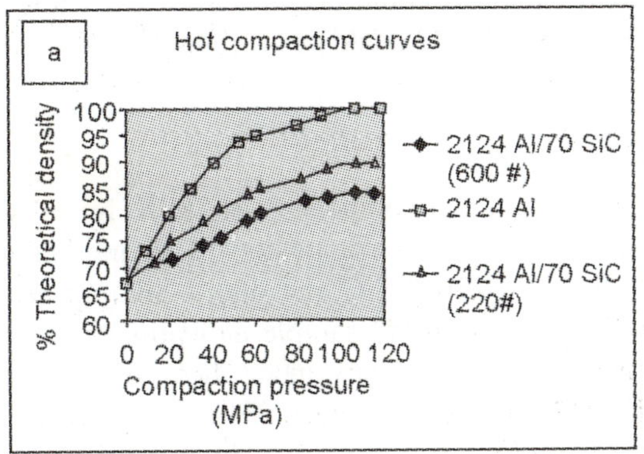

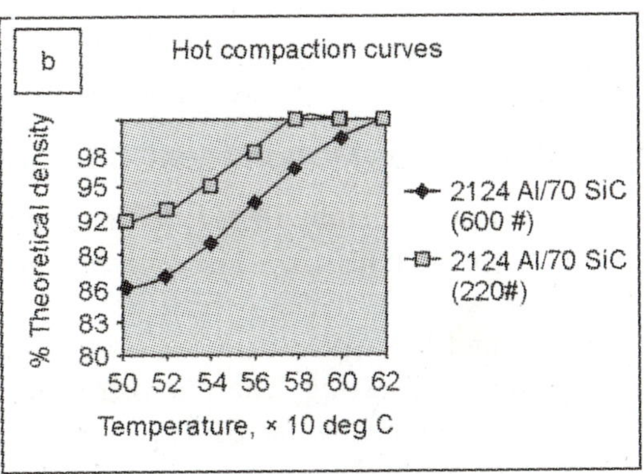

Fig. 2. (a) 1st stage and (b) 2nd Stage vacuum hot Pressing Curves.

Different composite samples, M1, M2, M3, B1, B2, B3 and T, were vacuum hot pressed at 570°C. Table 3 shows the densities of the different 2124A1/SiC_P compacts measured by Archimedes's principle and also normalized by 3.071 g/cc, the TD for 2124 Al-70SiC to get % TD. In general all the compacts could be compacted at 570°C to the densities better than 91% T.D. It also can be observed that the mono-sized samples (M1, M2 and M3) were not compacted to the full density. M3 sample was least compacted to 91% and MI sample is maximum compacted among the mono-sized samples. In MI as the reinforcement size is large, the interstitials to be filled are also large it is easy to fill these interstitials with plastic flow of matrix material. So they obtain more density during first stage it self. Need for further densification during second stage is also less and the same is also easy due to the larger pores to be filled during second stage. They will be filled with less viscous matrix prevailing at lower temperature. Where as in case of M3, reinforcements are finer so are the interstitials. The number fraction of the finer reinforcements are large when compared to coarse reinforcements, so they offer maximum resistance to the plastic flow of the matrix. This results in lower densification during 1st stage. During 2nd stage, as the needed volume of densification is more and the size of the interstitials to be filled are also small, the temperature at which low viscous matrix in required quantities available is more. As the 570°C is not adequate to provide the necessary liquid phase, the densification is lowest in case of M3. As the SiC particle size (220 #) in M2, is in between the those of M1 and M3, its compaction behavior is also in between those of the M1 and M3.

Table 3: Effect of Vacuum Hot Pressing at 570°C on the density of the MMC compacts.

Sample	M1	M2	M3	B1	B2	B3	T
Density (g/cc)	2.97	2.91	2.80	3.05	3.04	3.01	3.04
% Theoretical Density	96.7	94.8	91.1	99.3	98.9	97.9	99.1

The bi- modal and tri-modal composites B1, B2, B3 and T could be densified above 98% T.D. which is better than the % T.D. of mono-sized samples M1, M2 and M3 during hot pressing at 570 °C. Increased packing density of the starting powders in case of bi-modal and tri-modal samples when compared to the mono-modal samples may be responsible for better densification in case of earlier samples. Samples B, B2 and T were densified above 99% where as sample B3 was densified to 98% only. Presence of 36 # SiC in B1, B2 and T samples must be the reason for better densification as the number fraction of 36 # SiC is lower and interstitial size is more. The lower sized SiC present with in these interstitials also contribute to differential densification as could be seen from the Table 3 even though the differences are marginal. Relatively lower density in B3 could be attributed to the inadequate temperature as explained in case of M3.

Microstructure of Mono Modal Composites

Microstructural study of M1, M2 and M3, the mono modal composites, at 200X. revealed fairly uniform microstructures (10). The 36 # SiC particles present in M1 composite are having defects like porosity, micro-cracks. Such defects are not evident in the 220 # SiC and 600 # SiC present in the samples M2 and M3. Some finer SiC particles are also evident in M1 sample indicating fracture of 36 # SiC during hot pressing. The microstructures also show the porosity in the form of dark regions. Figs. 3 a, b, c and d show the SEM pictures of M1, M2 and M3. The derivatives, of the order of 10-20 microns, of the broken 36 # SiC can be observed from Fig. 3(a). Fig. 3(b) shows no such evidence of particle breakage in M2. Micrograph of M3 sample in Fig. 3(c) consists of three regions *i.e.,* fully infiltrated region showing uniform distribution of 600 # SiC, the dark regions of matrix devoid of particles and uninfiltrated regions. Fig. 3 (d) shows the uninfiltrated region at high magnification , which is highlighted by presence of loose SiC particles. Infiltration is totally absent in these regions and the particles are in fully packed and mechanically interlocked. Due to this interlocking they are not falling out during cutting and polishing. Inadequate liquid matrix as a result of lower processing temperature looks to be the sole reason for un-infiltration of these regions. However, presence of entrapped gasses due to inadequate degassing could not be ruled out. This region is mainly responsible for getting the lower densities in the as VHP condition. The darker region of Fig. 3 (c) is found to be a Silicon rich phase (80%) with multiple micro cracks similar to solidification crack.

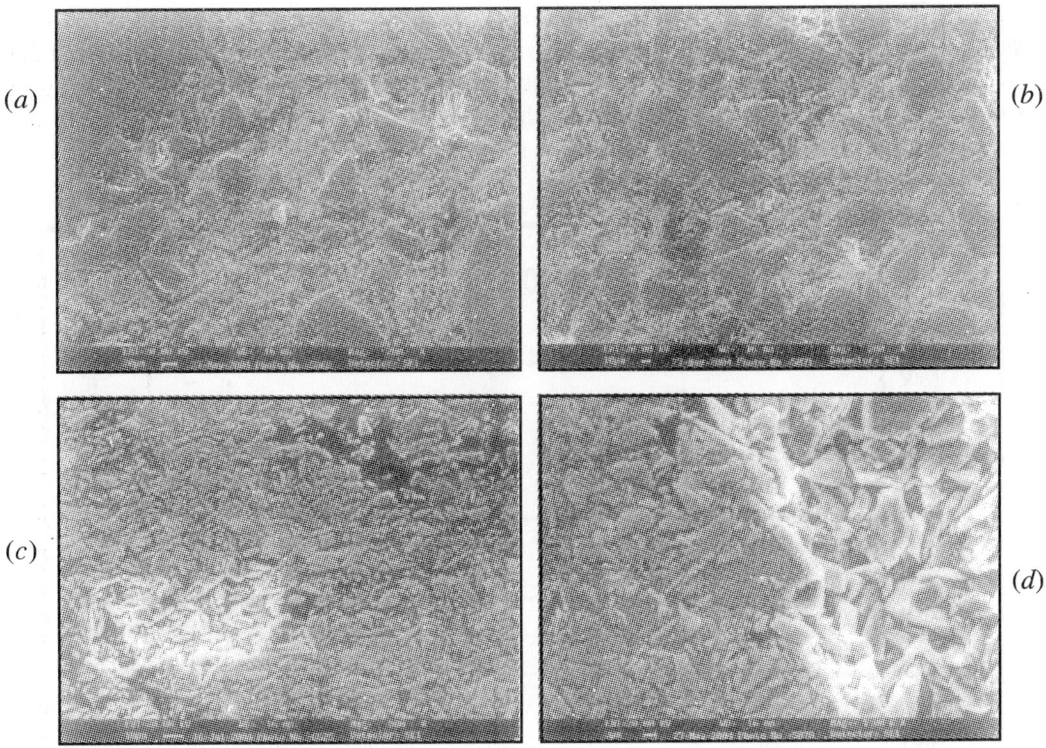

Fig. 3. SEM micrographs of (a) M1 at 200x, (b) M2 at 200x, (c) M3 at 500x and (d) M3 at 1000x.

Microstructure of Multi Modal Composites

The bi-modal and Tri-modal composites were studied for their microstructure using SEM.

Figs. 4 (a) and (b) show the micrograph of B1 sample at 50x and 500x. Uniform distribution of 36 SiC is evident from Fig. 4 (a). The inter particle spacing between the 36 # SiC got filled with unresolved gray phase. The area fraction of 36 # SiC and unresolved gray regions are calculated to be 50% each. When observed at high magnification (500X) the gray region resolved to be composite of Al /600 # SiC where Al and 600 # SiC are in the ratio of 60:40. As the volume of the material to be compacted to full density is confined to smaller regions and the volume fraction of SiC is lower (40% against 70% in M3) the B1 'sample could be compacted to full density even though M3 could not be compacted to full density at 570 °C. Better compactability of B1 over M1 can be attributed to the better initial packing density in B1 powder. Fig. 4 (c) shows the micrograph of B2 sample again showing uniform distribution of 36 # SiC. However, the inter particle spacing between the 36 # SiC is larger when compared to B1 (Fig. 4 (a)) as the volume fraction of 36 # SiC is lower in B2. The inter particle spacing between the 36 # SiC fully got resolved to be a composite with 220# SiC at high magnification (200X). Figs. 4(d) and (f) show the micrograph of B3 sample at 50x and 1000x. From the first micrograph uniform distribution of 220 # SiC could be observed. However, the inter particle spacing between unresolved gray phase. When observed at 1000x the 600 # particles are fully resolved and some porosity is visible in 220 # particles. Fig. 4(e) shows the micrograph of the tri-modal sample T again at 50x. The low magnification micrograph revealed the uniform distribution of 36 # SiC. The interparticle spacing of 36 # SiC is filled with grayish phase containing uniformly distributed 220 # SiC. However, the second stage inter particle spacing between the 220 # SiC is unresolved at this magnification. When observed at high magnification (500X) the 600 # particles present in the second stage inter particle spacing between the 220 # SiC fully got resolved.

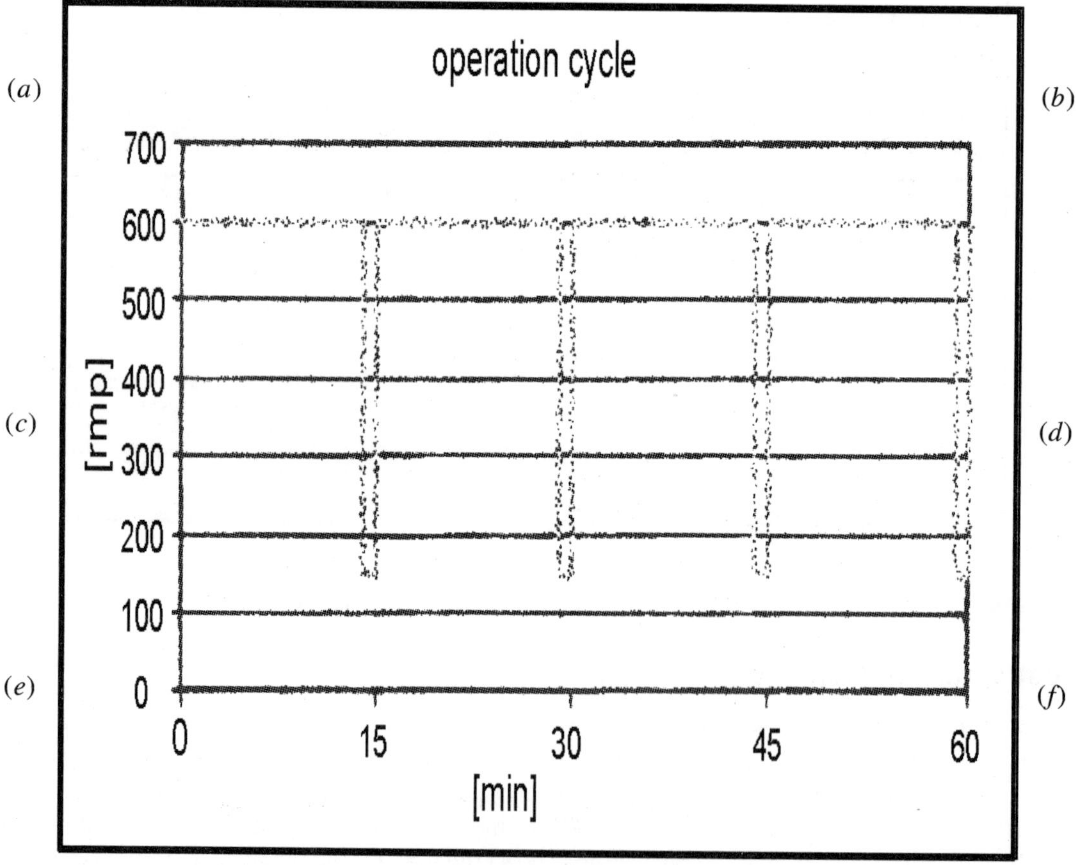

Fig. 4. Microstructures of (a) B1 at 50x & (b) at 500x (c) B2 at 50x (d) B3 50x (e) T at 50x & (d) B3 at 1000x.

Flexural Strength of Composites

The as-VHP composites M1, M2, M3, B1, B2, B3 and T are tested for their flexural strength using 3 point bend test. All the samples have span length of 40 mm. Table 4 shows the flexural strength for different samples. Among the mono-sized samples M2 reinforced with 220 # SiC resulted higher strength than M1 and M3, even though its density is lower than M1. Generally in MMCs as the reinforcement particle size decreases the strength increases due to decreased inter- particle spacing in the composites reinforced with finer particles. In case of coarser reinforcements like 36 # SiC, the reinforcements naturally have inherent defects which are responsible for the lower strength of the corresponding composites. Due to these defects the composites fail prematurely before complete load transfer from matrix to the reinforcements. The observed lower strength in M1 is consistent with the reported values else where in case of MMCs with 10-20 vol.% of coarser SiC (12). In these composites the fracture occurs due to the fracture of reinforcement particle along the preexisting defects, while the interface is intact. The 3.3% porosity present in the M1 sample also might have contributed to some extent in lowering the strength.The lowest strength observed (122 MPa) in case of M3 is not inline with the above argument as both the factors *i.e.,* reinforcement particle size and inter particle spacing are not favourable to result in low strength material. The 12 μm SiC comes under fine reinforcements category and the resultant MMCs fail due to failure of interface. This should result in effective load transfer from matrix to reinforcement. Secondly interpaticle spacing is also smaller when compared to M1 and M2. Even though both the factors are favouring stronger composite, the M3 showed very low strength. This is primarily due to extensive porosity (8.9 %) present in the sample M3 and the reasons for this porosity are already discussed. The microcracks, present in particle free regions and the porosity present in uninfiltrated regions act as crack nucleation sites and the sample fails prematurely.

The bi-modal composites B1 and B2 and the tri-modal composite T exhibited the almost similar flexural strength levels (244, 253 and 255 MPa respectively). Where as the B3 sample exhibited exceptionally high strength value of 459 MPa. The strength values of B1, B2 and T are better than the mono-modal composite M1 and lower than M2 and lies in between. Even though these multi modal samples have better density than mono-modal composites their flexural strength values are at par with the mono-modal composites, with an exception of B3 composite. This is due to presence of 36 # SiC in all the composites B1, B2 and T. where as 36 # SiC is absent in B3 sample. As explained in case of fracture of MI, the 36 # SiC because of its large size possesses inherent defects like microcracks and is prone to cracking during processing also. In all the cases where 36 # SiC is present the fracture is a premature one and is due to the fracture of 36 # SiC particles. In these samples the particles fail prior to completion of load transfer through the interface. Where as in case of B3 the larger fraction of the reinforcement is 220 # SiC and they are free from defects and thereby resulting complete load transfer. Due to this reason the strength of the M2 was also higher than M1. More over the minor fraction of the reinforcement *i.e.,* 600 # SiC further strengthens the matrix (effectively 2124A1/30 % SiC) in which the 220 # SiC particles are embedded. Due to this the B3 samples have exhibited consistently higher strength value.

Table 4. Flexural strength for vacuum hot pressed composites

Sample	*M1*	*M2*	*M3*	*B1*	*B2*	*B3*	*T*
σ_{max} (MPa)	199–217	312–314	107–137	246–250	251–255	442–475	245–264
Av. σ_{max} (MPa)	208	313	122	248	253	458.5	254.5

All the fractured samples were studied under SEM and the typical fractographs are shown in Fig. 5. Fractograph at 30x indicates more or less uniform fracture surface with clearly revealing the 36 # SiC particles embedded in the well defined dimples. Particle fracture responsible for sample fracture is evident from Fig. 5 (a) viewed at 100x. The well-defined dimples have formed around the fine SiC particles, which have generated due to particle fracture during processing. Fractograph of M2 sample at low magnification (100 X) clearly revealed an uniform fracture surface with the 220 # SiC particles embedded in the well-defined dimples. Both the interfacial fracture and the particle fracture are evident in the Fig. 5 (b) viewed at 1000x. The interfacial fracture is responsible for higher strength in M2. However, the dimples are shallow. Fig. 5 (c) shows the fractograph of M3 sample at low magnification (30 X) indicating non-uniform fracture surface exhibiting porous regions as dark and rest as bright. The fracture surface also exhibited extensive level difference. When observed at 1000 X (Fig. 5d) the bright regions revealed purely interface failure along with shallow dimples. At same magnification the darker regions were found to be un-infiltrated regions where no evidence of matrix is found. In the fractograph also the particle are held together due to mechanical interlocking.

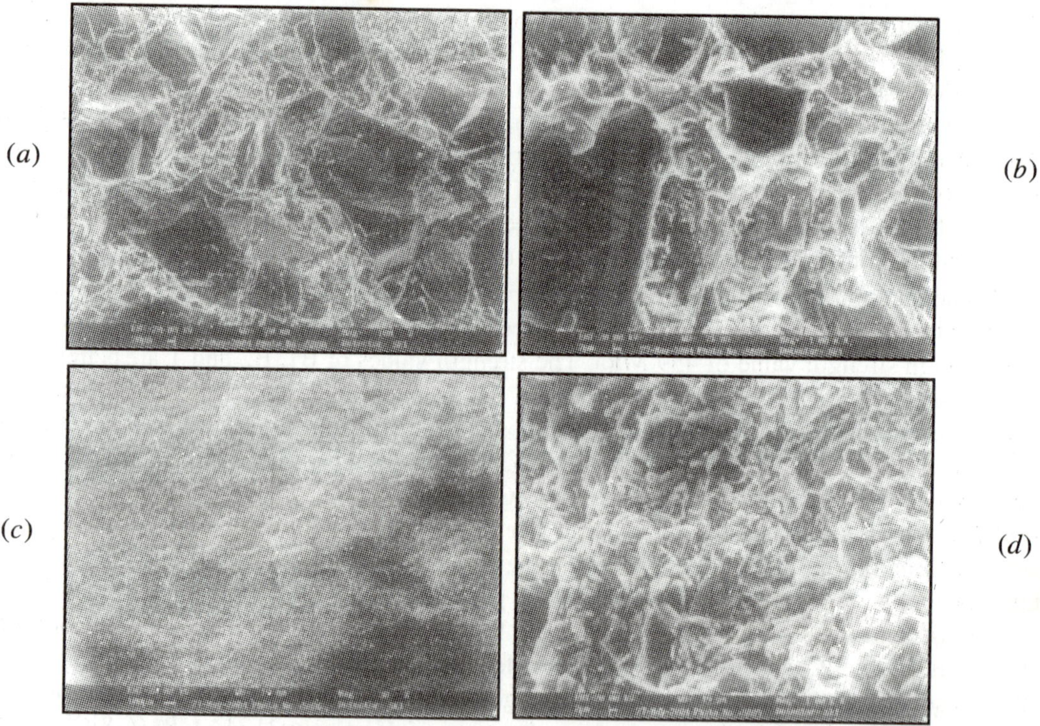

Fig. 5. Fractographs of (a) M1 at 100x, (b) M2 at 1000x, (c) M3 at 30x and (d) M3 at 1000x.

Fig. 6 a, b, c, d shows the fractographs of the broken bend test samples for the composites BI, B2, B3 and T. As explained earlier fracture of coarse 36 # SiC particles are evident in case of BI, B2 and T. The higher volume fractions of the exposed 36 # SiC fractured surfaces when compared to their original volume fraction in the respective composites is also an indication of the failure due to fracture of these particles. At observed magnifications involvement of the minor fraction reinforcements is not clear. Where as in case of B3 sample interfacial cracking at 220 # SiC and well-developed dimples around the fine 600 # SiC particles, indicating the overloaded fracture. The fractographic observations are in consistence with the property data obtained.

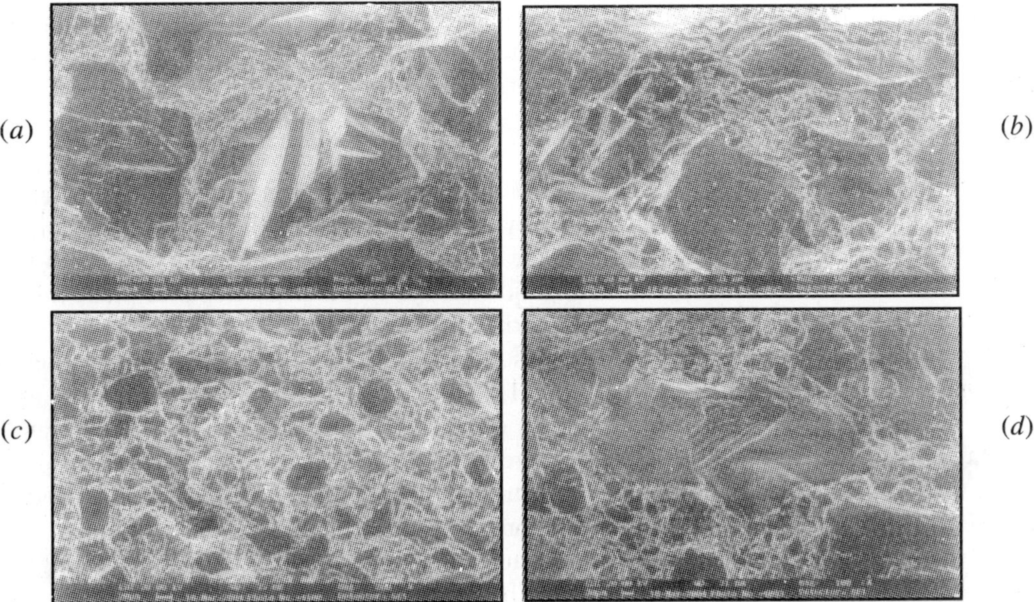

Fig. 6. Fractographs of (a) B1 at 100x, (b) B2 at 100x, (c) B3 at 200x and d) T at 100x.

Elastic Modulus

The elastic modulus values are measured by the Elasto-sonic testing equipment for sample B3. The average E value obtained for B3 is 219 GPa. This value is 212 % higher than that of the matrix Al alloy. The rise in the modulus is attributed to the higher modulus value of SiC reinforcement (480GPa). Effective load transfer from the matrix to reinforcement due to perfect interface is also responsible for the modulus enhancement in B3. The measured modulus value of 219 GPa lies in between the values calculated by Tsai -Halpin model (260GPa) and iso-stress model (175 GPa) (13). The improved modulus imparts necessary stiffness to the electronic packages even though the section thickness is reduced.

Table 5. Coefficient of Thermal Expansion of MMCs at different temperature intervals

Temperature intervals, °C	*CTE in * 10^{-6}/°C*			
	B1	*B2*	*B3*	*T*
30-100	10.06	9.28	7.86	9.00
30-200	10.10	9.41	9.41	9.59
30-300	10.70	10.00	10.20	10.27
30-400	10.78	10.01	10.68	10.37

Coefficient of Thermal Expansion (CTE)

CTE values of the packaging material should be closer to those of device materials such as GaAs and Si to minimise the thermal stresses and to enhance the life and performance of the chips. Table 5 shows the CTE values obtained for the samples B1, B2, B3 and T. In all the cases CTE is lowest at low temperature range *i.e.,* 30-100°C. As the range increases the CTE is also increasing. The CTE of the sample B3 is more dependent on the temperature range than that of the B1. It also can be noted that CTE of B3 is lowest in the temperature range of 30-100°C, which is practically significant range for packaging application.

Solution Treatment and Aging of B3

B3 sample was homogenised at 500°C, followed by solutionising and artificial ageing at 180°C for different lengths of time. Another sample was naturally aged for 30 days. All the samples were tested for their Vickers hardness at 30 kg load and the results were reported in Table 6. As the VHP was carried out at 570°C at which the alloy is in semi-solid state due to which Cu rich precipitate form. Due to the depletion of the Cu from solid solution the matrix is soft and as VHP hardness is low at 260 VHN, Upon homogenisation the Cu will get back into the solid solution and upon subsequent solutionisation solid solution super saturated with Cu will form. This will have higher hardness than the as-VHP sample. The peak hardness was obtained at 8 hours of ageing. However there are some serrations at the initial stages of ageing. The natural aged sample resulted similar hardness as in case of the peak age sample. The B3 sample was tested for the flexural strength in solution treated (ST), under aged (UA), peak aged (PA) and over aged (OA) conditions, along with the natural aged (NA) samples. And the results are given in Table 6 the solution treatment resulted in improvement of the average strength from 458 MPa to 493 MPa which can be attributed to the redissolution of alloying element in to the matrix. The quenched in dislocations formed at the interface due to CTE mismatch between the Al and SiC might have also contributed to the strengthening. The UA condition did not resulted higher strength and the trend is similar in case of hardness also. Where as the PA condition resulted in increasing the average strength from 458 MPa to 474 MPa this can be attributed to the precipitation of fine $CUAl_2$. Occurrence of peak ageing at 8 hours in composite is much faster when compared to 16 hours required in case of the matrix alloy. The faster kinetics can be attributed to the dislocation network created at the interface due to CTE mismatch. OA condition resulted in average strength drop to 414 MPa, which can be attributed to the coarsening of the precipitates. NA resulted in marginally higher strength than the PA sample.

Table 6. Vickkers Hardness values of as VHP, ST and aged B3 sample

Sample	*As VHP*	*ST*	*Aging time at 180C (hrs)*						*NA*
			2	*4*	*6*	*8*	*10*	*12*	
VHN	260	340	336	334	357	422	356	358	423
σ_{max} (MPa)	460	–	–	460	–	474	–	414	484

Dense Mono-modal M3

The mono-modal M3 has been processed at high enough (*i.e.,* 620°C) to get the 99 % TD. In as VHP condition the sample has yielded 570 M Pa flexural strength which is the highest value obtained in the present study. Fig. 7 (a) and (b) shows the micrograph and fractograph of the dense M3. Absence of porosity and decohesion at interface which are characteristic features of a strong composite can be observed from this figure. From the coarse Cu rich precipitates found in the microstructure it can be assumed that the strength value may further go up upon subsequent heat treatment. Studies are underway for evaluating fracture toughness, Elastic modulus, heat treatment and CTE. Even though, the fully dense M3 is a promising material with attractive strength values, the processing temperature is high for the H11 steel dies used for vacuum hot pressing the M3.

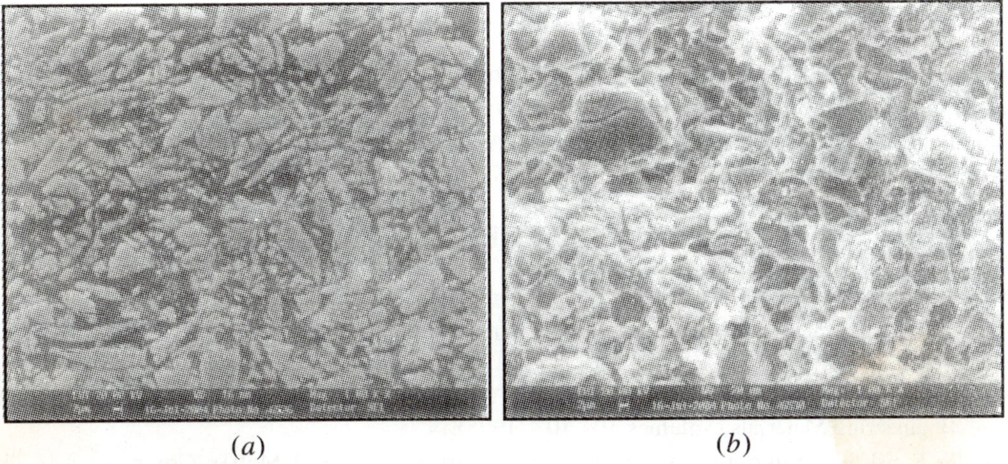

(*a*) (*b*)

Fig. 7. (a) Microstructure and (b) Fractograph of M3 sample hot pressed at 620°C.

4. CONCLUSIONS

1. 2124 Al/70% SiC composites have been synthesised by PM route using a two stage vacuum hot pressing technique. Hot pressing gives an edge over infiltrated electronic packages composites in terms of high strength and freedom in using different alloys of Al as matrix.
2. While mono-modal composites are reinforced with any one sized SiC *i.e.,* (36 #, 220 # or 600 #), bi-modal composites consists of combination of any two types of particles of SiC and trimodal composites consists of all the three types of SiC.
3. Mono-modal composites hot pressed at 570°C resulted in lower densities (91% TD in M3, 95% TD in M2 and 97% TD in M1). While lower densification in M1 is due to poor packing of reinforcements, lower densification in M3, reinforced with 600 # SiC, is attributed to inadequate infiltration of matrix alloy due to lower processing temperature.
4. The bi-modal and tri-modal composites (B1, B2, B3 and T) hot pressed at 580°C have been densified close to TD (>98 %) due to better packing of SiC in multi-modal composites.
5. The mono-modal composites resulted in lower bend strength due to improper densification. Where as, the multi-modal composites B1, B2 and T resulted in lower strength (245-264 MPa) due to presence of inherent defects in 36 # SiC. In as-VHP condition the maximum strength of 459 MPa obtained in B3 (reinforced with 220 # SiC and 600 # SiC).
6. Rise of VHP temperature to 620°C in case of modified M3 has resulted in higher density (99% TD) and highest strength of 570 MPa.
7. Multi-modal reinforcements bring down the required processing temperatures for complete densification by at least 50°C when compared with that in case of mono-modal composites.
8. The lowest CTE of 7.86×10^{-6}/°C in RT-100°C range makes B3 composite suitable for electronics packages applications as the obtained CTE is close to that of GaAs.
9. Elastic modulus of B3 composite (219 GPa) is three times to that of the matrix Al alloy.
10. Controlled CTE coupled with high modulus, low specific gravity and good thermal conductivity make this material suitable for electronic packaging applications.

ACKNOWLEDGEMENTS

The authors thankfully acknowledge the financial support received from DRDO, Govt. of India in order to carry out the present research study. We are grateful to Director, DMRL for his constant encouragement and permission to publish this work. Authors from NIT are thankful to Head, Metallurgical engineering Department, NIT, Warangal for permission to present this work.

REFERENCES

1. K.K. Chawla: Composite Materials, Springer-Verlag, NewYork, USA (1987).
2. W. S. Miller and F.J. Humphreys, Strengthening mechanisms in particulate metal matrix composites. Scripta Metall 25 (1991), pp. 33-38.
3. G. M. Newas, H. Neber-Aeschbacher, F. H. Wohlbier, : Metal Matrix Composites, Part 1 and 2, Key Engineering Materials, Volumes 104-107 (1995).
4. W.R. Mohan Vukobratovich, Recent applications of Metal Matrix Composites in Precision Instruments and Optical Systems, J. Mat. Eng., 10, (3) (1988), 225-235.
5. C. Zweben, Metal-Matrix Composites for Electronic Packaging, JOM, Vol. 44, No. 7, (1992), pp. 15-23
6. M. A. Occhionero, R.W. Adams, R.A. Hay and K.P. Fennessy: AlSiC for thermal management solutions and functional packaging Designs, Proceedings of the Annual IMAPS conference, San Diego CA, October, (1998)
7. M. K. Premkumar, W.H. Hunt, Jr. and R.R. Sawtell, Aluminium Composite Materials for Multichip Modules, J. Metals, 44 (7) (1982) 24-28.
8. Aghajanian, Keck, Rocazella: The fabrication of Metal Matrix Composites by pressureless infiltration technique, Journal of Material Science, 26, (1991), 447-454.
9. Vikram Jayaram, A.V. Pathak and K. J. Patel, : Fabrication of High Volume SiC Reinforced Aluminium Alloys for Microwave Integrated Circuit Substrate, Space technology cell, May (2002).
10. D.J. Lloyd, Particle reinforced Al and Mg matrix composites. Int. Mater. Rev. 39 (1994), pp. 1-23.
11. Asish Rao Narkale: Synthesis and characterization of 2124Al/70% SiC MMes for electronic packages, M.Tech Thesis, NIT, Warangal (2004).
12. V.V. Bhanu Prasad: Powder Metallurgy Processing of Silicon Carbide Particulate Reinforced Aluminium Matrix Composites, Ph.D. Thesis, IIT-Bombay (1998).
13. J.C. Halpin and S.W. Tsai: Environmental Factor Estimation in Composites Materials Design, AFML TR 67-423 (1967).

Unit–V

Powder, Processing and Applications

1

Milling Map for Mechanical Activated in-situ Synthesis of $MoSi_2$ from ($Mo + Si_3N_4$) Powder Mixture

Sheela Singh
Department of Metallurgical and Materials Engineering, Indian Institute of Technology, Kharagpur-721 302, India

R.V. Krishnarao
Defence Metallurgical Research Laboratory, Kanchanbagh, Hyderabad-500058, India

B.S. Murty
Department of Metallurgical and Materials Engineering, Indian Institute of Technology, Chennai, 600 036, India

M.M. Godkhindi
Department of Metallurgical and Materials Engineering, Indian Institute of Technology, Kharagpur-721 302, India

ABSTRACT

Attempts have been made to develop the milling map for mechanical activated *in situ* synthesis of $MoSi_2$ during the production of Si_3N_4 or $MoSi_2$ based composites. Efforts have been made to study the effect of milling energy and grinding media materials on the mechanical activation of the process of $MoSi_2$ synthesis through a reaction between molybdenum and silicon nitride. Powder mixtures of molybdenum (Mo) and silicon nitride (Si_3N_4) in the molar ratios of 1:3, 1:2 and 1:1 were ball milled using WC, steel, and ZrO_2 grinding media for mechanical activation. In order to evaluate the results obtained after pyrolysis of these milled powders, milling parameters has been converted to energy parameters viz. impact energy of the ball and total energy of milling. The optimum impact energy required for mechanical activation of ($Mo + x\ Si_3N_4)_x$ = 3, 2, 1 powder mixtures was found to be (0. 145-0.173J) leading to a reduction of pyrolysis temperature by (100-200°C). Samples milled with both higher and lower impact energy than the optimum show the presence of phases like $CoSi_2$, $ZrSi_2$, Fe_3Si and Fe_5Si_3 without any significant reduction in pyrolysis temperature required for $MoSi_2$ synthesis. Milling for prolonged period more than Shours does not further led to reduction in pyrolysis temperature but can led to the production of ultrafine $MoSi_2$ particle.

Key-words: Mechanical alloying, Nano particles, Pyrolysis, Intermetallic alloys, Composites.

1. INTRODUCTION

The properties, which have made $MoSi_2$ a well known structural material, are low density (6.31 $g.cm^{-3}$), high melting point (2030°C) and excellent oxidation resistance at elevated temperatures [1]. It ($MoSi_2$) however suffers from certain limitations such as room temperature brittleness, low high temperature strength and pesting (400-600°C). $MoSi_2$ is thermodynamically stable with respect to structural ceramics such as SiC and Si_3N_4 at elevated temperatures [2, 3, and 4]. The addition of Si_3N_4 to $MoSi_2$ improves low temperature fracture toughness and high temperature strength as well as eliminates pesting [5]. In turn the additions of $MoSi_2$ to Si_3N_4 have a potential to produce composites with improved fracture toughness, oxidation resistance and easy machining [6]. The most recent applications of $MoSi_2$-Si_3N_4 composites are ceramic glow plug in diesel engine [7].

These composite materials can be manufactured in several ways such as conventional hot pressing or hot isostatic pressing, reaction bonding, infiltration and combustion synthesis [8]. The preparation of particulate composites by conventional powder processing tends to generate large agglomerates and hence limits the degree of microstructural homogeneity. High reaction temperatures and longer holding time during sintering lead to the formation of larger particle sizes [9]. The presence of unreacted/unbonded reactants limits reaction bonding and melt infiltration process. Production of composites by the in situ synthesis methods could reduce cost and process steps in addition to providing fine size reinforcements [10]. Major advantages of the *in situ* processing are good/unique micro structural controllability, the short processing time, the decrement of processing temperature and the cost reduction. Composites containing *in situ* $MoSi_2$ in Si_3N_4 matrix can be produced by reacting ($Mo+Si_3N_4$) powders [11, 12].

The recently developed mechanically activated synthesis (MAS) process has often been found to decrease the time and temperature of the reaction, and could generate nanostructure at low energy consumption. The MAS also could increase the production rate in comparison to other techniques [13]. There are several reports on the production of $MoSi_2$ by mechanical alloying (MA) [14-16]. A new approach of $MoSi_2$ synthesis by mechanical activated annealing process (MA2P) was proposed by Gaffet et al. [17], which combines short duration ball milling followed by low temperature isothermal annealing. Gras et al. [18] proposed a mechanically activated self-propagating high temperature synthesis (MASHS) process, which included high-energy ball milling followed by self-sustaining reaction. They applied this method for the production of $MoSi_2$ using elemental powder mixture of (Mo + Si). MAS process is suitable for reactions that have a positive free energy [9] of formation at ambient temperature. One of the advantages of the MAS process is shorter milling time than that required for the *in situ* reactions in high-energy reaction milling. However, there are several factors such as rotation speed, ball-powder weight ratio (BPR), milling ball medium (Grinding medium) and milling model, which influence the milling process. It is however uncertain whether milling conditions have any influence on the mechanism of synthesis [19], though one should expect the ball milling energy should basically enhance the rate of synthesis.

There have been several reports on evaluation of the energy transferred during milling [for example Burgio et al. [20,] Magini et al [21] and Murty et al. [221], for various specific systems. Zhang et al. [19] developed the milling energy map of Mo-Si system. The present paper attempts to investigate the effect of milling energy and grinding media on mechanical activated synthesis of $MoSi_2$ using ($Mo + Si_3N_4$) powders in different molar ratios. For precise discussion, details regarding $MoSN_2$ composition is only furnished in the present study.

2. MATERIALS

Sinterable grade molybdenum (1-3 μm) powder obtained from former Soviet Union was used in the present study. The EDS analysis of the powder shows it to contain 99.9% Mo and 0.1% oxygen. Si_3N_4 powder (<1.5 μm) grade Silconide U95, having ($\beta/\alpha + \beta$) < 3 wt% was procured from Permascand AB, Sweden.

3. EXPERIMENTAL DETAILS

Mixtures of powders of Mo and Si_3N_4 in the molar ratio of 1:3, 1:2, and 1:1 ($MoSN_3$, $MoSN_2$ and MOSNI respectively) were chosen to produce Si_3N_4 *in situ* $MoSi_2$ composites. Powder mixtures were milled with ZrO and WC grinding media in a Fritsch (P5) planetary ball mill. Milling with steel grinding media was done in an Intel planetary ball mill. At a speed of 300 rpm. milling was carried out for various lengths of time (10-100h). The details of the milling conditions are described in Table 1. The angular velocity of the vial incases of Fritsch and an lntel mill was 351 rpm. Internal and external dimensions (diameter and height) in mm of WC, ZrO_2 and steel vials were (75, 70; 100,83), (100, 73; 130, 85), and (100, 74; 128, 100) respectively. Ball to powder ratios (BPR) used in this study were 10:1 and 15:1. A toluene bath was used for milling. To evaluate milling energy, the linear velocity of balls was calculated according to the procedure given by Burgio et al. [23]. The samples after every 5 h of milling were taken out and analyzed by X-ray diffraction (XRD).

Table 1. Experimental conditions during Milling of ($Mo+Si_3N_4$) powders.

Composition	*Sample Identity*	*Milling intensity (rpm)*	*No. of Balls*	*BPR*	*Wt of each ball (g)*	*Dia of Ball (mm)*	*Milling media*	*Milling hours (h)*
	M1	300	50	15:1	5.685	8.14	WC	0–70
$MoSN_1$	M2	300	50	10:1	5.61	8.1 1	Wc	0–30
	M3	300	100	15:1	2.25	8.0	ZrO_2	0–30
	M4	300	50	15:1	3.323	8.0	Steel	0–30
$MoSN_2$	M5	300	50	15:1	7.668	9.0	WC	0–100
	M6	300	50	15:1	5.569	8.08	WC	0–10
	M7	300	50	10:1	6.084	8.33	WC	0–30
	M8	300	50	15:1	6.62	8.56	WC	0–100
	M9	300	50	10:1	5.54	8.07	WC	0–30
$MoSN_3$	M10	300	100	15:1	2.88	8.50	ZrO_2	0–30
	M11	300	50	15:1	3.367	8.03	Steel	0–30

The unmilled samples and samples obtained after every 10h of milling were compacted in a cylindrical steel die of 10 mm diameter at a uniaxial pressure of 598 MPa for 60 seconds. The compacts were subjected to pyrolysis under argon atmosphere at different temperatures (1200-1600°C) in a horizontal tube furnace (Carbolite, U.K. equipped with a Eurotherm 818P temperature controller). A beating rate of 20°C min-1 in the temperature range of (1000-1600°C), and holding time of 0.5-2h followed by furnace cooling were employed. The pyrolised samples were analysed through a Philips diffractometer (PW 1840) using CoKα radiation with an iron filter was used. The microstructure of the samples was studied by a JEOL JSM 5800 scanning electron microscope (SEM). Energy dispersive X-ray microanalysis (EDS) of samples was carried out to estimate the phase composition.

4. RESULTS

4.1. XRD Analysis

Figs. 1 (a and b) shows the XRD patterns of $MoSN_2$ samples as a function of milling time, for BPR of 10: 1 and 15:1 with WC grinding media. The intensifies of peaks of Mo and Si_3N_4 decreased to a great extent after milling for 10h. A peak of WC also appeared after 10h of milling. Its intensity increased with increase in Si_3N_4 content ($MoSN_3$). Further increase in milling time causes a decrease in peak intensity and an increase in peak width of Mo and Si_3N_4. After 30h of milling Mo reduced to fine (nano) size and its peak disappeared from the XRD pattern for $MoSN_2$ (BPR-10:1).At higher BPR (15:1) Mo peak was still observed after 50 hours. The difference in Mo peak behavior could be attributed to the difference in milling energy and composition (Table 2) and will discussed later. But some trace of Mo peak and a clear peak of Mo were still observed in $MoSN_3$ and in $MOSN_1$ after 30h incase of samples milled with BPR 10:1 and after 60 hours incase of samples milled with BPR of 15: 1.

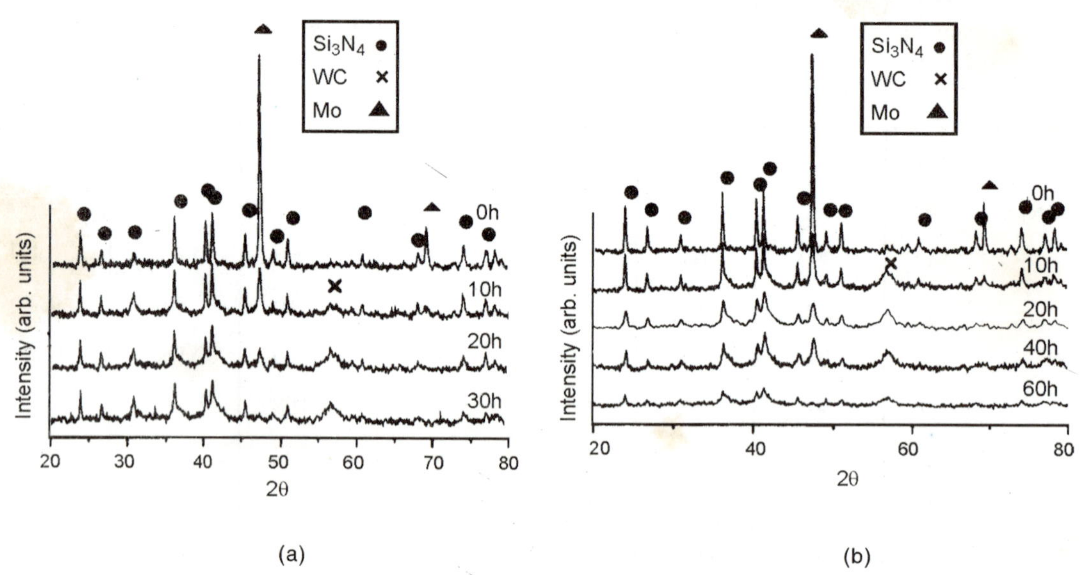

Fig. 1. shows the XRD patterns of $MoSN_2$ samples as a function of milling time, for BPR (a) 10:1 and (b) 15:1 with WC grinding media.

In Fig. 2(a) the effect of steel balls (grinding media) with BPR = 15:1 on XRD pattern of $MoSN_2$ is shown. Decrease in peak intensity of Mo and Si_3N_4 Was observed with increase in milling period (10-30h). However a high amount of iron contamination was also observed. After 30h of milling iron appeared as the major phase with low intensity peaks of Mo and Si_3N_4. Similar results were obtained for $MoSN_3$ and $MOSN_1$. But the iron contamination was relatively low. In Figure 2(b) the effect Of ZrO_2 balls (grinding media) with BPR = 15:1 on XRD patterns $MoSN_2$ is shown. Peak intensities of Si_3N_4 and Mo were found to decrease with increase in milling period. With 10h of milling a large decrease in the intensity of Mo peak was observed. After 30h of milling the peak of Mo completely disappeared. Large amount of ZrO_2 (T) contamination was observed within 10h of milling. Similar results were obtained for $MoSN_2$ and $MOSN_1$ samples, only the level of ZrO_2 (T) pick up in the samples decreases with decrease in Si_3N_4 content in the reactant. Abrasive wear of grinding media due to high wear resistant Si_3N_4 could lead to high contamination of iron and ZrO_2.

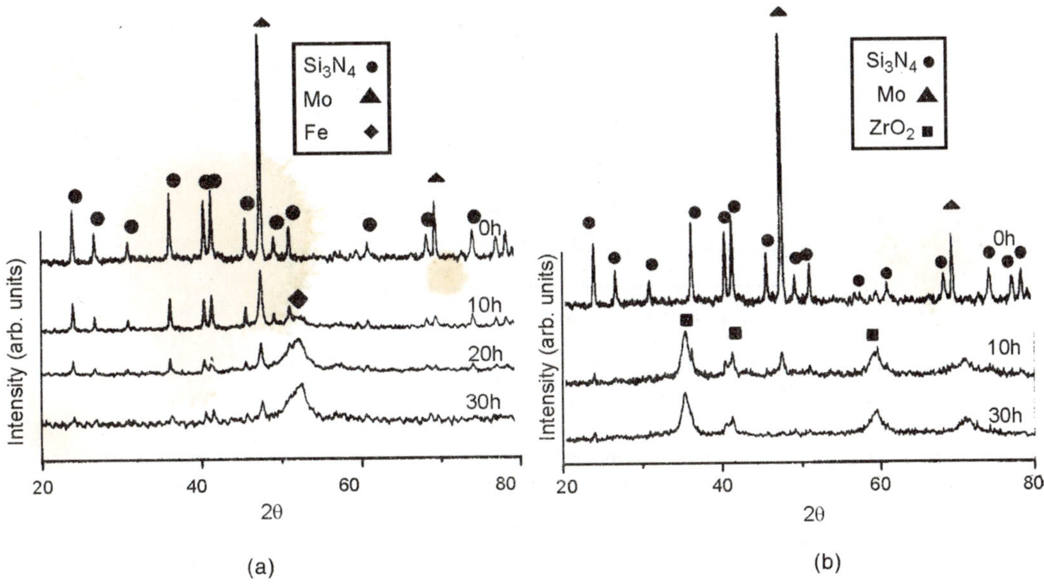

Fig. 2. Plot of XRD patterns-milling time of $MoSN_2$ samples milled with (a) steel balls and (b) ZrO_2 balls with BPR of 15 : 1.

4.2. Effect of Pyrolysis

Fig. 3 shows the XRD spectra of 100h milled and pyrolysed $MoSN_2$ samples (with WC balls, BPR= 15: 1) as a function of temperature. Peaks of molybdenum were absent in case of 100h asmilled $MoSN_2$ samples. Milled samples after pyrolysis at 1200°C show the presence of Mo, Si_3N_4 and Mo_5Si_3 Phases. With increase in temperature to 1400°C, Mo_5Si_3 was observed to be the major phase along with the presence of phases of Si_3N_4 and minor amounts of SiC. With further increase in temperature to 1500°C, $MoSi_2$ (T) becomes the major phase along with presence of minor phase of $MoSi_2$ (H), Si_3N_4. $MoSi_2$ (H) is not a stable phase at higher temperatures and was absent for the samples pyrolised at 1550°C. A new phase corresponding to $CoSi_2$ was observed at 1550°C and phases observed at this temperature were only $MoSi_2$ (T) and $CoSi_2$. At this temperature intensities of peaks of $MoSi_2$ reached a maximum. Further increase in temperature to 1600°C causes a decrease in $MoSi_2$ intensity. $CoSi_2$ remains stable up to a temperature of 1600°C. Complete conversion to $MoSi_2$ (T) without any phases of lower silicides occurs at 1550°C. In the unmilled samples complete formation of $MoSi_2$ could be achieved at 1600°C. Thus the mechanical activation of $MoSN_2$ sample decreased the pyrolysis temperature required to form complete $MoSi_2$ by about 50°C. Reduction in pyrolysis temperature observed after pyrolysis of $MoSN_1$, $MoSN_2$ and $MoSN_3$ samples were milled with different impact energy is shown in Table-2. $MoSN_2$ samples milled with ZrO_2 and steel shows the formation of $ZrSi_2$ and $FeSi_2$ because of heavy contamination pick up in the samples.

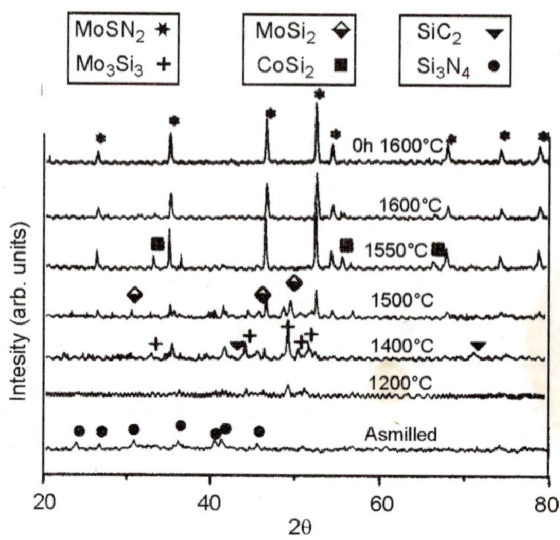

Fig. 3. XRD patterns - pyrolysis temperature of $MoSN_2$ samples milled for 100h with WC grinding media at ball to powder ratio 15:1.

5. DISCUSSION

5.1. Impact Energy

All the above results can best be understood if one considers the energy transfer during milling. The energy of each ball during the milling process is given by

$$E_b = \frac{1}{2}(m_b v_b^2) \qquad (iii)$$

Where E_b the energy of each ball is, m_b. is the mass of one ball v_b is the linear or absolute velocity of the balls. v_b can be calculated following the process given by Burgio et al. [23]. The impact energy of each ball can be given as $E_b' = \Phi_b E_b$, where Φ_b is the correlation factor for different degree of the filling of the vial. The total energy transferred per unit weight of powders for a given time can be expressed as

$$E_t = \frac{\nu E_b' n_b t}{W_p} \qquad ...(iv)$$

where E_t = the total energy transferred per unit weight of powders, n_b = No. of balls, t = Milling time (hours), W_p = Weight of powders. The frequency of impact (ν) according to Burgio et al. [23] is given by

$$\nu = \frac{K(\omega_p - \omega_v)}{2\pi} \qquad ...(v)$$

K is the proportionality constant and approximately equal to 1, ω_p=Speed of disc, ω_v =Speed of vial. All the milling parameters are accounted for, in two parameters *i.e.* the impact energy of each ball (E_b')

and the total energy transferred (E_t) to the powder after certain duration of milling. These parameters for different BPR and grinding media have been evaluated and shown in Table 2. In the table the impact energy of the ball E'_b and total b energy E_t transferred to the samples help in further understanding of the results obtained during milling and pyrolysis of the milled samples.

Table 2. Effect of milling parameters on mechanical activation of (Mo+Si_3N_4) powders milled at 300 r.p.m. with WC grinding media.

Compo sition	BPR	Sample name	E'_b (J)	Milling time (h)	Mo peaks			Reduction pyrolysis (h)	Product phases upon pyrolysis
					30h	60h	70h		
$MoSN_1$	15: 1	M 1	0.149	10–70	p	p	p	100	$MoSi_2$, Si_3N_4
	10:1	M2	0.147	10–30	p	–	–	100	$MoSi_2$, Si_3N_4, $MoSN_2$
$MoSN_2$	15: 1	M5	0.200	10–100	p	NP	NP	50	$MoSi_2$, Si_3N_4, $CoSi_2$
	15: 1	M6	0.146	10	–	–	–	100	$MoSi_2$,Si_3N_4
	10:1	M7	0.159	10–30	NP			100	$MoSi_2$, Si_3N_4
$MoSN_3$	15: 1	M8	0.173	10–100	p	P	NP	200	$MoSi_2$, Si_3N_4, SiC
	10:1	M9	0.145	10–30	PLD	–	–	100	$MoSi_2$, Si_3N_4

*P = Present, NP = Not present, PLD = Present but less distinct.

5.2. Milling Map

Fig. 4(a) shows impact energy of the samples milled with different grinding media as a function of milling time. It can be observed that the samples milled with ZrO_2 and steel have the lowest impact energy and an abrasive wear of grinding media occurs during milling due to the presence of Si_3N_4. During milling of samples with low impact energy the contamination pick up in the sample is sufficient enough to dilute molybdenum peaks and molybdenum peaks. In ZrO_2 milled sample (M3 sample). The peaks of molybdenum were not observed and were not due to mechanical activation but due to enormous contamination pick up in the samples. So milling with lower impact energy of balls (0.067-0.10J) does not seems beneficial.

The samples milled with WC grinding media having impact energy of (0.145-0.2J) undergo mechanical activation and show reduction in the temperature of pyrolysis required to form complete $MoSi_2$. The region of mechanical activation is shown by dotted contour in Fig. 4(b). The effect of milling parameters on mechanical activation is shown in Table 2. For mechanical activation, milling with impact energy of ball in the range of 0.145-0.173J appears beneficial both in terms of product quality and pyrolysis temperature. The degree of mechanical activation depends on impact energy of milling. Samples milled with lower impact energy of 0.145-0.16J have led to the reduction in temperature of 100°C and product phases are $MoSi_2$ and Si_3N_4. Samples milled with impact energy of 0.173 J led to the reduction in pyrolysis temperature of 200°C, but an additional phase of SiC was observed. Sample milled with higher impact energy of 0.2 J have shown a large amount of contamination and only 50°C decrease in the pyrolysis temperature. For a given total energy, for mechanical activation of Mo + Si_3N_4 powder mixtures samples milled with higher impact energy seems beneficial.

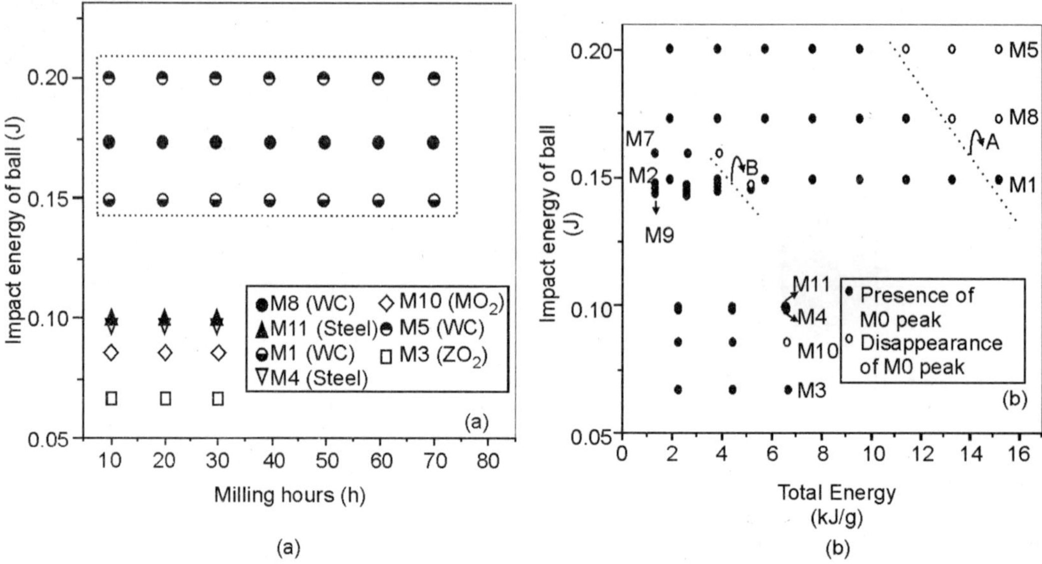

Fig. 4. Impact energy of the (Mo+Si_3N_4) samples milled with different grinding media using BPR 15:1 with (a) milling time (b) total energy. (For sample details refer table-1). [Region A: BPR 15:1, Region B: BPR 10:1]

6. CONCLUSIONS

Considerable degree of mechanical activation could be achieved in (Mo+Si_3N_4) powders of different compositions (Mo: Si_3N_4 = 1: 1/2/3) through milling with WC grinding medium. Use of steel or zirconia grinding media, however fails to achieve any activation. There appears an optimum impact energy range (in the present case 0.145-0.173J), which helps in achieving mechanical activation. With the impact energy lower than the optimum failing to generate sufficient stored energy, and energy larger than the optimum causing extensive wear of the grinding media. The subsequent high contamination has been observed to act as a diluent for the reaction. Further it has been observed that impact energy of a ball seems to be a more important parameter than the total energy transferred for the process of mechanical activation. A maximum of 200°C decrease in the synthesis temperature could be achieved in for $MoSN_3$ mixture with complete conversion to $MoSi_2$ and without any undesirable accretion to the product. A uniform distribution of submicron SiZe $MoSi_2$ particle was observed after pyrolysis of the mechanically activated powder mixtures.

REFERENCES

1. John J.Petrovic, Maria I.Pena, H Harriett Kung: Fabrication and microstructure of $MoSi_2$ reinforced Si_3N_4 matrix composites. J. Am. Ceram. Soc. 80 [5], 1111 (1997).
2. D.A Hardwick: Composites based on molybdenum disilicide: Progress and prospects. (Mater. Res. SOC. Symp. Proc. 350,), p. 165, 1994
3. Machester P.J., Schwartz S.S., JOM, 41, 52-55,1989

4. A.K.Vasudevan, and J.J Petrovic: A comparative overview of molybdenum disilicide composites. Mater. Sci. Eng. A-155, 1 (1992).
5. Tateoki Jizuka, Toshihiro Murao, Hiroyuki Yarnaimoto and Hiedeki Kita: Microstructure and properties of Mo_5Si_3 particle reinforced Si_3N_4 matrix composites. J. Am. Ceram. Soc. 85 (4), 954 (2002).
6. John J. Petrovic and A.K. Vasudevan: Key development in high temperature structural silicides, Mater. Sci. Engg A261, 1 (1999).
7. K. Yamada and N kamiya, High temperature mechanical properties of Si_3N_4-$MoSi_2$ and Si_3N_4-SiC composites with network structure of second phases, Mater. Sci. Eng A261, 270 (1999).
8. Nicholas J. Welham, Paul E. Willis and Tony Kerr: Mechanochemical formation of metal ceramic composites, J. Am. Ceram. Soc. 83 [1], 33 (2000).
9. Ruiming Ren, Zhenguo Yang and Leon L. Shaw, Synthesis of nanostructured sic through an integrated mechanical and thermal activation process. J. Am. Cerarn. Soc. 85 (4), 1 (2002).
10. Huang QiLiang, Cai Juan, Pan Wei, Chen Jian and Lian Jie: Insitu processing of Tin-Si_3N_4 composites by Ti- Si_3N_4 solid state reactions. Materials Letter 31, 221 (1997).
11. R.V. Krishnarao and J. Subrahmanyam: Formation of $MoSi_2$-Si_3N_4/SiC by reacting molybdenum with silicon nitride, Maters. & Manu. Process, 18 [4], 609 (2003).
12. R.V. Krishnarao and J. Subrahmanyam: Sintering of $MoSi_2$ by reacting (Mo+Si_3N_4) compacts, Mater. Sci. Eng. A 352 [1-2], 340 (2003).
13. Leon L. Shaw, Zhenguo Yang, Ruiming Ren: Dependance of SiC product morphologies on the mechanical activation, J. Am. Ceram. Soc. 85 (3), 1 (2002).
14. B.K. Yen, T. Aizawa, and J. Kihara: Synthesis and formation mechanism of molybdenum silicides by mechanical alloying. Mater. Sci. Eng. A 220, 8 (1996).
15. Pee-Yew Lec, Tzeng-Rong Chen, Ju-Lung Yang and Tsung-Shune Chin: Synthesis of $MoSi_2$ powder by mechanical alloying. Mater. Sci. Engg. A 1 92/193, 556 (1995).
16. B.B. Bokhonov, I.G. Konstanchuk and V.V. Boldyrev: Sequence of phase formation during mechanical alloying in the Mo-Si system. J. Alloys Comp. 218, 190 (1995).
17. E. Gaffet and N. Malhouroux-Gaffet: Nanocrystailine $MOSi_2$ phase formation induced by mechanically activated annealing. J. Alloys Comp. 205, 27 (1994).
18. Ch. Gras, D. Vrel, E. Gaffet and F. Bernard, Mechanical activation effect on the self-sustaining combustion reaction in the Mo-Si system. J. Alloys. Comp. 314, 240 (2001).
19. Houan Zhang and Xinyu Liu: Analysis of milling energy in synthesis and formation mechanism of molybdenum disilicides by mechanical alloying. International Journal Of Refractory Metals and Hard Materials 19, 203 (2001).
20. Burgio N., lasonna A., Magini M, Colloq Phys, 5 I, C4-265, 1990.
21. M. Magini and A. Lasonna: Energy transfer in mechanical alloying (overview). Mater trans JIM 36(2), 123 (1995).
22. B.S. Murty, M. Mohan Rao, S. Ranganathan: Milling Map and Amorphization during mechanical alloying. Acta Metall. Mater. 43 (6), 2443 (1995).
23. N. Burgio, A. Lasonna, M. Magini, S. Martelli and F. Padelia: Mechanical alloying of the Fe-Zr system: Correlation between input energy and end products, Nuoyo cimento 13, 459 (1991).

2

Processing and Characterization of Ruthenium Aluminide Alloys by Powder Metallurgy and Solidification Techniques

A. Srinivasan
Department of physics
Indian Institute of Technology,
Guwahati, India 781039.

Anil Borah
Department of Mechanical Engineering,
Indian Institute of Technology, Guwahati,
India 781039.

Ajit L. Mujumdar
Department of Mechanical Engineering,
Indian Institute of Technology, Guwahati,
India 781039.

P.S. Robi
Department of Mechanical Engineering,
Indian Institute of Technology, Guwahati,
India 781039.

ABSTRACT

RuAl alloys exhibit a combination of properties like high melting point (≈2050°C), high strength at room and elevated temperatures, high resistance to envirnment corrosion in some very severe chemical media as well as useful room temperature ductility and toughness. In spite of these promising properties, the processing of ruthenium aluminide still remains a challenge. Powder metallurgy (P/M technique is a viable method of processing these alloys so as to circumvent these drawbacks. The present paper discusses the processing of binary RuAl alloys and ternary Ru-Al-Ni alloys by P/M as well as solidification technique. The extent of mechanical alloying during attrition milling is characterized by X-ray diffractometer (XRD). The final microstructure obtained for these alloys processed by P/M technique is compared with that obtained by casting technique. Dense pellets with homogeneous structure were obtained for the P/M processed alloys after sintering. The cast and heat treated alloys showed segregation and non-homogeneous structure with a high amount of porosities. It is concluded that P/M technique is a better processing route compared to the solidification technique.

1. INTRODUCTION

Intermetallic materials like nickel and titanium aluminides are characterized by high elevated temperature strength, high oxidation and corrosion resistance, and high melting point. However, their applications for structural use are impeded due to their poor room temperature ductility and toughness. Recent studies revealed that intermetallic ruthenium aluminide (RuAl) alloys exhibit the unusual combination of very high melting point (≈2050°C), high strength at room and elevated temperatures, high resistance to environmental corrosion in some very severe chemical media, as well as useful room temperature ductility and toughness [1-5]. Though the properties obtained for ruthenium aluminide are promising for high temperature structural applications, the processing of these materials still remains a challenge. The high thermodynamic stability of RuAl imposes a number of restrictions for processing these alloys by conventional solidification routes [4-5]. Most of the refractory crucible linings tend to get attacked by the high temperature melt. Aluminum vaporizes at temperature above 1900 °C leading to high chemical inhomogeneity, deviation from the stoichiometric composition and porous material during casting. RuAl alloys can be processed by powder metallurgy (P/M) techniques and have the following advantages over the casting technique [6] (*i*) The ruthenium source material is converted directly from the powder, which is the primary product of the refining process. (*ii*) The intrinsic enthalpy of reaction is the thermally efficient driving force for densification, circumventing the high furnace temperatures necessary for melting or sintering. (*iii*) The material can be processed at low (sintering) temperatures, whereby volatilization of aluminum is prevented, resulting in a chemically homogeneous and porosity free material. (*iv*) The technique has the unique advantage of successfully processing several alloys and compounds, which cannot be prepared by conventional melt techniques, especially in systems containing elements with remarkably different melting points. (*v*) The material processed by P/M technique is free of coring and segregation which are observed in the products produced by melting technique. (*vi*) Grain size distribution is narrow in materials prepared by P/M route. (*viii*) Near net shaped components can be obtained ensuring minimum material wastage and circumventing the need of machining process.

Mechanical alloying process is a method for producing composite metal powders with controlled fine microstructure. This can be achieved by milling elemental powders necessary for the formation of the alloy under controlled atmosphere in an attrition mill. For proper mechanical alloying, it is essential to have minimum two types of powders viz. one ductile material and the other brittle material. In an attrition mill, due to the very high energy impact of the powder mixtures and hard balls, the ductile powder undergoes plastic deformation resulting in extensive work hardening. The brittle powder fragments into a fine powder and gets embedded into the ductile phase. The hardened ductile powders are repeatedly cold welded, fractured and re-welded leading to mechanical alloying [7]. It is imperative to strike a balance between the cold welding and fracturing processes in order to achieve efficient mechanical alloying. Attempts aimed on producing mechanically alloyed RuAl alloy powders have been reported in the literature [8-10]. Attempts were also carried out to characterize the structure of RuAl alloys processed by solidification techniques [1, 3-4, 11].

The high cost of ruthenium severely limits the use of these materials in structural applications. However, these can be used as a high temperature coating material for various applications such as turbine blades and shafts. One method of reducing the cost of these materials is by substituting ruthenium with elements like nickel, cobalt, etc. The mechanical properties of the material depend on the microstructure obtained and are having a strong influence on the processing route adopted. For justification of use of these materials in the above applications, a systematic processing followed by characterization of structure and properties is required. A comparison of the microstructure obtained for the RuAl alloys processed by different techniques is, therefore necessary. The present paper discusses the processing of binary RuAl alloys and ternary Ru-Al-Ni alloys by P/M as well as solidification technique. The extent of mechanical alloying during attrition milling is characterized by XRD and discussed. The final microstructure obtained for these alloys processed by P/M technique consisting of attrition milling, cold compaction followed by sintering is compared with that obtained by arc melting technique.

2. EXPERIMENTAL PROCEDURES

Elemental powders of Ru (purity of 99.4%, average size 4 μm), Al (purity of 99.9%, average size 7 μm) and Nickel with (purity of 99% and size ~10 μm) were used as the starting materials. The processing iof the materials was carried out by two techniques: (*i*) powder metallurgy technique, and (*ii*) solidification route. The elemental powders of stoichiometric compositions $Ru_{47}A1_{53}$, and $Ru_{38.5}Al_{16.5}Ni_{45}$ were used as the starting material.

Powder metallurgy processing was carried out by milling the elemental powders for 50 hours in an attrition mill using 1 wt% stearic acid ($C_{18}H_{36}O_2$) as the process control agent. Milled powders were taken out at regular time intervals and the structural analysis of the milled powder was carried out by XRD, using Cu Kα radiation (λ = 154Å) on SEIFORT XRD 3002 PJ/T powder X-ray diffractometer. The extent of alloying was monitored from the relative intensities of the different peaks from the XRD data. X-ray peak profile from a single reflection could be approximated to a Lorentzian curve. Multiple (upto 3) Lorentzians were used for analysing overlapping peaks. From the peak broadening data the particle sizes and strain of Ru, RuAl and Ni phases present in the as-milled powder samples were estimated from the full width at half maximum (FWHM) of the respective XRD peaks using the Williamson-Hall method [12]. This was followed by degassing, cold compaction of the milled powders at a pressure of 500 MPa to obtain 10 mm diameter X 10 mm height pellets, and then sintering at 1500°C for 24 hours. The details regarding the extent of mechanical alloying and the compaction procedure are explained elsewhere [8].

For processing by solidification route, the powders were mixed homogeneously in a mortar and pestle. To avoid the loose powder from flying off during the arc melting, the powder mixture was compacted at a pressure of 200 MPa in a die and punch assembly to obtain a cold compact rectangular rod of size 50 mm × 12 mm × 10 mm. This was kept in the water cooled copper mold of a laboratory level are furnace and evacuated to a pressure of 10^{-5} mbar. This was followed by purging with argon gas. Melting was done by striking the arc between the tungsten electrode and the cold compacted alloy. The arc was struck 2-3 times along the length of the sample to ensure proper melting of the entire rod. The arc was stopped and the sample was allowed to solidify in the water cooled copper furnace hearth. A part of the solidified alloy rod was annealed at 1400°C for 24 hours in an argon atmosphere to study the structural evolution taking place during the heat treatment.

The cast samples, cast and heat treated samples as well as the powder metallurgy processed samples were sliced using a precision saw and polished for microscopic observation following the standard Metallographic specimen preparation techniques. The microstructure of the samples were studied using a Leo make 1430 VP model scanning electron microscope (SEM) attached with OXFORD energy dispersive X-ray analysis (EDS) system.

3. RESULTS AND DISCUSSION

Figs. 1 (a) and (b) shows the XRD patterns of milled powders of Stoichiometric composition $Ru_{47}A1_{53}$ and $Ru_{38.5}Al_{16.5}Ni_{45}$ respectively. In the powders with $Ru_{47}A1_{53}$, no evidence of formation of RuAl phase was found up to 2 hours of milling. Peaks corresponding to Al were not seen after two hours of milling. Peaks corresponding to RuAl phase were observed after 7 hours of milling indicating the mechanical alloying. With further milling, peak broadening was observed for Ru and RuAl phases indicating a continuous size reduction and presence of residual stress in the milled powders. A detailed analysis of the size reduction and residual stresses in this material is reported elsewhere [8].

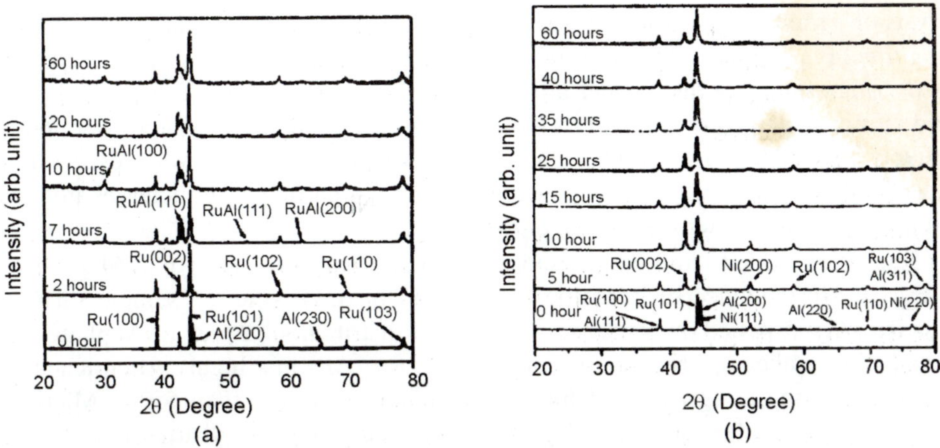

Fig. 1. XRD pattern at various milling times for (a) $Ru_{47}Al_{53}$ and (b) $Ru_{38.5}Al_{16.5}Ni_{45}$.

Fig. 1(b) shows XRD pattern for $Ru_{39.5}Al_{16.5}Ni_{45}$ system and does not reveal the formation of any binary or ternary phases during the milling process. Intensity of Al peaks disappeared after 5 hours of milling. Chakravorty and West [13] have reported that the continuous solid solution across the Ru-Al-Ni pseudo-binary range was interrupted by a miscibility gap. They proposed a constitutional diagram for Ru-Al-Ni system consisting of an extensive three phase field of β-RuAl, Ru solid solution and Ni solid solution (γ). The disappearance of the Al peaks with milling time might be an indication of the formation of a disordered γ-phase and Ru solid solution formation during the milling process.

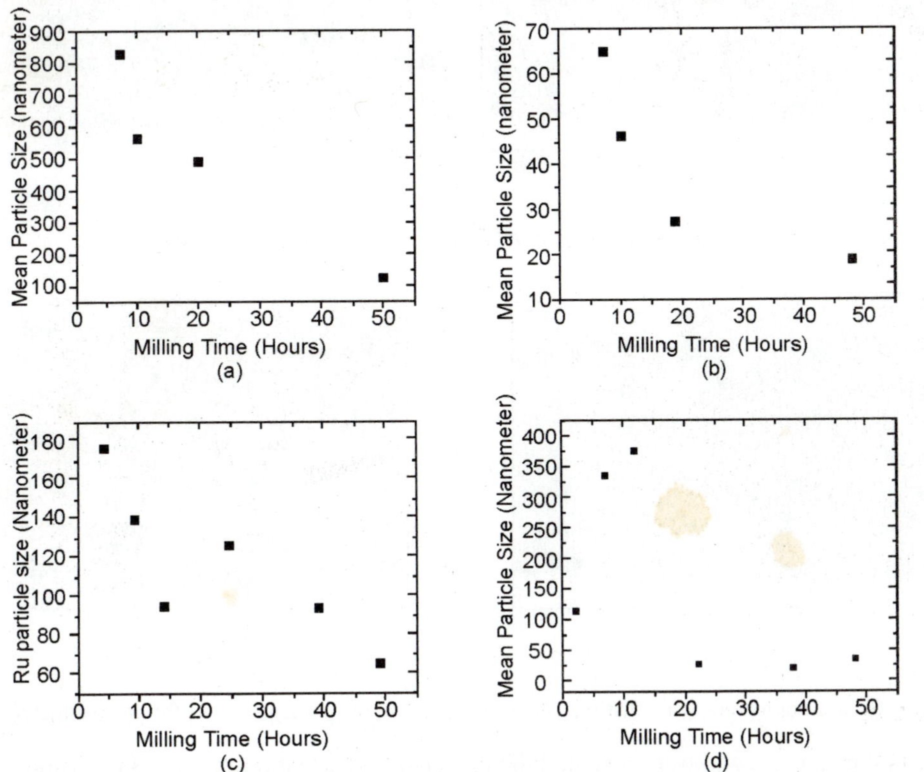

Fig. 2. Particle size variation with milling time for (a) Ru phase in $Ru_{47}Al_{53}$, (b) RuAl phase in $Ru_{47}Al_{53}$, (c) Ru in $Ru_{39.5}Al_{16.5}Ni_{45}$, and (d) Ni phase in $Ru_{39.5}Al_{16.5}Ni_{45}$.

Analysis of various peaks from the XRD data revealed continuous reduction in the powder size with milling time. Fig 2 shows the results of the crystallise sizes for various phases with milling time for the above compositions of the powder mixture. The sizes of Ru and RuAl in $Ru_{47}Al_{53}$ system were 120 nm and 17 nm respectively and that of Ru and Ni in $Ru_{39}.sAl_{16}.sN_4s$ system were 66 rim and 25 nm respectively after 50 hours of milling. The variation of Ru particle size with respect to the milling time is expected due to the brittle nature of Ru. However, Ni particle size increases initially up to 15 hours of milling time, thereafter decreasing rapidly.

SEM images of milled, cold compacted and sintered $Ru_{47}Al_{53}$ and $Ru_{39.5}Al_{16\cdot5}Ni_{45}$ alloys are shown in Fig. 3. No evidence of porosities was found in these sintered samples. The alloy with composition $Ru_{47}Al_{53}$ revealed the presence of a uniform distribution of grey RuAl phase (identified as $Ru_{56}Al_{44}$) and white ruthenium rich phase (identified as $Ru_{95}Al_5$) in a highly Al enriched matrix. The grain sizes of these phases (grey and white) were found to be less than 5 μm. Microstructure of sintered $Ru_{39.5}Al_{16.5}Ni_{45}$ alloy, shown in Fig. 3(b), revealed a uniform distribution of a white phase (identified as $Ru_{67}Ni_{33}$) and a black phase (identified as $Ru_3A1_{61}Ni_{34}Fe_2$).

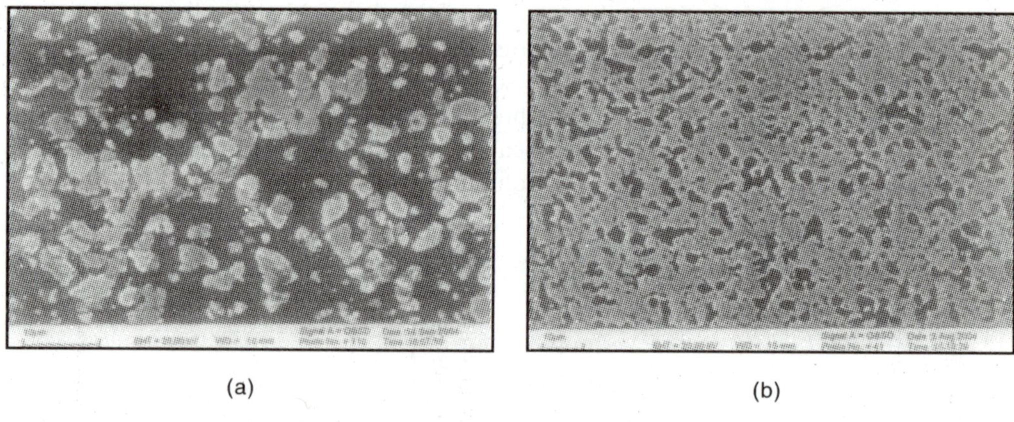

(a) (b)

Fig. 3. SEM images of sintered (a) $Ru_{47}Al_{53}$, and (b) $Ru_{39.5}Al_{16.5}N_{45}$

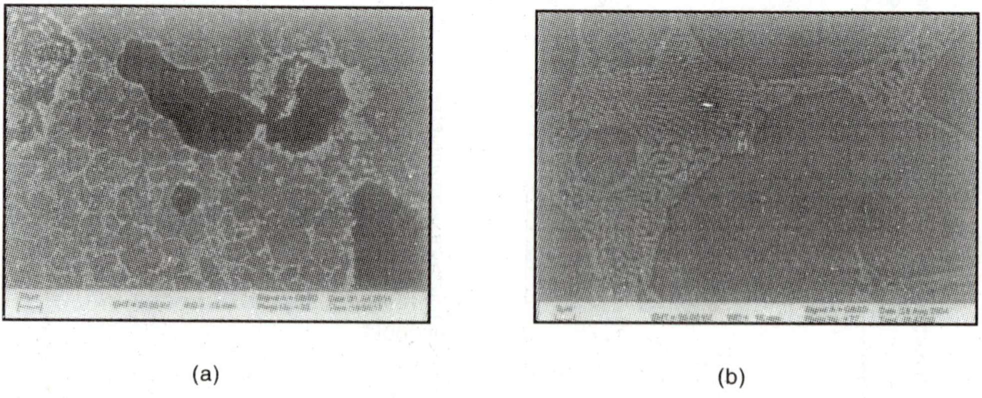

(a) (b)

Fig. 4. SEM p hotograph of as cast $Ru_{47}Al_{53}$ alloy showing (a) porosities, and (b) micro-structure.

Alloys in the as-cast condition, revealed the presence of large amount of gas porosities as well as rnicro-porosities. The gas porosities observed is due to the vaporization of aluminum due to the high temperature during the melting stage. The analysis of the alloy with a starting composition of $Ru_{47}Al_{53}$ revealed aluminum loss which might have occurred during the melting stage. The structure of this alloy shown in Fig. 4 revealed a continuous network of eutectic mixture ($Ru_{88}Al_{12}$ + $Ru_{63}Al_{37}$) around the pro-eutectic ($Ru_{56}Al_{44}$) phase. The microstructure of the cast alloy after heat treatment is shown in Fig 5. Fine needle shaped second phase precipitates were observed in the pro-eutectic RuAl phase. From this observation as well as from the binary phase diagram [14], it can be concluded that this alloy exhibits precipitation hardenable behavior.

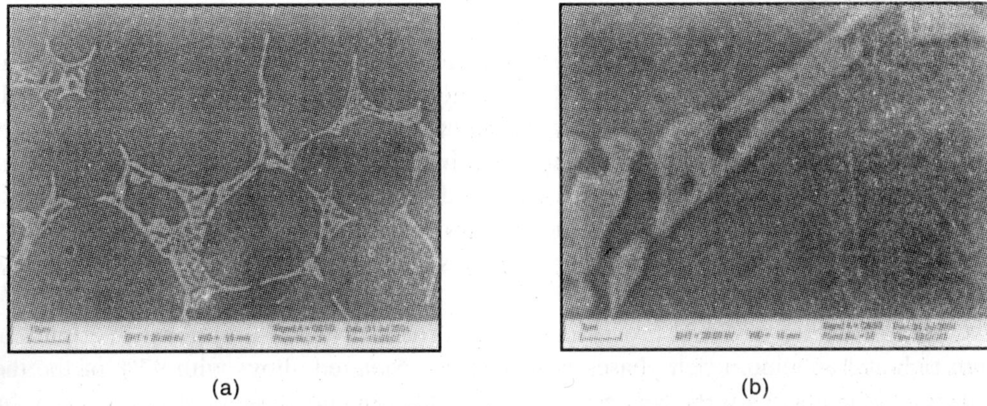

(a) (b)

Fig. 5. SEM photograph of as-east and heat treated $Ru_{47}Al_{53}$ alloy showing presence of fine needle shape second phase precipitates (a) at low magnification and (b) at high magnification.

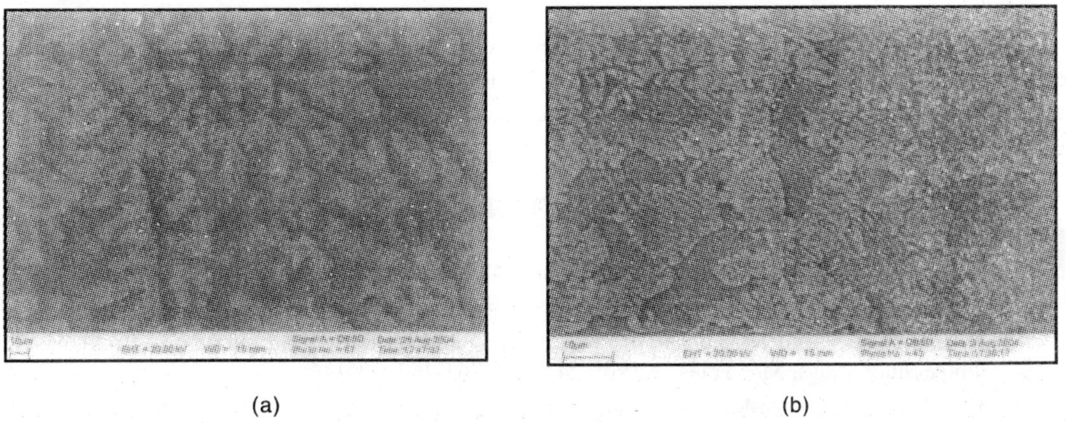

(a) (b)

Fig. 6. SEM micrographs of $Ru_{39\cdot5}Al_{16.5}Ni_{45}$ alloy (a) in as-east condition and (b) cast and annealed.

SEM photograph of as-cast $Ru_{39.5}Al_{16.5}N_{45}$ alloy is shown in Fig. 6(a). The cast structure revealed dendritic structure at low magnification. The micro-structure consists of two phases: a primary white phase (identified as $Ru_{66}Al_7Ni_{27}$) and a dark phase (identified as $Ru_{28}Al_{15}Ni_{57}$). Overall composition of the alloy was $Ru_{38}Al_{13}Ni_{48}$. The overall composition of the alloy showed a decrease in aluminium and increase nickel concentration. A structural evolution was found to occur when the cast alloy was annealed at 1400°C for 24 hrs, shown in Fig. 6 (b). During the annealing process, the morphology of both the phases changed to a laminar type structure. It appears that this morphological change occurring in these two phases is due to a diffusion assisted reaction. Due to the rapid cooling of the molten metal inside the water cooled copper mold, the solidification of the alloy occurs under a non equilibrium condition resulting in the two phases deviating from the equilibrium composition. During the annealing process, individual atoms diffuse to obtain equilibrium composition resulting in a laminar structure as shown Fig. 6 (b). It has been proposed by researchers that the development of laminar structure can lead to the improvement in the ductility and toughness of intermetallic alloys [15].

4. CONCLUSIONS

RuAl and Ru-Al-Ni alloys were prepared by two different routes, viz., solidification and P/M. Microstructural studies of the alloy prepared by casting revealed loss of aluminum due to volatilization resulting in gas porosities. Micro-porosities occurred due to shrinkage during solidification of the alloy. Microstructure of binary RuAl alloy consisted of primary dendrites of RuAl phase surrounded by a continuous network of eutectic lamellae. Heat treated cast samples showed a precipitation reaction resulting in needle shaped precipitates in the primary RuAl phase. Ternary Ru-Al-Ni alloy showed dendritic structure consisting of two phases, which resulted in a laminar type of structure during the annealing process.

The microstructure of RuAl processed by P/M route revealed a uniform distribution of fine ruthenium rich and aluminum rich phases in the matrix. Sintered alloys with 97% of the theoretical density devoid of porosities were obtained. Absence of a eutectic network structure, very fine grain size, and uniform distribution of phases are the advantages of the PIM processed alloy. These observations suggest that the P/M processing of RuAl alloys is superior compared to the solidification technique.

ACKNOWLEDGEMENTS

The authors acknowledge Indian Space Research organization (ISRO) for the funding of this work as per the sanction No. 9/1/(1)/99-II. The authors also thank Mr. K. Susheelan Nair, Vikram Sarabai Space Centre, Thiruvananthapuran, India, for his valuable suggestions and technical support during the course of this project.

REFERENCES

1. R.L. Fleischer, R.D. Field and C.L. Briant, Mechanical properties of high temperature alloys of RuAl, Met. Trans. A, Vol. 22A, pp. 404-414, (1991)
2. R.L. Ficischer and McKee D.W., Mechanical and oxidation properties of RuAl based high temperature alloys, Met Trans. A, Vol. 24A, pp. 763-769, (1993).
3. I.M. Wolff and G. Sauthoff, Mechanical properties of Ru-Ni-Al alloys, Metall. Mater. *Trans.* A, Vol. 27A, pp. 1395-1400, (1996).
4. I.M. Wolff and G. Sauthoff, Role of intergranular phase in RuAl with substitutional additions, Acta Mater., Vol. 45(7), pp. 2949-2969, (1997)
5. I.M. Wolff, Towards a better understanding of ruthenium alunlinide, JOM, pp. 34-39, (1997).
6. I.M. Wolff, Synthesis of RuAl by reactive powder processing, *Metall. Mater. Trans*., A, Vol. 27A, pp. 3688-3699, (1995).

7. L. Lu and M.O. Lai, Mechanical Alloying, Kluwer Academics, Boston, (1998).

8. Anil Bora et. at., Powder metallurgy processing of ruthenium aluminium alloys, J Material Processing Technology, Vol. 153-154, pp. 952-957, (2004).

9. K.W. Liu and F. Mucklich, Thermal stability of nano-RuAl produced by mechanical alloying, Acta Mater., Vol. 49, pp. 395-403, (2001).

10. K.W. Liu and F. Mucklich High-temperature stability of nano-grained B2-structured RuAl-based intermetailics by mechanical alloying, Intermetallics, In Press, Available online, (2004).

11. N. Ilic' et.al., Properties of eutectic Ru-Al alloy produced by ingot metallurgy, Materials Science and Engineering A, Vol. 329-331, pp. 38-44, (2002).

12. Yu. Rosenberg, V. Sh. Machavariani, A Voronel, S. Garber, A. Rubshtein, A. I. Frenkel and E A Stern, Strain energy density in the X-ray powder difftaction from mixed crystals and alloys, J Phys.: Condens. Matter, Vol. 12, pp. 8081-8088, (2000).

13. S. Chakravorty and D.R.F. West, The constitution of the Ni-Al-Ru System, J. Mater. Sci., Vol. 21, pp. 2721-2730, (1986).

14. T. B. Massalski (ed.), Binary Alloy Phase Diagrams, ASM International, USA, 2nd edition, pp. 204, (1990).

15. I.M. Wolff and G. Sauthoff, High-Temperature behavior of precious metal base composites, Metall. Trans., Vol. 22A, pp. 2642-2652, (1996).

3

Effects of Precursors and Calcination Temperatures on the Morphology of ZnO Nanopowders

Ashish Anand
Department of Metallurgical Engineering and Materials Science, IIT Bombay, Powai, Mumbai 400076, India

Om Prakash
Department of Metallurgical Engineering and Materials Science, IIT Bombay, Powai, Mumbai 400076, India

T. K. Gundu Rao
SAIF, IIT Bombay, Powai, Mumbai 400076, India

ABSTRACT

Zinc oxide powders, as aggregates of nano- and sub-micron particles, were synthesized using/ (or preparing) various precursors (mostly the metal-organic chelates viz. Zn-acetate, Zn-citrate, Zn-oxalate), which on decomposition in air between 350 and 800°C yielded ~15 to ~90 nm size crystallites with very narrow size distribution. Crystallite size progressively grew with increasing calcination temperature. The particle size (as assembly of nanocrystallites) and shape were influenced by the choice of Zn-precursor. With decreasing crystallite size, the lattice (hexagonal, $P6_3$) volume marginally decreased (<1%) to that of bulk lattice. Soft agglomerates of ZnO as observed by SEM and TEM warrant further work with appropriate dispersants surfactants to realize free-flowing nano particles.

INTRODUCTION

Nanopowders (*i.e.,* constituent powder-particles controlled to nanocrystalline size between (5 nm and 100 nm) are being actively considered (from applications point of view) for their enhanced properties arising from large specific surface area, increased defects density and metastable characteristics. Zinc oxide (Wurtzite structure), as a fine chemical, in nanopowder form is almost set to find many applications in electronic industry, bio- and environment friendly technologies [1-4]. In order to realize custom make ZnO powders, with appropriate chemical, electrical, and/or optical properties specific to the end applications, controlled morphology, nature and extent of agglomeration, as well as chemical purity are important parameters that need to be considered. Presently, various shapes of ZnO particles *viz.* spherical, prismatic, ellipsoidal, bi-pyramidal and nano-rod have been reported in literature [2,4-6].

Nanopowders of zinc oxide have been prepared by various methods using different precursors, which, on calcination, often, give rise to different powder morphologies. The shape, size, increased surface area and powder activity can be gainfully exploited for varied applications. In the present paper, syntheses of nano-size zinc oxide powders from decomposition of different, mostly chelate, Zn-precursors are reported. The effects of an organic dispersing medium (ethylene glycol-EG) have also been looked into. Dispersion of precursor in EG is expected to be effective in decreasing crystallite and particle size. The various ZnO powders, thus synthesized, were characterized by XRD, SEM and TEM techniques.

EXPERIMENTAL

Two approaches have been followed for the preparation of zinc oxide nanopowders viz. (*i*) various zinc oxide precursors were prepared and subsequently thermally decomposed, and (*ii*) zinc acetate was dispersed in EG and decomposed at different temperatures. In the first approach, zinc hydroxide, zinc citrate and zinc oxalate precursors were prepared from Zn- acetate using aqueous and organic solvents at appropriate reactor parameters [7], and subsequently decomposed in air at temperatures around 400°C (taking guidelines from our DSC experiments with precursors [7]). In the second approach, zinc acetate and EG, in the mol ratio of 1:4, were mixed thoroughly by sonication and subsequently calcined at different temperatures from 400 to 800°C to obtain different discrete size nano powders of ZnO.

Powder XRD patterns (using X'pert pro XRD unit), and micrographs using SEM (FEI Quanta-200) and TEM (Philips CM-200), of the calcined powders were collected to see the ZnO phase formation/ crystal structure, and view the powder morphology. ZnO crystallite sizes were computed from the XRD peaks using Scherrer equation. Here, however, no background corrections (expected to be < 2 nm) were made to the computed crystallite sizes.

RESULTS AND DISCUSSION

The XRD (Cu Kα) patterns of the ZnO powders, synthesized from decomposition, at 400°C, in air, of Zinc acetate samples: (*i*) with EG and (*ii*) without EG are shown in Fig. 1.

It is seen from Fig. 1 that EG dispersed zinc acetate, on decomposition, yields relatively smaller (34 nm) crystallite size ZnO powder compared to that without EG (60 nm). The effect of calcination temperature on the XRD peaks widths (see Fig. 2) reveals progressive increase in ZnO crystallite size with increasing temperature. The lattice parameters for ZnO are given in Table 1.

Table 1. Effects of calcination temperature on lattice parameters and lattice volume of ZnO (hexagonal, P6$_3$ mc), prepared by decomposition of EG dispersed Zn-acetate

Calcination Temp. (°C)	*Particle Size (nm)*	*Lattice Parameters (Å)*	*Lattice Vol. (Å)^3*
300	32	$a = 3.3186, c = 5.1764$	49.3705
400	34	$a = 3.2791, c = 5.1944$	48.3684
500	47	$a = 3.2749, c = 5.2014$	48.3338
600	60	$a = 3.2742, c = 5.1960$	48.2389
800	70	$a = 3.2699, c = 5.1933$	48.0873

For ZnO (hexagonal, P6$_3$ mc) bulk lattice a =3.2495 Å, c = 5.2069 Å, Lattice vol. 47.6149 (Å).

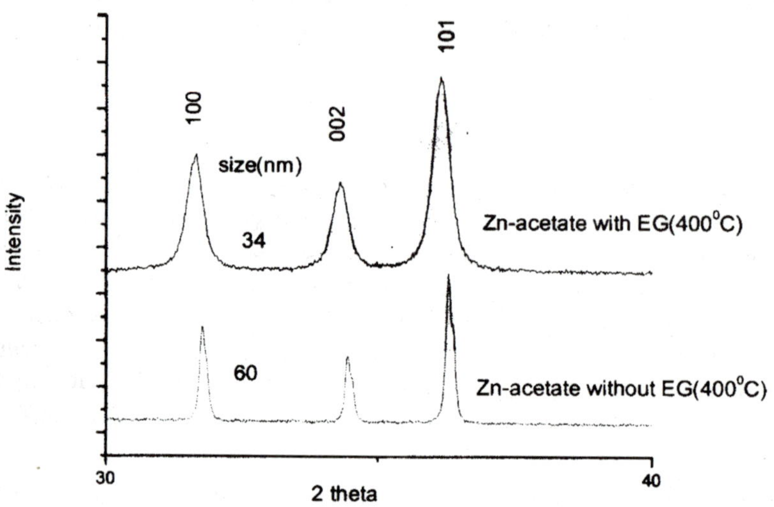

Fig. 1. XRD pattern of the ZnO powders showing the effect of sonically dispersed Zn acetate in EG on the peak widths and thereby the crystallite size of ZnO.

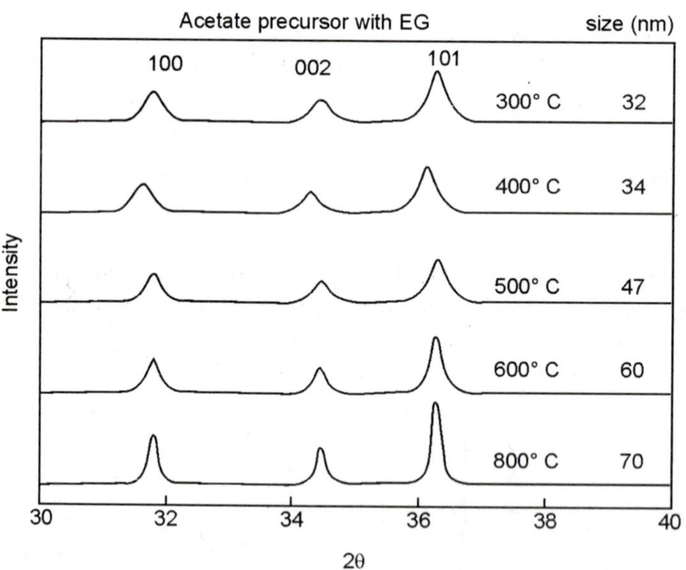

Fig. 2. XRD patterns of ZnO powders showing the effect of calcination temperature on the peak widths and, thereby, revealing a progressive increase in crystallise size with increasing temperature.

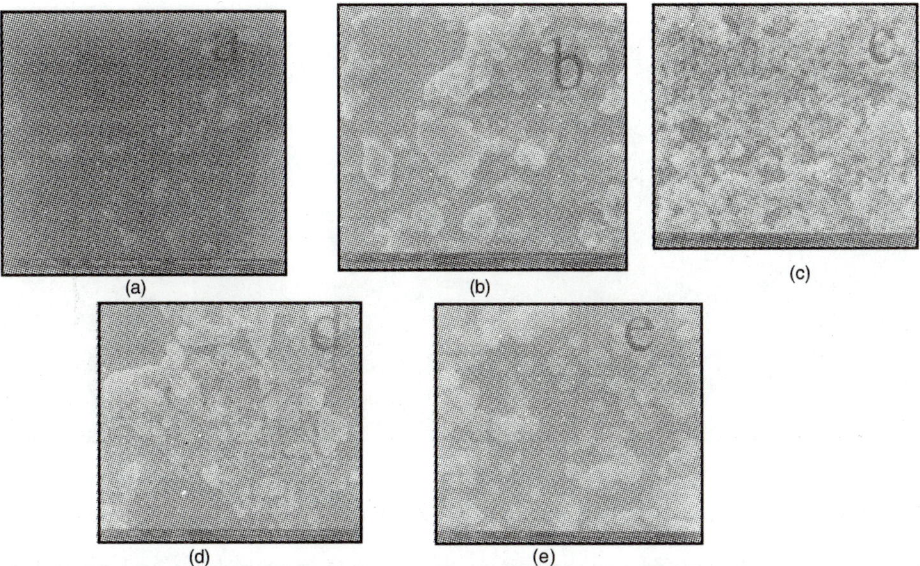

Fig. 3. SEM photo micrographs of ZnO powders prepared from Zn-acetate dispersed in EG medium and calcined at: (a) 300°C, (b) 400°C, (c) 500°C, (d) 600°C and (e) 800°C (Mag. 1 cm = 2 μm).

It is observed that even as crystallite size increases, with increasing calcination temperature, tendency of forming bigger agglomerates decreases as shown by the SEM micrographs (Fig. 3). This is attributed to development of well faceted crystallite at progressively higher temperatures. The average particle size (Fig. 3(e)) for the 800°C calcined ZnO powder is ~200 nm which is essentially an agglomerate of ~70 nm size crystallite (revealed by XRD peaks and TEM). The van der Waals forces at this length scale are quite effective due to considerable surface charge and defects. Hence, there is a need to develop appropriate surfactants to realize free flowing nanosize particles of the length scale of crystallite.

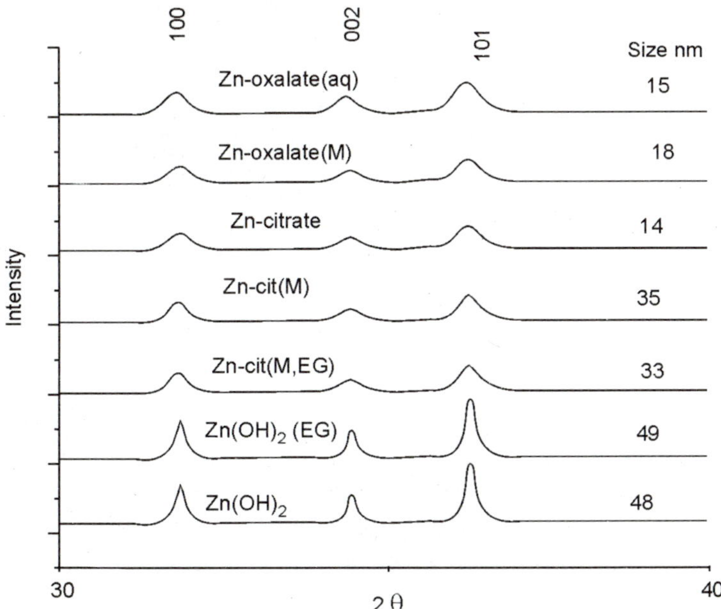

Fig. 4. XRD of ZnO obtained through calcination and effect of precursors on crystallise size.

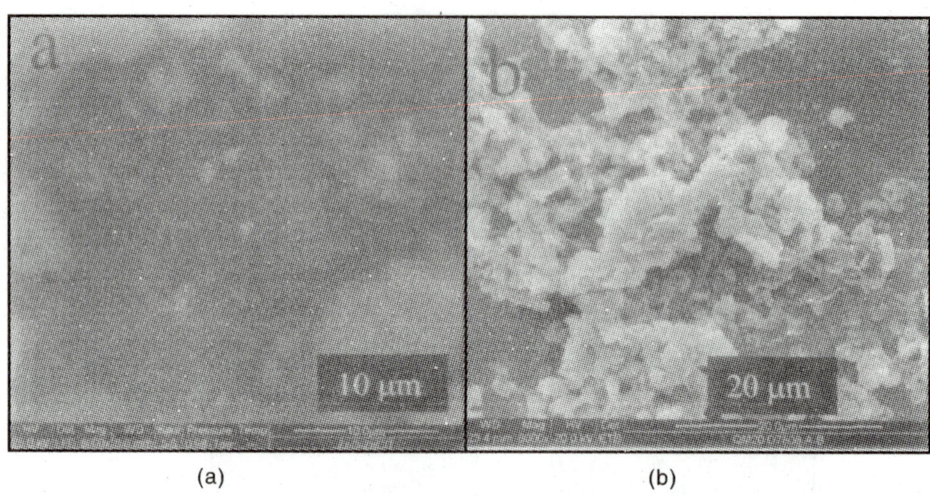

(a) (b)

Fig. 5. SEM micrographs of ZnO prepared form: (a) Zn-citrate and (b) Zn-oxalate, in aqueous medium.

The effects of different Zn-precursors on the particle morphology and crystallite size of the thermally decomposed zinc oxide powder are highlighted in Figs. 4-6. It is seen from XRD peaks/ widths (Fig. 4) that Zinc oxalate and Zinc citrate precursors yield finer ZnO crystallites (~14-18 nm) compared to that from $Zn(OH)_2$. TEM micrographs of these powders also validate this finding. SEM (Fig. 5) and TEM (Fig. 6) micrographs of a few ZnO powders clearly reveal that both the particles and the crystallites are markedly different in size for different precursors. Further, these may also depend upon reactor parameters [7].

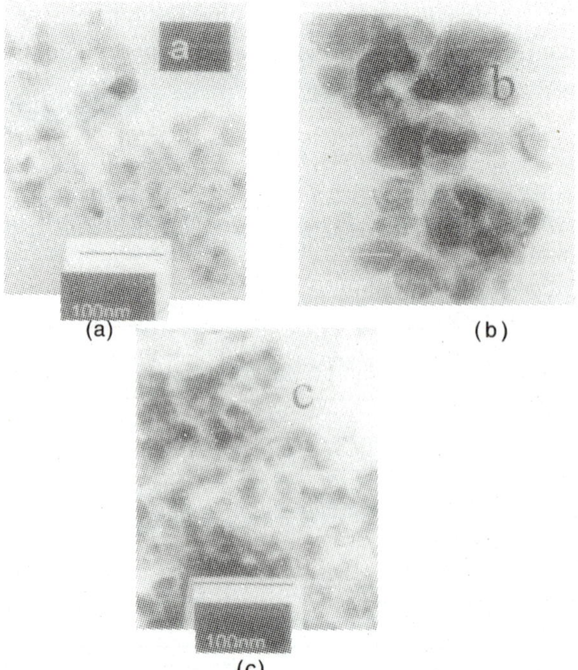

(a) (b)

(c)

Fig. 6. TEM micrographs of zinc oxide powders prepared from (a) Zn-citrate, (b) Zn-hydroxide and (c) Zn-oxalate.

CONCLUSIONS

Nanocrystalline (20-80 nm) zinc oxide powders were successfully synthesized from different ZnO precursor salts. The ZnO powders, due to invariable agglomeration, however, showed particles of different sizes ranging between ~100 nanometers to about a micron. Zn-acetate dispersed in ethylene glycol, on decomposition, produced ZnO powder consisting of smaller crystallite and particle sizes vis-a-vis EG free Zn-Acetate.

The formation of sub-micron/micron size agglomerations require further work with surfactants to realize free flowing nano particles of crystailite-length scale.

REFERENCES

1. Y. Jakewon, K.Y. Hun. Kim, Characterization of ZnO powder synthesized by the polymerized complex method via an organochemical route, Journal of Ceramic Research, Vol. 3, pp. 1 46-149, (2002).
2. C. Lu, C. Yen, Influence of hydrothermal conditions on morphology and particle size of zinc oxide, Ceramic.International, Vol. 26, pp. 351-357, (2000).
3. O. Yamamoto, Influence of particle size on antibacterial activity of zinc oxide, International Journal of Inorganic Materials, Vol. 3, pp. 643-646, (2001).
4. J.Q. Hu, X.L. Ma, Z.Y. Xie, N.B. Wong, C.S. Lee, S.T. Lee, Characterization of zinc oxide crystal whiskers grown by thermal evaporation, Chem. Phys. Lett. Vol. 344, pp. 97-100 (2001).
5. D. Jezequel, J. Guenot, N. Jouini and F. Fievet, Preparation and morphological characterization of fine, spherical, monodisperse particles of ZnO, Materials Science Forum, Vol. 152-53, pp. 339-342.
6. K. G. Kanade, B. B. Kale, A. V. Murugan and B. K. Das, Effect of solvents on synthesis of nanosize zinc oxide, Inorganic Materials, Editors: D. Bahadur, S. Vitta and Om Prakash, Narosa Publishing House, New Delhi, pp. 177-180, (2004).
7. Ashish Anand, Zinc oxide nanopowder synthesis and its characterization, Dual Degree Project Dissertation, IIT Mumbai, pp. 34-40, (2005).

4

Electroless Ni Plating of WC

Sükrü TALAS, Ayhan EROL,and Fatih ÇOLAK
Afyon Kocatepe University,Technical Education Faculty, Afyon,Turkey

ABSTRACT

In this study the production of ceramic metal composites using WC powders plated with Ni by electroless method were attempted and their properties were also investigated. Transient liquid phase sintering was used to produce high ceramic to metal ratio composite. Microstructural properties in specimens before and after sintering process were characterized using SEM, DTA-TG and Microhardness tests.

1. INTRODUCTION

Metal matrix composites are widely utilized in various industrial areas and also have potential applications in aerospace industry. The advantages of metal matrix composites over the conventional metals and their alloys are the superior mechanical, corrosion resistance, high strength-to-weight ratio and wear properties at higher operational temperatures [1-4]. Electroless metal coatings on ceramic powders [5,6] have been practically applied to produce intricate shapes from powder mixtures, allowing the use of high ceramic to metal ratio within the composite.

Ceramic reinforced composites were produced by which ceramic powders, WC, plated with metal, Ni, were sintered to make final product. Metal plating is applied to ceramic powders by means of electroless method and hardness tests are also presented.

2. EXPERIMENTAL PROCEDURES

Composites of Co, Cr, Al and Ni base metals reinforced with WC find commercial applications. For this reason Ni coating of WC powders by electroless method was used to assess the formability of final product. The plating method was easy to use as well as less costly. Water soluble Nickel Chloride was mixed with Hydrazine Hydrate heated to a temperature at which the reaction is maintainable. Following the electroless plating process various materials characterization techniques were used to define microstructure and phases formed during the process if exist.

Electroless plating of WC powders with Ni is accomplished in a bath of which its control is dependent on two main factors: bath temperature and pH. WC to Ni ratio was arranged to 7 to 3, therefore it is high in ceramics than metal components. The acidity of bath was kept at pH 9 which is of slightly acidic nature and operational temperature was kept at 95°C throughout specimen preparation. The adjustment of pH is done through ammonia addition at 95°C. Process parameters are given in Table 1.

Powders plated with Ni were compacted using a hydraulic press at a pressure of 140 Bar and sintered at temperatures of 1050°C, 1150°C and 1250°C with or without Ar gas shroud. Sintered specimens were characterized using Leo 1430VP scanning electron microscopy attached with RONTEC EDX detector at Afyon Kocatepe University, Technical Research and Development Center.

Table 1. Chemical bath components for electroless coating of WC with Ni

Components	*Additions*
Tungsten Carbide (WC)	14 (by weight)
$NiCl_2.6H_2O$ (NIckel chloride)	24 (by weight)
Hydrazine Hydrate ($N_2H_4.H_2O$)	30 vol.
Water	70 vol.

Hardness measurements were made using Shimadzu HMV 2L micro tester. DTA data was obtained in air up to 1150°C maximum temperature with 59 grs of Ni plated WC powders using LINSEIS DTA-TGL-81 equipment.

3. RESULTS AND DISCUSSION

SEM images of Ni plated WC particles are given in Fig. 1a and 1b. Figure shows EDS analysis of Ni plated WC powders, indicating that a shell of Ni exists around WC powders.

Fig. 1. (a) SEM image of Ni plated WC powders (b) EDS analysis of Ni plated WC powders.

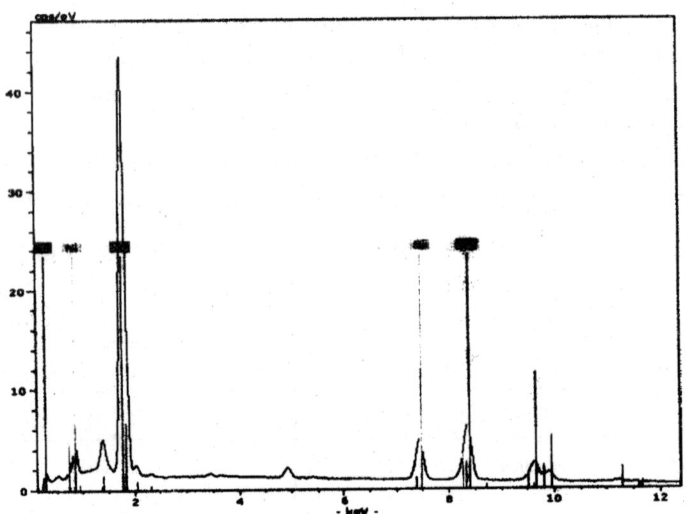

Fig. 1. (b) EDS analysis of Ni plated WC powders.

Average particle size of WC powders was measured to be 1.1 μm, however, average particle size of plated powders were 1.33 μm after electroless plating process. It can be concluded that 0.23 μm of Ni was deposited onto WC powders by electroless plating process. Such thickness of Ni may help form bonding during the sintering process. It is conclusive that plating is successful however DTA data (air, 1150°C max) of Ni plated WC powders shows a weight increase around 500°C due probably to fact that oxidation of Ni takes place. A complete reaction up to 1150°C is given in Fig. 2 below.

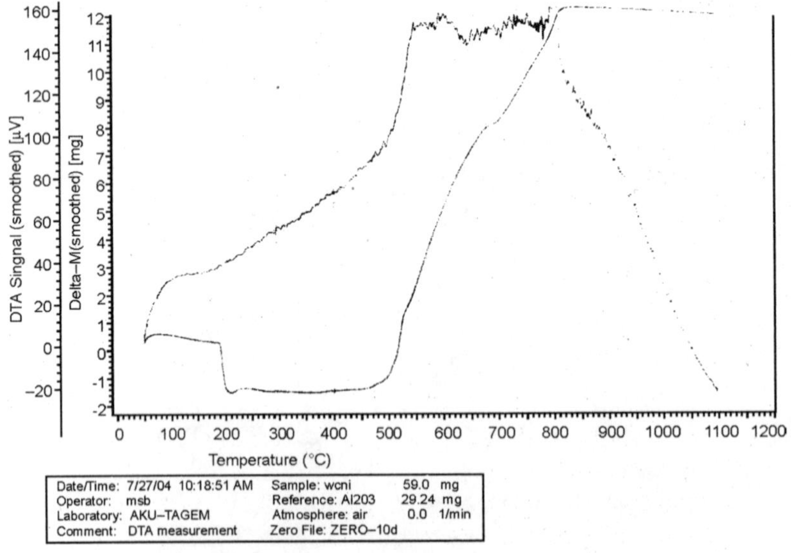

Fig. 2. DTA analysis results obtained from Ni plated WC powders.

According to these results an increase in weight on TG line at 500°C is due to oxidation of powders from air, which was expected on the course of heating. It is also possible to see exothermic reactions occurring in the temperature range 540°C and 860°C and some unexpected endothermic reactions were also detected. Between these two points it is possible to say that reactions lead to the formation of an intermediate phase of WC-Ni system, WC_xNi_y, which can be proposed from W-Ni phase diagram. It is however possible that some oxides of Ni may also be formed.

Using the data obtained from Literature [6] and DTA results above two sets of experiments were arrange three sintering experiments were carried out. First set of experiments were carried out in the air with three different temperatures of 1050°C, 1150°C and 1250°C and heating rate of 10°C per minute and 1 hour holding time. Second set of experiments were carried out under Ar atmosphere at temperatures of 1050°C and 1150°C (See Figs. 3 a and b).

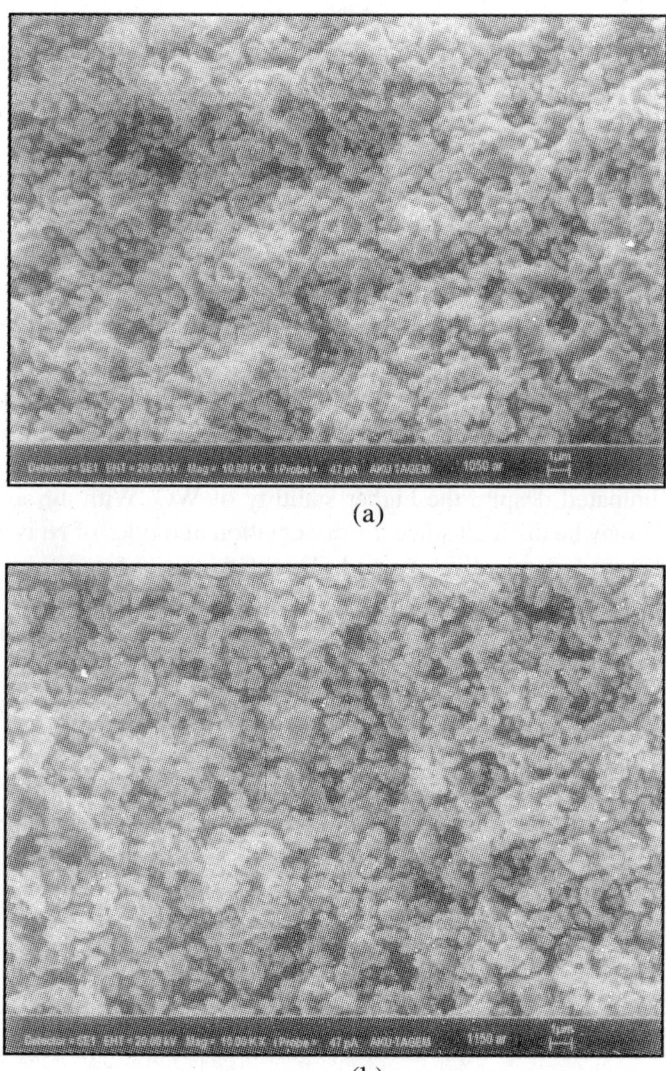

(a)

(b)

Fig. 3. SEM images of the specimens sintered at (a) 1050°C (b) 1150°C under Ar atmosphere (10.000X).

Results showed that, specimens sintered at temperatures of 1050°C and 1250°C did not produce good results however specimen treated at 1150°C was successfully sintered. Specimens treated under

Ar atmosphere were also successfully sintered. Temperature 1250°C was not attempted because this temperature is not feasible since one of the purposes of this work was to reduce the cost of application.

Electroless plating of WC powders were used by many workers however the bath composition for plating was different. For example, Sharma and co workers used attempted electroless Ni plating in a bath containing P which resulted in the formation of Ni_3P phase [5]. Considering the adverse effect of P compounds, the presence of Ni_3P may be undesired. However the use of $NiCl_2.6H_20$ and Hydrazine Hydrate ($N_2H_4.H_2O$) prevented undesired effects of newly formed compounds during the process of sintering by effectively forming Ni layer around WC particles. Such a layer is supposedly helped the formation of liquid phase which advanced by a diffusion driven capillary forces and sintered WC particles as a result of such action. Products of reactions are thought to be Ni derivatives with a substantial amount of solute C; Gidikova *et. al.* [7] proposed a W rich compound of Ni_2W_4C, which is probably the case in this study. EDX analysis indicates that the presence of such elements are confirmed in the structure.

From SEM images in Figure 3 a and b, it can be seen that in specimen treated at 1150°C under Ar atmosphere WC powders are in close contact with each other and formed a continuous agglomerated structure. This suggests that bonding between the plated particles occurred with a mechanism by which a liquid phase formed and advanced by capillary force. Particle size was found to be increased from 1.4 mm to 4 mm due to agglomeration. Sintering in air did not result in extensive oxidation degradation because the formation of Ni oxide may have prevented further diffusion of oxygen through the oxide layer formed on the WC plating.

However the degraded appearance of sintered Ni plates WC powders sintered in air given in Fig.4. also is subject of discussion. At such a high temperatures the formation of Ni oxides is highly possible according to Ellingham diagram [8]. The activity of C is also higher and therefore the formation of CO may not be eliminated despite the higher stability of WC. With air sintering experiments the thickening of Ni oxides may be difficult since the dissociation of oxides of Ni is easily attainable and the formation of CO is thermodynamically favoured. Therefore the O freed from the dissociation of Ni oxides may be taken by C and lead to the formation of CO and hence corroded WC and freed some Ni for sintering. It is evidential that there is no O peak observed in EDX analysis result of air sintered specimens.

Hardness measurement on sintered specimens were carried out following a metallurgical preparation of specimens mounted on bakelite and Vickers scale was used for hardness. The average of five measurements was taken for more accurate readings which are given in Table 2 below.

Table 2. Average micro hardness values of specimens sintered at various temperatures.

Sintering Conditions	*1050°C Argon Atmos.*	*1150°C Argon Atmos.*	*1150°C Air*
Hardness (HV)	113,6	159,8	239,8

Using the recipe described in experimental procedures section and data obtained from literature and DTA experiments temperatures of 1050°C, 1150°C and 1250°C were chosen for sintering for both air and Ar atmosphere. As a general view the formation strong and uniform bonds between the particles occurred at 1150°C for both atmospheres. An increase in hardness value when specimens treated in air is probably because of O forming hard phases of oxides with Ni.

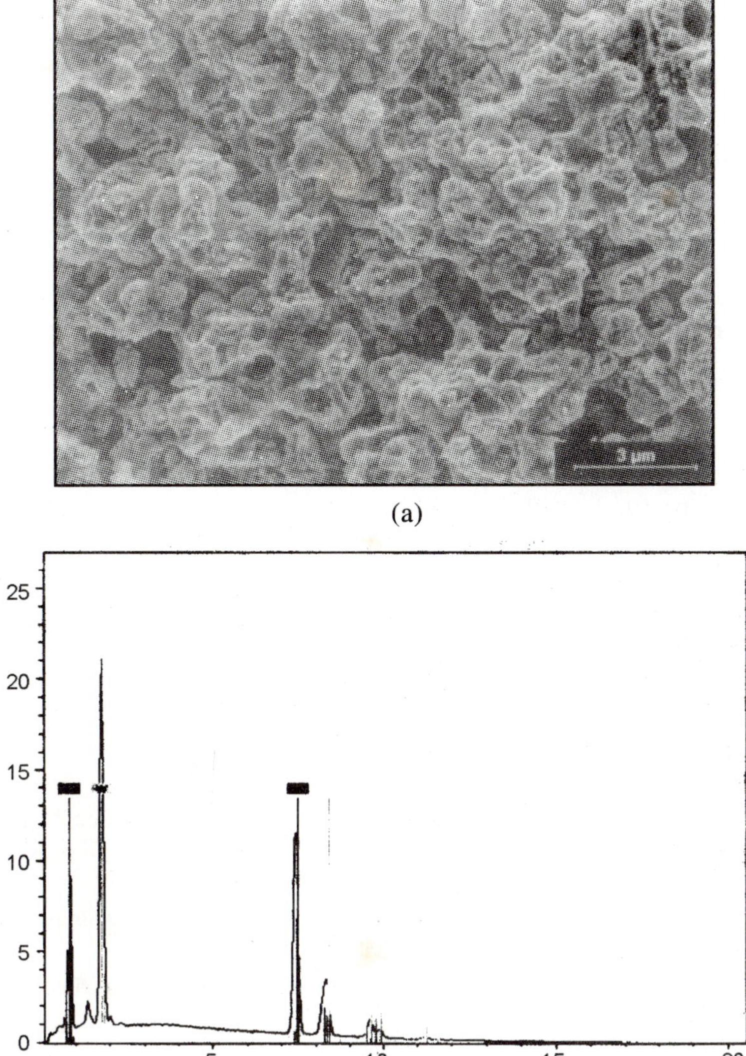

(a)

(b)

Fig. 4. SEM image and EDS analysis of specimens sintered at 11500°C in air.

4. CONCLUSIONS

Results suggest that electroless plating of WC powders are useful for plating Ni onto particles, producing high purity Ni layer. Therefore results below can be given:

- WC powders are successfully Ni plated by clectroless method.
- Optimal temperature for sintering is proposed to be 1150°C for both Ar atmosphere and air sintering.
- As a result of sintering the formation of an intermediate phase of Ni_2W_4C is proposed from DTA and EDS results.

- A decrease in hardness with respect to sintering temperature is due probably to the effect of Ni, forming agglomeration. This is also affected by the porous structure of composite since the deformation is easily attained through unconnected WC particles.
- The presence of Ni coating on WC powders enhanced the bonding strength between particles by forming intermetalics and uniformly distributing Ni layer between the particles of WC.

ACKNOWLEDGEMENT

This work was supported by Scientific Research Projects Unit (Research Project No. 031.TEF.30) and Institute for Graduate Studies of Pure and Applied Science of Afyon Kocatepe University.

REFERENCES

1. P. T. Wang, A constitutive model for consolidation of granular materials, Powder Technology, Vol. 54, Issue 2, pp. 107-118 (1988).
2. A. Erol, Ph. D Thesis, Bradford University, (2000).
3. A. Grosjean, M. Rezrazi, J. Takadoum and P. Berçot, Hardness, Friction and Wear Characteristics of Nickel-SiC Electroless Composite Deposits, Surface and Coatings Technology, Vol. 137, Issue 1, pp. 92-96 (2001).
4. Malecki, A. and Micek-Ilnicka, A., Electroless Nickel Plating From Acid Bath, Vol. 123, Issue 1, pp. 72- 77 (2000).
5. Sharma, A., K., Suresh, M., R., Bhojraj, H., Narayanamurthy, and H., Sahu, R., P., Electroless Nickle Plating of Magnesium Alloys, Metal Finishing, pp. 10-18 (1998).
6. M. Velez, H. Quinones, A.R. Di Giampaolo, J. Lira and I. C. Grigorescu, Electroless Ni-B coated WC and VC powders as precursors for liquid phase sintering, lnt. Jour. Ref. Met. Hard Mater., 17, pp. 99-102 (1999).
7. Gidikova, N., Simenova, Y., Assenova, E., Dannev, K., R., Proceedings of the Conference World Tribology Congress, London, 1997, p. 490.
8. Lars Erik Svensson, Control of Microstructures and Properties in Steel Arc Welds, CRC press, 1994 p. 123.

5

Nano and Fine Processing of High Temperature Ceramics and in situ Composites by SHS Process

S. K. Mishra

National Metallurgical Laboratory, Jamshedpur 831 007 (India).

ABSTRACT

Processes have been developed for the fabrication of various advanced high temperature monolithic and composite materials namely ZrB_2, TiC, ZrB_2-Al_2O_3, TiB_2, TiB_2-TiC, ZrB_2-TiC etc. by self-propagating high-temperature synthesis (SHS) technique. The studies are focused on the SHS processing, mechanism of phase formation, microstructure evolution, sintering studies of SHS produced powder and in-situ formation of advanced materials and composites. The SHS powder showed better sinterabilities compared to the conventionally prepared powder. SHS techniques were successfully used for the fabrication of in-situ composite powder and alsq attain high densification through SHS dynamic compaction. The present talk is focused towards monolithic borides and their in-situ composite with alumina.

Materials index: Zirconium diboride, Titanium dihoride, alumina.

Key words: SHS, High Temperature Material, Borides, Carbides, Composites.

1. INTRODUCTION

The self propagating high temperature synthesis (SHS) is a well-known method for the fabrication of several high temperature ceramics, intermetallic and nano-composite materials in a single-" of processing. The SHS technique has inherent advantages over the other methods that require high temperature furnaces and longer processing times. Materials produced by the SHS process have advantages such as high purity of the products, low energy requirements and relative simplicity. The exothermic heat generated during SHS reaction is used for the completion of chemical reaction and once the reaction is initiated at a particular ignition temperature (T_i), the combustion front propagated within the reactants at a particular combustion velocity (V_c) to yield the final product. The feasibility of synthesising a given compound or composite through tins route and sustainability of the wave front propagation mainly depends on the extent of heat liberated by the exothermic reaction(s) and the heat dissipated from the system to the surrounding. The adiabatic temperature, T_{ad} the theoretically calculated maximum rise in temperature in adiabatic condition, for a particular set of reaction is important for self-sustainability[2, 3] of the reaction. Since most of the exothermic reactions are initiated at a temperature (T_i) above the room temperature, the total enthalpy generated by the exothermic reaction is assumed to be consumed in increasing the temperatures of the products, and there is no heat loss to the surroundings. The total change of enthalpy, ΔH_{Ti} at any temperature T_i can be given by

$$\Delta H_T = \int_{T_i}^{T_{ad}} \sum \Delta C_{P_i} \, dT \qquad \text{...(1)}$$

where, ΔH_T is the enthalpy change associated with the exothermic reaction, ΔC_{pi} represent the difference between the specific heats of product i and Reactants at Constant pressure. For simplicity, various phase transformations of the reactants and products are not considered in equation 1. The adiabatic temperatures for the different chemical reactions used for the synthesis of various products and possible reaction products are presented in Table 1.

Table 1: Adiabatic Temperatures for different reactions adopted here to synthesize different advanced composities

Reaction equation	*Adiabatic temperature (T_{ad})K*
$Ti + C = TiC$	3200
$ZrO_2 + B_2O_3 + 5Mg = ZrB_2 + 5MgO$	2100
$3ZrO_2 + 10Al + 3B_2O_3 = 3ZrB_2 + 5Al_2O_3$	2200
$TiO_2 + B_2O_3 + 5Mg = TiB_2 + 5MgO$	2445

It is well known that various high temperature materials could be prepared by the SHS techniques and a few reviews are available in literature[4,5,6]. Amongst the various high temperature materials, zirconium diboride (ZrB_2) and TiB_2 are emerging as potential advanced ceramic materials because of their excellent properties- high melting point, hardness, elastic modulus and electrical conductivity, excellent chemical resistance to HCI, HF and other non-ferrous metals, cryolite and non-basic slags. Excellent properties make these borides ideal candidates for high temperature application requiring wear, oxidation and corrosion resistance. They have several applications, like cathodes for electrochemical processing of aluminium (Hall-Heroult process), evaporation boats, and crucibles for handling molten metals, thermowells, and thermocouple sleeves for high temperature uses, wear parts, nozzles, armour, cutting tools and as dispersoid in metal and ceramic matrix composites to improve mechanical properties. Several techniques have been established to prepare zirconium diboride powders. These are: synthesis from elements by melting, sintering or hot pressing, borothermic reduction of metal-oxides and boric oxide, reduction of the metal oxide with carbon or boron carbides, metallothermic reduction of metal oxides and self propagating high temperature synthesis (SHS) from elemental powder of boron and zirconium or titanium. Even though the SHS process has advantages, the use of elemental powder makes the process expensive. Hence, it is observed that in all the above processes, the processing times are quite long and they require high temperatures in the range of 2273 to 2473 K or pure elemental powders as starting raw material to achieve the purity of 95% in the final product. The use of high temperature furnaces for longer time or the high purity metal powder makes the process costly.

Besides the powder processing, sintering of Ti/Zr diboride is also known to be difficult due to their high melting temperatures and covalent nature. Most often, Zr/Ti diboride powders are sintered at 2273-2473 K by hot pressing and large grain growth limits the densification[7]. Pressureless sintering studies are very limited and different low temperature additives have been used to sinter them, which ultimately degrades the high temperature properties. Though some reports on sintering of titanium diboride are available, very few reports pertaining to the sintering of zirconium diboride powder are available in literature[8]. Various binders such as Fe, Cr, CrB_2, Cu, Ni, NiB were used to get dense titanium diborid compacts[9-11].

Recently, the alumina based ceramic matrix composites with the dispersion of hard phase material such as borides and carbides have shown promise as advanced materials for high temperature application because of their excellent refractoriness, low susceptibility to oxidation and good mechanical properties at elevated temperatures. Besides the conventional powder metallurgical route, in-situ processing of composite is gaining momentum because of certain inherent advantages such as reduced number of processing steps, lesser segregation of the dispersoid and homogeneous dispersion of second phase in the matrix. Some of the important in-situ composites that we have studied are ZrB_2-Al_2O_3 and Al_2O_3-ZrB_2-TiB_2. The alumina (Al_2O_3) –zirconium/titanium diboride composite is an important technical ceramic for use at high temperatures requiring wear, oxidation and corrosion resistance[12-13]. The addition of zirconium diboride to alumina is expected to have a better strength, higher hardness, fracture toughness and greater tolerance to impact loading as compared to non-reinforced alumina. Very few reports are available on zirconium/titanium diboride-alumina composite in which composite has been prepared by taking particulates of alumina and zirconium diboride[14]. Aluminothermic reduction in a resistive heating furnace has been used by Ray et al.[15], these composites were not only found to enhance the mechanical strength but also alumina matrix was found to become conductive at 20 Vol% addition of zirconium diboride. The ZrB_2 addition is also advantageous over titanium diboride because it does not have many stable phases like titanium dibode. In this manuscript, the different process developed for the fabrication of ZrB_2, nano-size TiB_2, and in in-situ composites Al_2O_3/TiZrB_2, are discussed.

2. SHS PREPARATION OF ZrB_2 POWDER AND SINTERING BEHAVIOUR

For the synthesis of zirconium diboride (ZrB_2) powder by the SHS process, stoichiometric amount of zirconia (99% purity, particle size < 5 mm), boron oxide or boric acid (99% purity, particle size < 40 μm) and Mg (99% purity, particle size < 100 μm) were taken and mixed by a planetary ball mill (PM400- Retsch, Germany) for 30 min at 300 rpm. The mixture was ignited from the top by an electrical arc in argon atmosphere. The reaction was carried out in a closed chamber flushed and filled with IOLAR-1 argon. After the completion of SHS reaction, the reacted product consisting of ZrB_2 and MgO, Mg was leached out. Dissolving in dilute HCl and washing with distilled water removed the reaction by-product, MgO. No milling was required to yield the fine ZrB_2 powder. The XRD analysis of the final products showed only the presentce of ZrB_2 in the leached powder. The presence of Mg and its compounds could not be detected in the XRD pattern and also by the EDX and ICP analysis. The detailed process was described elsewhere[19]. The SEM of the synthesized powder revealed the formation of sub-micrometer sized particles and the TEM showed the generation of ~ 100 nm powders (Fig. 1a, b).

It was established that SHS product ZrB_2 powder has more defect concentration compared to the powder prepared by non-SHS process. Fig. 2 shows the TEM micrographs of the ZrB_2 powders prepared by SHS and non-SHS carbothermic process. It is clear that the powder prepared by SHS technique has very large defect density compared to that of non-SHS carbothermic processed powder. These studies gave a first experimental manifestation to the theoretical prediction that SHS powder should have more defects owing to very high rate of heating and cooling during SHS reaction. The detailed work is reported elesewhere[20].

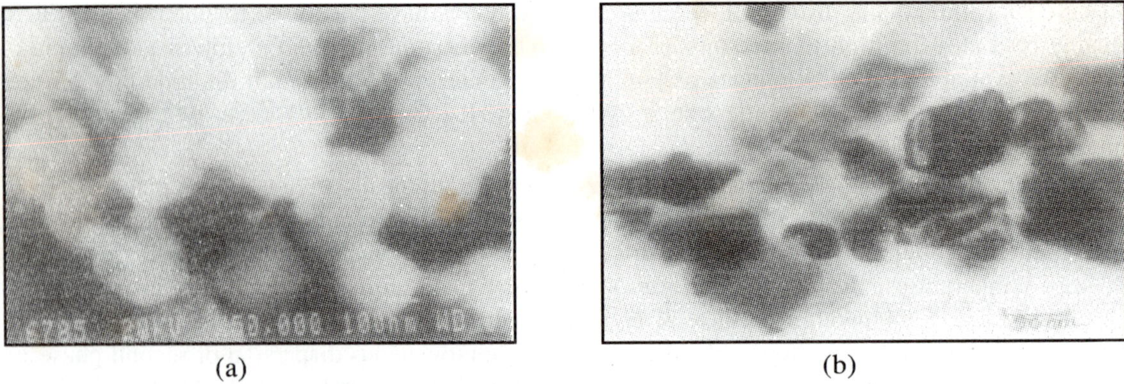

(a) (b)

Fig. 1. Micrographs of SHS producted ZrB_2 powder : (a) SEM and (b) TEM.

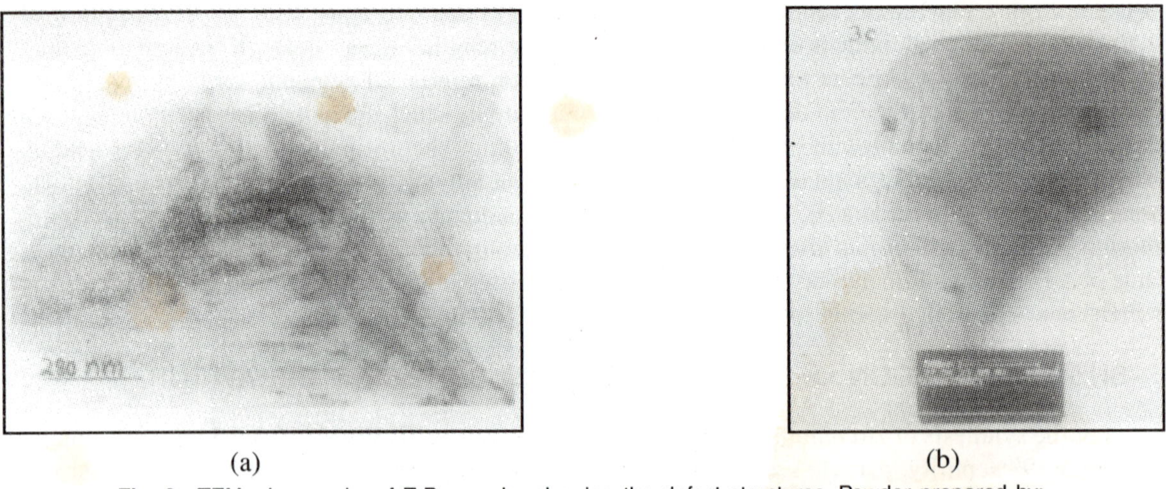

(a) (b)

Fig. 2. TEM micrographs of ZrB_2 powder showing the defect structures. Powder prepared by: (a) SHS process and (b) carbothermic process.

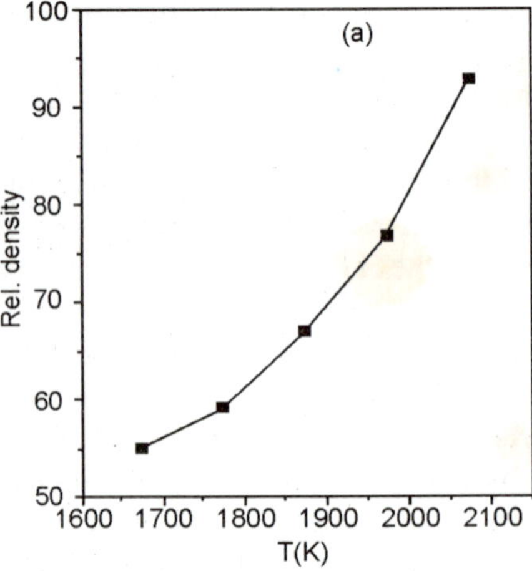

Fig. 3. Density versus sintering temperature curve for ZrB_2 powder prepared by SHS process.

The zirconium diboride powder could densify at low temperatures as low as ~1773 K and a maximum densification of 94% of the theoretical density could be achieved at 2073 K for 1h. the sintered pellets were having high electrical conductivity and had good mechanical properties. During sintering, formation of zirconium oxide to the surface was detected. The microstructures showed faceted growth at few places, which indicated the evaporation condensation mechanism during sintering. The evaporation condensation mechanism in the boride systems is well known. However, the densification above 90% indicated the mass transfer through volume diffusion process. The Variation of density with the sintering temperature is plotted in Fig. 3.

The excellent sinterability was primarily due to high surface energy of the ultra fine powder produced by the SHS technique and the high defect concentrations in the synthesised powders. From the sintering equation[21], where the rate of densification at temperature T are expressed as:

$$\frac{d\rho}{dt} = \frac{K}{T} e^{-\frac{\Delta E}{kT}} \quad ...(2)$$

where, K is Boltzmann constant P is the density value at any, temperature T (K), t is the sintering time and k is rate constant. The Arhenius plot yielded the value of activation energy, $\Delta E = 247.7 \pm 4$ kJ/mole. Densification above 90% suggested this activation energy possibly represented the mass transfer process of volume diffusion through grain boundary. The details of the sintering process have been discussed elesewhere[22, 23]. During sintering, except the formation of ZrO_2 at the surface no phase change could be detected in the sintered samples. The SEM image of pellet sintered at 2073 K for 1h revealed the formation of dense packed structure with grain size ~20 μm (Fig. 4). Extensive studies on the TEM of these sintered samples were investigated to find out the grain-boundary structures and the details are reported[23]. The pick up of Fe was observed in some cases which was mostly found at tripple points forming a complex boride of Zr and Fe as $(Zr, Fe)_2B$ with a = 5.09 A and c/a = 0.54 shown in Fig. 5 and the corresponding SAED.

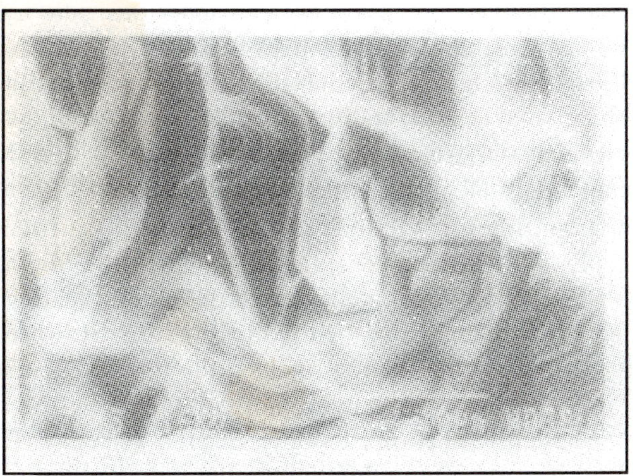

Fig. 4. SEM micrograph of a ZrB_2 sintered at 2073 K.

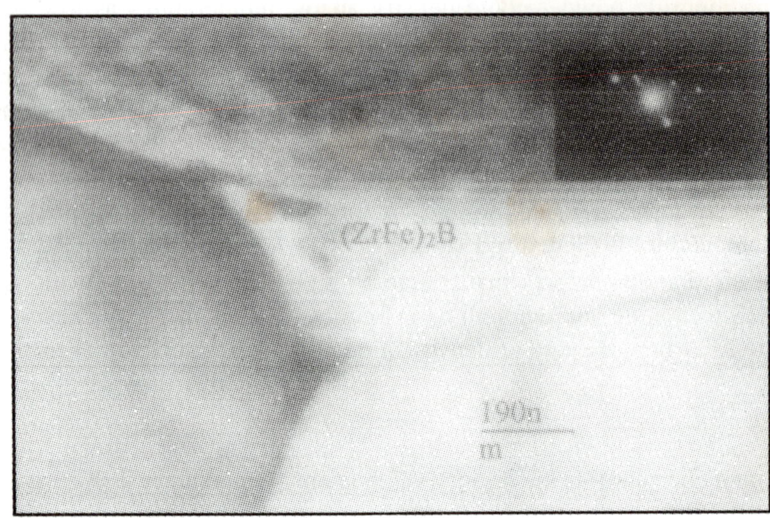

Fig. 5. TEM Micrograph of the sintered ZrB_2 sample showing the $(Zr, Fe)_2B$ at the triple point junction. Corresponding SAED pattern is shown in inset.

3. FORMATION OF NANO-SIZE TITANIUMN DIBORIDE (TiB_2) BY A NOVEL SHS PROCESS

For the SHS of nano TiB_2 powder, a novel processing route has been adopted using the stoichiometric mixture of Mg, titanium dioxide and boron oxide with SHS diluent NaCl that yielded nano-sized titanium diboride powder. This process uses the magnesiothermic reduction in the presence of NaCl during SHS reaction. The stoichiometric amount of the oxides and magnesium powders were mixed in a planetary ball mill with varying amounts of NaCl in the range between 0-20 wt% followed by SHS reaction. After the combustion synthesis, sodium chloride was washed away with distilled water and the byproduct MgO was leached out using dilute HCl. With the increase of NaCl from 0-20 wt% in the mixture, the adiabatic temperature (T_{ad}) was found to decrease from 2445 K to 2165 K.

From the TG-DTA plot ignition temperature of the reaction was found to be ~958 K. The thermograms of different samples with varying percentages of NaCl showed that the reaction initiation temperature remained independent of the NaCl addition and the rate of heating, which suggested that the NaCl content did not affect the mechanism of the (1738 K) present SHS process and the NaCl was remaining neutral. The insignificant role of heating rates on the reaction initiation temperatures also suggested that the present SHS process was possibly not controlled by the diffusion processes as observed by Pathak et al. in the case for the alumina-silicon carbide system[25]. The XRD patterns of the finally leached samples showed the presence of TiB_2 phase only. As the NaCl content was increased, extensive line broadening for the NaCl was observed, suggesting the generation of fine crystallises in the NaCl added samples. The crystallite, sizes calculated using Debye-Scherrer equation (Ignoring the defect or strain contribution) for the powders are presented in Table 2.

Table 2. Variation of crystallite size with different NaCl contents.

NaCl (wt%)	*Angle (°2θ)*	*Broadening (°β_0)*	*Instrument line broadening (°β_1)*	*Broadening (°β_c)*	*Crystallite size (nm)*
0	52.35	0.28	0.2	0.19	52
5	51.85	0.33	0.2	0.33	31
10	52.5	0.44	0.2	0.39	26
15	52.05	0.48	0.2	0.44	23
20	52.8	0.5	0.2	0.46	22

From the Table 2, Significant changes in the crystallite sizes could he seen with the increase of NaCl concentration. The SEM and TEM studies also showed the decrease of particle sizes with the addition of NaCl (Figure 6 a, b, c). During exothermic reactions, very high temperatures (> 2423 K) were achieved within a fraction of minutes. Since the combustion temperature was much higher than the boiling point of NaCl (1465°C), evaporation of NACI was expected and the presence of NaCl in the EDS analysis indicated that all the NaCl did not escape out from the mixture after the synthesis. This NaCl may have partially melted/vaporized and formed a coating on the TiB_2, particles, which helped in reducing the grain growth of particles and the particle sizes decreased. The decrease of adiabatic temperatures with the addition of NaCl may also have helped in reducing the grain growth.

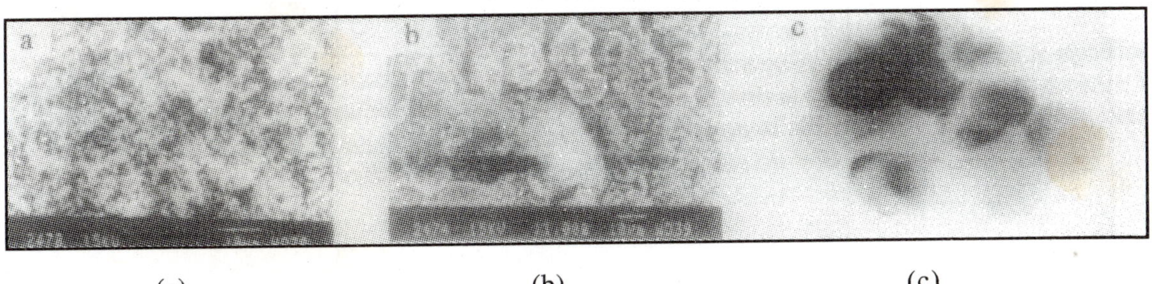

(a) (b) (c)

Figure 6. Micrograph of nanosize TiB_2 powder prepared by SHS process (a) without NaCl, (b) 20 Wt% added Nacl and (c) TEM images of 20 Wt% added NaCl.

4. SHS OF IN-SITU ZrB_2-Al_2O_3 COMPOSITES

The raw materials, zirconium oxide, fine Aluminium (< 5μm in size) powder and boron oxide (<200 μm) powders, were mixed together in a ball mill. The mixture was compacted in the form of 20 mm diameter and 40-60 mm long cylindrical pellets. The SHS reaction was carried out under argon atmosphere (0.5 MPa) in a closed reaction chamber. The pellets were ignited with an electrically heated tungsten coil placed on the top surface of the green pellets. Once the ignition of pellet started the ignition source was switched off. A thermocouple (W-Re 5/20, 100 μm thick) was inserted from the bottom of the sample to 5 mm from the bottom end to measure the combustion temperature and the combustion velocity. The reactions were carried out for different amount of aluminium content (near the stoichiometric regime (28-32 wt%/) in the mixture. The ignition temperature of the reaction for stoichiometric green mixture was measured as 1548 K from the DTA analysis. From the time-temperature plot (Fig. 7) obtained during SHS, the combustion temperature and velocity were measured as ~ 2327K and ~ 5cm/sec respectively. The XRD patterns of the reacted powder showed only the presence of ZrB_2 and Al_2O_3 phase. SEM Microstructual investigations showed the presence of two types of phases in the synthesised product, the large grains (10-20 μm) of Al_2O_3 and the small grains (0.5 to 3 μm) of ZrB (Fig. 8). At some places melt growth of Al_2O_3 was observed possibly due to attainment of high combustion temperatures at the reactioninterface or due to precipitation of alumina and subsequent fusion by absorbing heat of reaction during SHS reaction between solid ZrO_2, liquid Al and liquid B_2O_3. Similar observations were also made during SHS of TiC-Al_2O_3 in-situ composites by Feng et al.[26]. The details of the ZrB_2-Al_2O_3 composite fabrication process have been published[27].

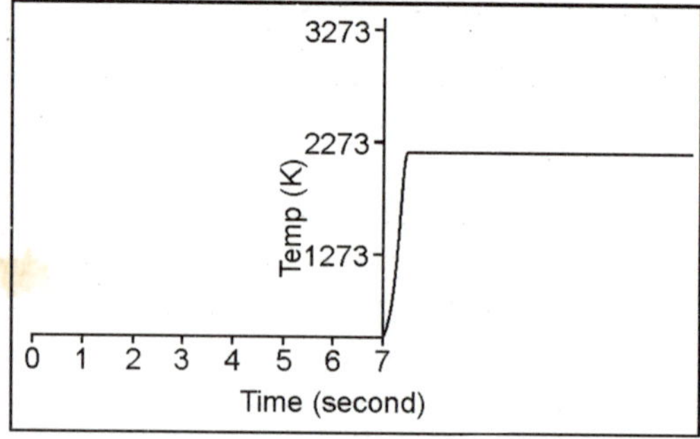

Fig. 7. Time-temperature curve during SHS synthesis of Al_2O_3-ZrB_2 composite.

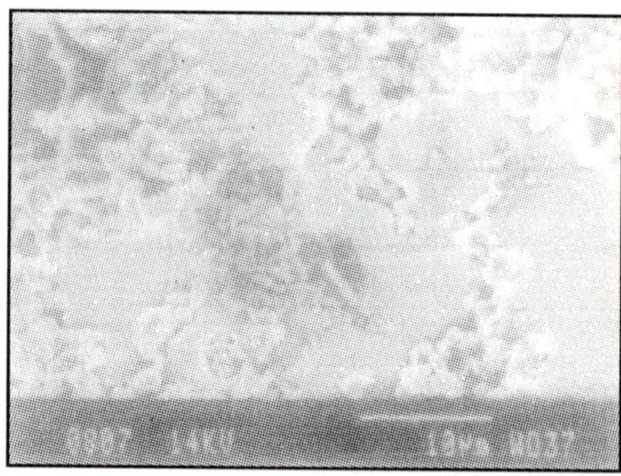

Fig. 8. SEM micrograph of the as synthesized Al_2O_3-ZrB_2 composite

A time-resolved X-ray diffraction study of the formation of a zirconium diboride-alumina composite from zirconium dioxide, aluminium and boron oxide by revealed that, during the formation of the composite, solid solutions of Al in ZrB_2, forms namely $(Zr,Al)B_2$ with a slight change in the lattice parameters of ZrB_2 before conversion to the ZrB_2-Al_2O_3 composite. During synthesis in air, oxidation is limited to surface only. Dendritic growth structures of ZrB_2 were observed (Figure 9) during synthesis of the composite, suggesting formation of a molten pool of the materials followed by solidification of the high melting point material ZrB_{28}. The details of the dynamic X-ray studies and microstructure evolution have been reported by Mishra et al.

Fig. 9. SEM microstructure of Al_2O_3-ZrB_2, composite showing the dendritic structure of ZrB_2 in alumina matrix.

These composite powders showed a dense microstructure of 94% density at 2073 K, which is significantly lower sintering temperature than the conventionally used 2273 K with hot pressing. Fig. 10 shows a SEM micrograph of the sintered alumina-ZrB_2 Composite. The diffusion of alumina into zirconium was detected by detailed EDS analysis but no diffusion of ZrB_2 was observed into alumina.

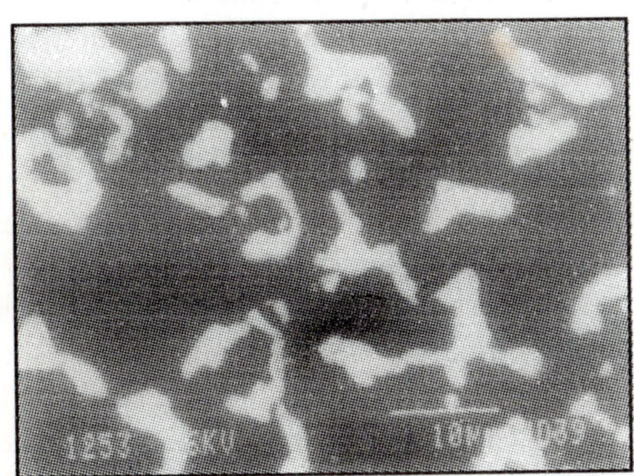

Fig. 10. SEM micrograph of the Al_2O_3-ZrB_2 composite sintered at 2073 K.

5. Al_2O_3-ZrB_2, Al_2O_3-Zr/TiB_2 IN-SITU DENSED COMPOSITE BY SHS DYNAMIC PROCESS

I. Al_2O_3-ZrB_2 in-situ Composite

The in-situ densified composite was also prepared by using novel SHS dynamic compaction using same raw materials. This process could control the grain growth and very fine grain sized composites having the grain sizes less than 5 μm (Fig. 11) with densification of 96% is achieved. It is clear that in situ composites were consisting of very fine grains compared to the sintered SHS prepared composite powder, where the alumina grains were coarser than boride grains (Fig. 10).

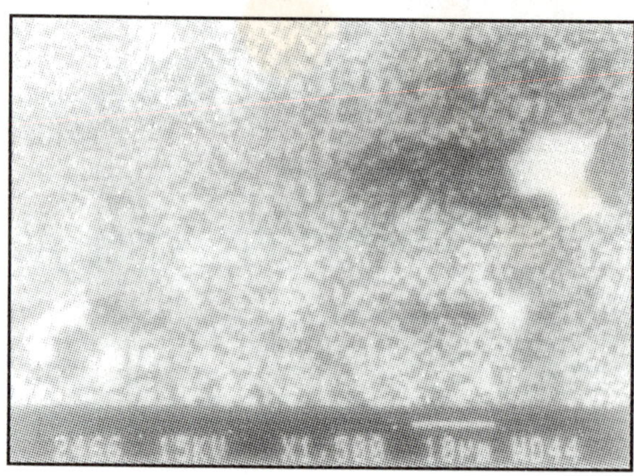

Fig. 11. SEM micrograph of the Al_2O_3-ZrB_2 composite prepared by SHS dynamic compaction

II. Al_2O_3-Zr/TiB_2 in-situ Composite and the Affect of Addition of Al_2O_3

Al_2O_3Ti/ZrB_2 composites were prepared by the SHS dynamic compaction process, where the raw materials zirconium oxide, fine aluminium boron oxide and titanium (99% < 200 μm) powder, were mixed together in a ball mill for an hour. Batches with stoichiometric composition as per the reaction I given in equation 3 and with 5-20 wt% of extra alumina as SHS were used.

$$3ZrO_2 + 2\,Ti + 5B_2O_3 + 14Al \rightarrow 3ZrB_2 + 2\,TiB_2 + 7Al_2O_3 \qquad ...(3)$$

The mixture was palletised in the cylindrical form of 60 mm diameter and 15 mm height. The SHS reaction and dynamic compaction processing was carried out in air and the loads were simultaneously applied for different processing times using a hydraulic press. The pellets were ignited with an electrically heated tungsten coil placed on the side surface of the green pellets. Once the ignition of pellet started, the ignition source was switched off and after prefixed synthesis time (T_s) between 1-60 seconds the sample was loaded with different pressure in the range of 70-100 atmosphere. The facility used was developed at ISMAN. The details of the process are described elsewhere[29]. The thermodynamic calculations for estimated percentage phase and adiabatic temperature is shown in Table 3.

The measured density was found to increase with the increase of diluent alumina content upto 10 Wt% addition but it decreased for addition of diluent by 15 and 20 wt%. The decrease in density is due to increase of alumina, a lesser density in the composite and also due to formation of lighter phases T_iO_{2-x}. The composite density will depend on the percentage phases present in the final product and their effective densities. However the microstructure was found to be highly densed. Porosity was hardly seen in the product. It was observed that, lower percentage additions of the Al_2O_3 diluent led to a homogeneous microstructure (Fig.12 a, b, c) and lower grain growth.

Samples with 20 wt% added alumina though showed at large scale homogeneous but has some segregation in the sample. This decrease in grain growth is due to the reason that addition of Al_2O_3 as diluent decreases the liquid alumina phase significantly as was estimated in thermodynamic calculations (Table 3). The adiabatic temperature T_{ad} decreased by 140 K by addition of 5 wt% alumina in the reaction but the liquid phase alumina remained similar and solid phase alumina started appearing. Further addition of alumina though did not change the adiabatic temperature but amount of liquid phase in the product decreased and solid phase increased significantly. It will lead to inhibit the grain growth because of the presence of more second solid phase and less liquid phase in the product.

Table 3. Thermodynamic Calculations with different % of alumina in reaction 3

Alumina diluent (%)	*Products*	*Number of moles*	*wt%*	*Tad (K)*
	Al_2O_3(L)	7	59.91	
0	TiB_2(S)	2	11.67	2468.797
	ZrB_2(S)	3	28.42	
	Al_2O_3(L)	7.5842	61.82	
5	TiB_2(S)	2	11.12	2340.0979
	ZrB_2(S)	3	27.06	
	Al_2O_3(L)	6.5299	50.81	
10	Al_2O_3(S)	1.6384	12.75	2326.7912
	TiB_2(S)	2	10.61	
	ZrB_2(S)	3	25.83	
	Al_2O_3(L)	5.2804	39.30	
15	Al_2O_3(S)	3.4721	25.84	2326.8791
	TiB_2(S)	2	10.15	
	ZrB_2(S)	3	24.71	
	Al_2O_3(L)	4.0336	28.77	
20	Al_2O_3(S)	5.3031	37.82	2326.6254
	TiB_2(S)	2	9.73	
	ZrB_2(S)	3	23.68	

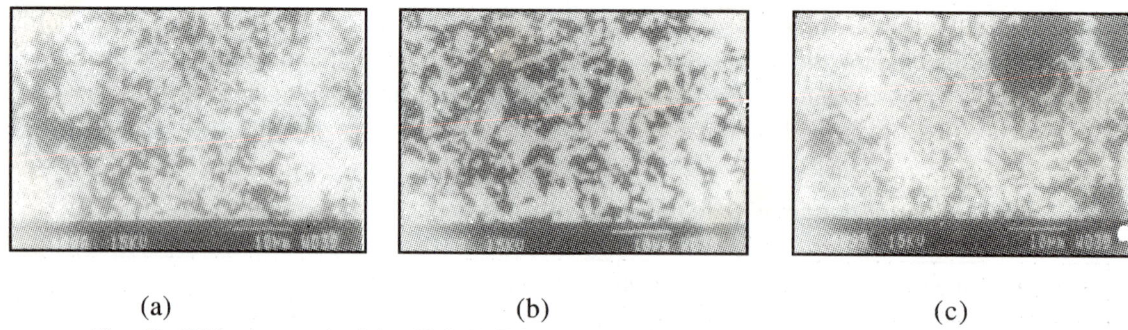

(a) (b) (c)

Fig. 12. SEM micrograph of the Al_2O_3-Zr/TiB_2 composite prepared by SHS dynamic compaction: (a) 0%, (b) 5 wt% and (c) 20 wt%

The micro hardness were measured with a Vickers indenter, five indents were taken for each sample on a polished surface at 200 gf (1.96 N) respectively. The microhardness varied between 2485 HV to 1880 HV. The values obtained are shown in table 4. The microhardness values were found to decrease for 5 and 10% addition and then it increases for 15 and 20 wt % additions. This may be due the fact that there are two things, which is happening with the addition of alumina. Firstly, the grain size is decreasing with the addition of alumina so it should increase the hardness, as they are inversely proportional empirically. Secondly, with the addition of alumina the soft phase is increasing hence there will be decrease in resultant hardness. Both the reasons contribute towards hardness and hence the resultant hardness can vary accordingly. Further studies are under progress.

Table 4. Micro Hardness samples made by SHS dynamic compaction with different % of Al_2O_3 as diluent using reaction 3

Diluent (%)	*Hardness values (Vickers) measured at 200g load.*
0	2485 HV
5	2436 HV
10	1888 HV
15	2095 HV
20	2239 HV

CONCLUSIONS

Using SHS, ultra fine ZrB_2 and nano sized TiB_2 powders were prepared using cheaper raw materials. The detailed study of mechanism vis a vis microstructure evolution and correlation was carried out to establish the SHS parameters and mechanisms involved during the different reactions. It was found that SHS produced powders are more smterable compared to the powders developed by non-SHS processes. The sintering temperature could be brought down by 200-300 K compared to the temperatures reported 'in literature, which is very important for sintering of high temperature materials. The processes have been successfully used and controlled to fabricate dense in-situ and near net shape composites of $Zr/TiB_2-Al_2O_3$, for varyig percentages of the phases in the composite with controlled grain-growth. A hard composite of 28 Gpa has been fabricated.

ACKNOWLEDGEMENT

The authors wishes to acknowledge L.C. Pathak, S.K. Das, S. Das of National Metallurgical Laboratory, Acad. A.G. Merzhanov, D. Yu. Belov, V. Shcherbakov of institute of Mackrokinetics and Materials Science (ISMAN), Russian Academy of Sciences, Moscow, Russia, for their kind help and discussions during the above research work. They also acknowledge the financial support provided by the Department of Science and Technology, Govt. of India.

REFERENCES

1. Munir Z.A., Am. Ceram. Soc. Bull. 67 (1988) 342.
2. Novikov N.P., Borovinskaya J.P. and Merzhanov, A.G. In, Combustion processes in chemical technology and metallurgy, Ed. A.G. Merzhanov, 174 (1975), Chemogolovka.
3. Bowen C.R. and Derby, B. Br. Ceram. Trans. 96 (1997) 25.
4. Munir, Z.A.U. Anselmi-Tamburini, Mat. Sci. Rep., 3 (1989) 277.
5. Sheppard L.M., Adv. Mater. Proc., 2(2), 25-32. (1986).
6. A. Makino, C.K. Law, J. Am. Ceram. Soc. 77(3), 778-786 (1994).
7. Keeskes, J. Kottke.T. J. Am. Ceram. Soc. 73 (1990) 1274.
8. Cheng Y.M. and Gadalla, A.M., Synthesis and analysis of ZrB_2-based composite, Materials and Manufacturing Processes, 11(4), 575-587, (1996).
9. Einarsud M., Hagen E., Petersen G. and Granede T., J. Am.Ceram.Soc., 80(12), 3013 (1997).
10. Duabdesselam, Munir Z.A. J. Mater. Sci. 22, (1987) 1799.
11. Shim K. B., Edirisinghe M. J., and Ralph B., Brit. Ceram. Trans., 95, 1, 15-22, (1996).
12. Liu Jenq and ownby P. Darrell, J. Am. Ceramic society, 1991, 74 (1), pp. 241-243.
13. Postrach S., Potschke J., J. Euro. Ceram. Soc., 20 (2000) 1459.
14. Sundaram V., Logan K.V. and Speyer R.F., J. Mater. Res., 1997, 12, pp. 1681-1684.
15. Ray S.P.: Metall. Trans. A, 23A (1992) 2381.
16. Mishra (Pathak) S.K., Das S., Goel R.P. and Ramachandrarao P., J. Mater. Sci. 16 (1997) 965.
17. Kosolapova T, Ya., editor, Hand book of high temperature compounds, Ohemisphere Publishing Corporation, Taylor and Frands Group, 864-867, 1990.
18. (Pathak) Mishra S.K., Das S.K., Roy A. and Ramchandrarao P., J. Mater. Res. 14 (1999) 3594.
19. Ramachandrarao P. and Mishra S.K. - Zirconium diboride synthesis and sintering- Refractories and Furnaces-New options and new values. 31-39, 2000. Published by Allied Publishers.
20. Mishra S. K., Das S., Padiak L.C., Mater. Sci. Eng. A, under review
21. German R.M., Metal Powder Industries Federation, Prinecton, New Jercy, (1994) pp. 145.
22. Mishra (Pathak) S.K., Das S., Das S.K. and Raniachandrarao P., J. Mater. Res. 15 (2000) 2499.

23. Mishra (Pathak) S.K., Das S.K., Ray A.K. and Ramehandraraw P., J. Am. Cerm. Soc. Nov. (2002).
24. Mishra S.K., Das S., Ramachandrarao P., J. Mater. Res. Nov. (2002).
25. Pathak L.C., Bandyopadhyay D., Shrikanth S., Das S.K. and Ramachandrarao P., J. Am. Ceram. Soc. 84 (2002) 915.
26. Feng H.J. and Moore J.J., Met. and Mater. Trans. B, 26,(1995) 265.
27. Mishra S.K., Das S.K., Ramachandrarao P., Belov D., Mamyan S.S., Met. and Mater. Trans. A (in press).
28. Mishra S.K., Das S.K.. Ramachandrarao P., Belov D., Mamyan S. S., Phil. Mag. Lett. (in press).
29. Mishra S.K.,V. Shcherbakov An innovative process for in-situ fabrication of Al_2O_3-ZrB_2 composite, patent Indian, 1654DEL2004 dated 1/9/04 and US patent filed no awaited.

6

Effect of Copper Coated Tungsten Powders During Infiltration Densification of W-Cu System

Burak Özkal
Istanbul Technical University, Metallurgical & Materials Engineering
Dept. Maslak-Istanbul-80626 Turkey

Anish Upadhyaya
Indian Institute of Technology, Dept. of Materials & Metallurgical Engineering, Kanpur 208016, India

M. Lutfi Öveçoglu
Istanbul Technical University, Metallurgical & Materials Engineering
Dept. Maslak-Istanbul-80626 Turkey

Randall M. German
Penn State College of Engineering, P/M Lab., 16801 University Park, Pennsylivania, U.S.A.

ABSTRACT

Lack of solubility between tungsten and copper inhibits densification especially compositions with high volume fraction of tungsten. It is most probable that a homogenwous distribution of copper phase will lead higher degree of rearrangement during sintering with decreasing tungsten-tungsten particle contacts. In this study tungsten powders were coated with copper via electroless coating and densification behavior of these composite powders with further copper infiltration were examined. For this purpose 9OW-10Cu composition was selected as initial composition before infiltration. Compacts having same pre-composition prepared directly from coated powders and elemental mixtures were infiltrated to meet with target composition of 7OW-30Cu. Although both compacts have initially same composition, compacts prepared from copper coated powders were led totally different microstructural evolution by inhibiting tungsten-tungsten contacts by keeping pre-copper coating layer during infiltration.

1. INTRODUCTION

W-Cu binary system is an attractive field for many powder metallurgists since the mutual solubility almost zero between the components. Many efforts have been focused to this unique system to explore sole rearrangement process during liquid phase sintering [1-3]. With the combination of high welding and erosion resistance of tungsten and excellent electrical and thermal properties of copper, W-Cu composites have been used in high current circuit breakers for many decades and as a result of this wide usage "W-Cu" name almost have become identical with term of "electrical contacts" [4-5]. Today this composites are still find attractive application fields as a result of high thermal conductivity of Cu and low thermal expansion coefficient of W. Having thermal conductivity values starting at 50% higher than Mo and almost 10 times higher than Kovar and Alloy 42 without the high expansion rates of copper and gold, W-Cu composites have still attractive physical properties and found post modern application fields including microwave carriers, hermetic package bases and housings, ceramic substrate carriers, heat spreaders and sinks, GaAs and silicon device mount and Laser diode mounts in addition to their wide usage in electrical contacts, EDM electrodes and spot-welding electrodes [6-7].

Processing of W-Cu composites is usually done through copper infiltration of a porous tungsten compact or through liquid phase sintering of compacts pressed from mixed copper and tungsten powders [8]. For high tungsten contents liquid phase sintering generally results in some final porosity due to insolubility of tungsten in liquid copper which dramatically limits the densification by eliminating solution-recipitation stage [9]. For tungsten contents above % 80 (w/o) it is difficult to reach theoretical densities with only rearrangement support [9-10]. There are two main accepted solution for this problem one is use of submicron tungsten powders which is expensive and other one is addition of transition metal aids such as Fe, Ni or Co to promote activated sintering unfortunately which degrades thermal conductivity of alloy as a result of third phase addition [11-13].Therefore infiltration process still dominant technology to make W-Cu composite materials for many decades, although it produces large slugs which require extensive machining operations to make components with final dimensions.

It is most probable that a homogeneous distribution of copper phase will lead higher degree of rearrangement during sintering of W-Cu composite system [14-16]. Theoretically upper limit of this approach is to reach zero tungsten contiguity by eliminating every single tungsten-tungsten particle contact in microstructure. Recently, there have been interesting efforts made for production and performance evaluation of coated powders or micro-encapsulation concept. W-Cu composite powders with a unique phase distribution, each individual particle consists of a copper and a tungsten phase wherein the copper phase substantially encapsulates the tungsten phase or very interestingly vice versa where the tungsten phase substantially encapsulates the copper phase are commercially available in P/M market [17-18].

2. EXPERIMENTAL PROCEDURE

In this experimental work effect of coated powder properties during infiltration of W-Cu composites were investigated. In the background of this study, as coating procedure three different methods were evaluated and finally electroless coating was selected. Percentage of 5 -50 Cu-W composite powders were previously prepared and characterized by aid of SEM-EDS and XRD supported analysis [18]. In this study 90W-10Cu composition was selected for starting perform composition for infiltration. For this purpose two types of performs were prepared as comparison set. In first one while the tungsten powders were coated with copper via electroless copper coating technique to obtain 90W-10Cu composition, second perform was prepared by direct mixing of W-Cu powders in turbula blender. As it is mentioned in Table. 1 two types of W powders and one Cu powder used in this study. In the first perform coarser W1 powder were used, in the second perform relatively fine W2 powder and atomized Cu powder were used. As infiltrant Cu same atomized Cu powder was selected. Both perform were pressed under 150 MPa pressure as cylindrical compacts having the diameter of 1.24 mm and 10 mm height. Infiltrant Cu powders were also calculated and weighed for final 70W-30Cu target composition for both perform and pressed under same pressure with the diameter of 0.62 mm cylindrical compacts. Figure 1 illustrates the samples before the infiltration experiment. Infiltration experiment was realized under reducing H_2 atmosphere at 1250°C for 1 hour. As heating rate 10°C/min. were selected during experiment. After the sintering experiment, samples weighed and samples' macro photographs were taken and density measurements were realized according to Archimedes principle. Hardness tests were performed using Misawa Seiki Seisakusho Rockwell Hardness tester using Rockwell B scale. Electrical conductivity of infiltrated samples were measured using two different principle. In the first one Sigmatest 2.067 which works according to Eddy-Current principle. Measurements were done from top and bottom surfaces of infiltrated bodies according to Eddy-Current Principle. Also in order to measure electrical conductivity of samples with the direction of top to bottom a digital low resistance ohmmeter, Biddle Instrument™ were used. Standard metallographic procedures were followed to prepare samples to optical microscopy.

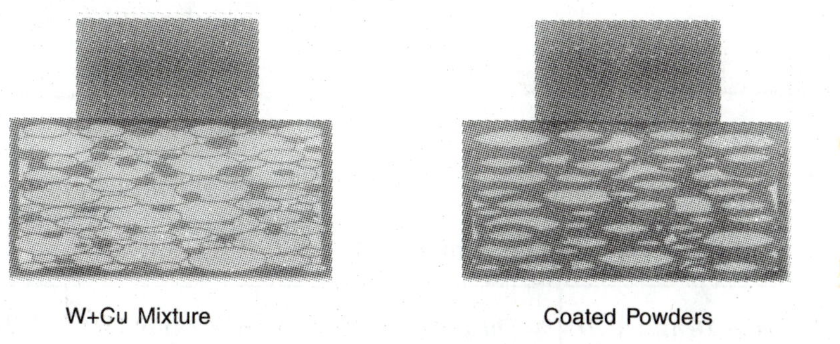

Fig. 1. Illustration of infiltration experiment.

Table 1. Initial powder properties used in experiments

Powder	*Particle size Distribution* (μm)			*Apparent Density* (g/cm³)	*Tap Density* (g/cm³)	*Picnometer Density* (g/cm³)	*Production Particle Shape*
	D_{10}	D_{50}	D_{90}				
Tungsten 1 (W1)	2.4	6.2	10.8	4.05	6.06	19.12	Reduction Polygonal
Tungsten 2 (W2)	0.9	2.9	6.1	3.96	5.92	19.8	Reduction Polygonal
Copper (Cu)	8.5	13.1	17.4	3.39	2.44	8.45	Atomization Spherical

3. RESULTS AND DISCUSSION

Macro photographs of samples before and affter infiltration experiment were given in Fig. 2. As is seen there is no certain copper bleedout problem occurred during infiltration. Also weighing of samples before and after infiltration experiment indicates that maximum of % 0.6 mass loss occurred in both samples. Measured green and sintered densities were summarized in Table 2. As it is seen from this table, both compacts have reached to acceptable high density values. Since relatively fine W powders were used in mixed sample it is logical to get higher density values. With respect to coarse W powders, fine powders creates more capillarity and also supports solid state sintering between W-W particles.

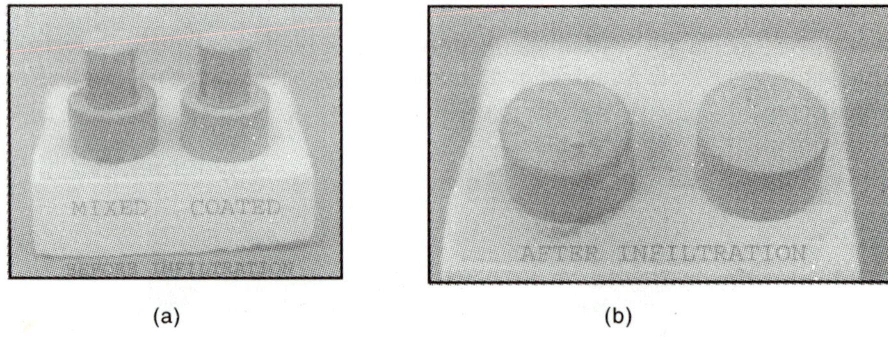

(a) (b)

Fig. 2. Macro photographs of samples (a) Before and (b) After Infiltration experiment.

Table 2. Green and Sintered Density of Samples.

Sample/Density	*Green Density (GD) (g/cc)*	*% Theoretical GD*	*Sintered Density (SD) (g/cc)*	*% Theoretical SD**
Mixed	10.46	60.4	14.42	100.28
Coated	10.18	58.8	14.05	97.71

* 14.38 g/cc was accepted as a theoretical density of alloy 70W-30Cu

At first glance, optical micrographs show totally different microstructures. If we evaluate both Cu and W phases within the microstructure from Figs. 3a and 3b; after infiltration, it is very obvious that samples prepared from premixed powders shows more W contiguity than precoated powders. While there are several isolated Cu lakes in premixed sample we see almost continuous copper network in precoated sample. Figs. 4a and 4b gives more details about samples microstructures with aid of higher magnification.

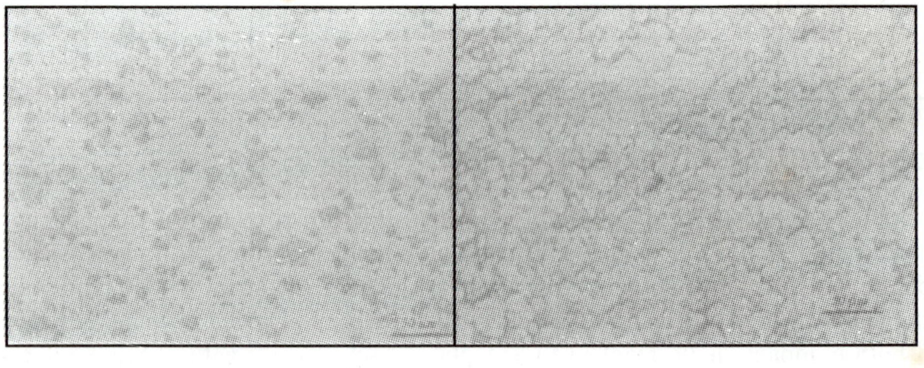

(a) Mixed (b) Coated

Fig. 3a and 3b. Optical Micrographs of samples after Infiltration (Lower magnification, micron marker indicates 50 μm).

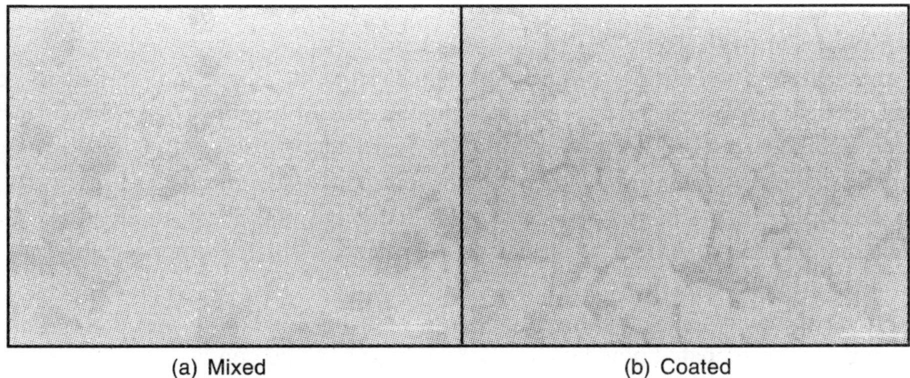

(a) Mixed (b) Coated

Fig. 4a and 4b. Optical Micrographs of samples after Infiltration (Higher magnification, micron marker indicates 10μm).

Table 3 summarizes electrical conductivity and hardness values of infiltrated samples. Electrical conductivity measurements from top surfaces where the Infiltration done shows much higher values compare to bottom surfaces. This difference is much lower in precoated samples where the Cu connectivity is much higher. According to electrical conductivity measurements based on Eddy current principle is it difficult to make an interpretation based on microstructural differences.Therefore, another set of measurements realized using low resistance ohmmeter. For this purpose top and bottom surfaces of samples machined and using special torchmeter same compression force applied during measurement to each sample. In this measurement sample prepared from coated powders showed % 8 higher electrical conductivity than mixed powders.

Table 3. Electrical Conductivity and Hardness Values of Infiltrated Samples

Sample/Property	*Electrical Conductivity* % (IACS) Bottom/Top*	*Hardness (Rockwell B)*
Mixed	18.84 – 26.41	86.4
Coated	22.53 – 25.35	79.8

* According to Eddy-Current Principle

Measured hardness values are amicable with measured density values. It is normal to expect higher hardness value from a sample which has a more rigiditiy due to W-W skeleton structure. In literature for the 70W-30Cu composites hardness values reported between 86-96 HR_B. Therefore, a measured hardness value like 79.8 HR_B is out of this range. But this low value supports the W-W particle separation observed in the microstructure; it should be more easy to move for a W particle or particle group in a ductile connected Cu matrix.

CONCLUSION

These results can be drawn from experimental work:

W-W contact formation during compacting and also sintering can be inhibited up to a certain level with use of Cu coated powders. But these observations should be supported with further studies in order to eliminate handicaps of using different initial W powders in this study.

Electrical conductivity measurements indicates lower values in compare to reported literature for 70W-30Cu alloys. Since the sintered densities are high enough, these low conductivity values should be related contaminations or oxides in initial powders have. This is supported by low picnometer densities of initial powders.

Electrical conductivity measurements realized between top and bottom surfaces of samples showed that sample prepared from coated powders has % 8 higher electrical conductivity than sample prepared from mixed powders. Since the appartures used in this experiment are not standar and open to some conductivity losses, it would not be possible to convert these values to % (IACS) scale for better comparison. It is believed that % 8 conductivity difference should be directly related with continious copper network within the microstructures. This idea is also supported by low hardness value which is a result of more W-W particle separation in the microstructure.

It is obvious that these results should be repeated by additional effort to prove which will involve with more number of samples and dimensionaly big samples to have a better statistically evaluation.

REFERENCES

1. W.J. Huppmann, H. Riegger, Modelling of Rearrangement Processes in Liquid Phase Sintering, Acta Metallurgica, V23, pp. 965-971, (1975).
2. P. M. Ossi, R. Roberti, G. Silva On the Rearrangement Mechanisms during Liquid Phase Sintering of a Model System, Scripta Metallurgica, V19, pp. 569-574, (1985).
3. V. Smolej, S.Pejovnik, A. Kaysser, Rearrangement during Liquid Phase Sintering of Large Particles, Powder Metallurgy International, V14, no. 1, pp. 34-36, (1982).
4. ASM Handbook Vol. 7, Powder Metal Technologies and Applications, pp. 870 (1998).
5. W. F. Wang, Effect on Tungsten Particle Size and Copper Content on Working Behavior of W-Cu Alloy Electrodes during Electrodischarge Machining, Powder Metallurgy , V40, no. 1, pp. 295-300, (1997).
6. http://www.spectramat.com/wcumat.htm
7. http://www.osramsylvania.com/chem/tungcopow.htm
8. J.L. Johnson, Densification, Microstructural Evolution and Thermal Properties of Liquid Phase Sintered Composites, Ph.D. Thesis, The Pennsylvania State University, University Park, PA, USA, (1994).
9. ASM Handbook V7, Powder Metal Technologies and Applications, pp. 870 (1998).
10. K.V. Sebastian, G.S. Teldolkar, High Density Tungsten-Copper Liquid Phase Sintered Composites from Coreduced Oxide Powders, The International Journal of Powder Metallurgy & Powder Technology, V 15, no. 1, pp. 45-52 (1979).
11. R.M. German, K.F. Hens, and J.L. Johnson, Powder Metallurgy Processing of Thermal Management Materials for Microelectronic Applications, International Journal of Powder Metallurgy, V30, no. 2, pp. 205-214 (1994).

12. J.L. Johnson, and R.M. German, Chemically Activated Liquid Phase Sintering of Tungsten- Copper, International Journal of powder Metallurgy, V. 30, no. 1, pp. 91-102 (1994).

13. I.H. Moon, and J.S. Lee, Activated Sintering of W-Cu Contact Materials, Powder Metallurgy, V. 22, no. 1, pp. 5-7, (1979).

14. C. Ghosh, B. Ozkal and A. Upadhyaya, Effect of Coating and Activators on Sintering of W-Cu Alloys, 11th International Metallurgy & Materials Congress, Proceedings E-Book, pp. 1688-1696 October (2002).

15. A. Upadhyaya, and R.M. German, Densification and Dilation of Sintered W-Cu Alloys, International Journal of Powder Metallurgy, V. 34, no. 2, pp. 43-55, (1998).

16. B. Ozkal, A. Upadhyaya, M.L. Ovecoglu, and R.M. German, Comparative Effects of Coated Powders on the Sintering Behavior of W-Cu Alloys, Proceedings 2nd Turkish National Powder Metallurgy Conference, S. Saritas (ed.), Turkish Powder Metallurgy Association, Ankara, Turkey, pp. 185-191. (2000).

17. C. M. Capus, Coated Powders Broaden Horizons for Particulates, Metal Powder Report, Vol. 53, no. 9, pp. 30-31, (1998).

18. B. Ozkal, Comparative Sintering Investigations of Copper Coated Tungsten Powders, Ph.D. Thesis, Istanbul Technical University, Maslak, Istanbul Turkey (in Turkish), (2002).

7

Sintering Mechanism Maps For Ultrafine Titanium Powders

B. B. Panigrahi
Department of Metallurgical and Materials Engineering, Indian Institute of Technology, Kharagpur 721302, India

K. Das
Department of Metallurgical and Materials Engineering, Indian Institute of Technology, Kharagpur 721302, India

T. R. Ram Mohan
Department of Metallurgical Engineering and Materials Science, Indian Institute of Technology, Bombay (Mumbai) 400076, India

V. V. Dabhade
Department of Metallurgical Engineering and Materials Science, Indian Institute of Technology, Bombay (Mumbai) 400076, India

P. Ramakrishnan
Department of Metallurgical Engineering and Materials Science,. Indian Institute of Technology, Bombay (Mumbai) 400076, India

M.M. Godkhindi
Department of Metallurgical and Materials Engineering, Indian Institute of Technology, Kharagpur 721302, India

ABSTRACT

Sintering mechanisms of micron size and attrition milled nanocrystalline titanium powders have been analyzed through the sinterig map model. In the micron size titanium powder two lattice diffusion mechanisms (both, from surface source and grain boundary source) are the operating mechanisms. Nano Ti shows a continuous change in mechanisms at various stages: (*i*) particle movements dislocation movement at lower temperature, (*ii*) grain boundary diffusion up to moderate temperatures and (*iii*) lattice diffusion and grain boundary rotation at higher temperatures.

1. INTRODUCTION

Sintering behavior of titanium powder has yet not been well understood due to its high reactivity, phase transformation, complex diffusion behavior and a large variation in reported activation energies of self-diffusion, etc. Sintering of P/M grade titanium powders were reported to be controlled by lattice diffusion and dislocation pipe diffusion mechanisms [1,2]. There have been no such reported studies on sintering behavior of nano size titanium powders. Sintering of ultrafine or nano powders suffers due to agglomeration, adsorbed gases, difficulties in compaction and grain growth, etc. [3]. Nano powders were reported to yield relatively low activation energy of sintering. The new densification mechanisms, such as, particle sliding or grain boundary rotation were proposed for sintering of nano powders [4,5]. There are several other investigators who see no difference in the mechanism of mass transport between nano and normal powders [6]. The present investigation attempts to analyze the effect of particle size on sintering mechanism of ultrafine titanium powders, *i.e.*, micrometric and attrition milled nanosize titanium powders. Efforts have been made to analyze the experimental data through the sintering diagrwn model of Ashby [7].

2. EXPERIMENTAL

High purity (~99.7%) titanium (mTi) powder (~5 μm), which was in the agglomerated form (~20 mm), was used. Nanocrystalline titanium (nTi) powder (0.020-0.040 μm) was produced through attrition milling for 40 hours. After milling the stearic acid (added as process control agent) was removed by heating the milled powder at a temperature of 250°C for 12 hours in a flowing high purity argon gas passed over heated (1000°C) titanium getter. Milled powder was found to pickup additional ~1.57% of transition element impurities (Fe, Cr, Ni, etc.). Powders were compacted (6.53 mm diameter) at a uniaxial pressure of 300 MPa, where both the powders showed similar green density of about 70% (theoretical). Dilatometric studies were carried out for a range of temperature (up to 1250°C, for an isothermal holding time of 60 min) under high purity treated argon atmosphere. Dimensional changes (linear) of the sample were continuously recorded on a strip chart recorder. Dimensional measurements, phase analysis (XRD analysis) and microstructural analysis of the sintered samples were carried out after each sintering.

3. RESULTS

Both mTi and nTi samples showed sintered density of about 83% (theoretical) at 1250°C. MTi samples showed isotropic shrinkage, whereas, nTi samples showed shrinkage anisotropy at higher temperature, *i.e.*, very high shrinkage in axial direction and smaller shrinkage in radial direction. The nTi samples show weight loss of ~0.2 to 0.4% during sintering. It was noticed that during polishing mTi samples showed mirror finish whereas nTi samples showed considerable grain pull out. Finely distributed Ti-Fe intermetallics were observed in SEM analysis of sintered nTi samples.

Dilatometric plots of the samples sintered at 1000°C are shown in Fig. 1. mTi compact showed normal linear expansion and shrinkage was found to start at about 650°C. A change in sintering rate at about 880°C, a temperature close to the $\alpha \rightarrow \beta$ transition in Ti could be seen. After isothermal heating for 60 min, during cooling sample shows continuous contraction.

The nTi samples show shrinkage to start at about 300°C and show discontinuity in shrinkage at about 500-575°C, followed by a rapid shrinkage. After switching-.off of the furnace, during cooling samples show a continuous contraction up to about 220°C and from about 220°C to room temperature, the samples show an unexpected expansion of about 0.05 to 0.1 %.

Activation energies for the rate-controlling step were calculated using the equation (1).

$$\ln\ r = n \ln C - \frac{nQ}{RT} + n \ln t \qquad ...(1)$$

where, r is shrinkage rate, Q is activation energy, R is universal gas constant, T is temperature, t is time, n is the time exponent and C is a constant. The value of 'n' is obtained from the plot of ln r vs ln t, whereas nQ/R is obtained from the plot of Inr vs $1/T$ (for 60 min of isothermal sintering).

Fig. 2 shows the plot of ln r vs 1/T for mTi and nTi samples. The data set shows a clear discontinuity at about 850-900°C in mTi samples and 700-800°C in case of nTi samples (Fig. 2). Two different Q values were obtained in nTi, 84 and 60 kJ/mol at low and higher temperature range respectively. Similarly for mTi also analysis was carried out separately in two parts, 184 kJ/mol (< 875°C) and 113 kJ/mol (for higher temperatures). Since, the mTi data in the beta range shows a curved dependence, the further analysis could reveal two Q values: 158 and 92 kJ/mol for lower and higher beta range respectively.

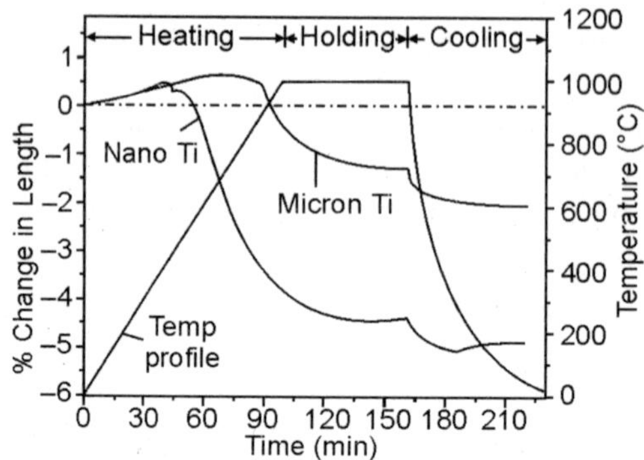

Fig. 1. Dilatometric plots of mTi and nTi samples, 1000°C.

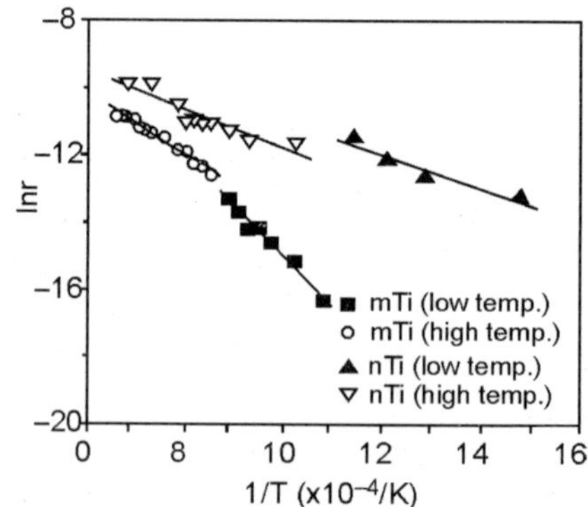

Fig. 2. Plot of lnr vs 1/T for mTi and nTi samples for estimating Q of sintering.

4. SINTERING MECHANISMS THROUGH MODEL STUDIES

Efforts were made to analyze sintering data through the sintering diagram approach of Ashby[7], using his proposed rate equations for various sintering stages and mass flow mechanisms. For the stage-1, the sintering mechanisms are: (*i*) surface diffusion from the surface source ($\dot{x}_1$,), (*ii*) lattice diffusion from a surface source ($\dot{x}_2$), (*iii*) vapour transport from a source ($\dot{x}_3$), (*iv*) grain boundary transport from sources on grain boundary, ($\dot{x}_4$) (*v*) lattice diffusion from sources on grain boundary ($\dot{x}_5$), (*vi*) lattice diffusion from dislocation sources ($\dot{x}_6$). The second stage of sintering was reported to be controlled by grain boundary diffusion ($\dot{x}_7$) and lattice diffusion ($\dot{x}_8$) for stage-3.

The theoretical sintering rates ($\dot{x}/a = dx/a.dt$) were predicted for above-mentioned modes (using equations proposed by Ashby). Where a is particle radius, x is neck radius and x is neck growth rate. The necessary physical data for the calculations were taken from literature [8] for both alpha titanium and beta titanium. As a first step the neck ratio (x/a) was obtained experimentally from the shrinkage data ($\Delta L/L_0$, where L_0 is initial length of sample and ΔL is change in length) using the formula [$x/a = (2\Delta L/L_0)^{1/2}$] This equation was used as a good approximation to estimate average neck ratio (x/a), because a reasonably accurate measurement of neck size through microstructure is nearly impossible when the particles are in ultrafine size and particles are irregular shaped. The experimentally measured shrinkage rates were converted to equivalent neck growth rates (change in the ratio of neck radius (x) to particle radius (a) in unit time). Next, using the calculated values of neck ratios and assuming various average particles sizes in the range of 1 to 30 gm for mTi and 0.003 to 0.045 μm for nTi, sintering rates were predicted. The rate controlling mechanisms were next predicted by comparing the calculated values against the experimentally determined sintering rates.

The predicted rates ($\dot{x}/a$) for each mechanism can be represented through a set of three-dimensional contour plots as a fomentation of particle size and homologous temperature. For example, Fig. 3(a) shows such plot for one mechanism ($\dot{x}_2$) for mTi. The assumed particle size (a) is plotted on x-axis. The homologous temperature (T_h) corresponding the experimental data is shown on y-axis and x/a values for each temperature are also shown on y-axis. The rates ($\dot{x}/a$) calculated for each particle size at various temperatures are shown by contour curve on z-axis (perpendicular to plane of paper) and the rate curves are labeled with actual values. The predicted rates, which are within one order of magnitude of experimentally observed rates, have been marked by a thick dotted line in the contour plot.

For mTi, when experimental rates were compared against the predicted rates of various assumed particle sizes, $\dot{x}_2$ and $\dot{x}_5$, seem to match within one order of magnitude. In the cc-range, $\dot{x}_2$ matches within the experimental data when particle size is assumed 9-13 μm (Fig. 3a) and $\dot{x}_5$, for particle size of 11-15 μm (Fig. 3b). Similarly in β-range, $\dot{x}_2$ matches within the experimental data when particle size is assumed 17-22 μm (Fig. 3a) and $\dot{x}_5$, for particle size of 21-28 μm (Fig. 3b).

For nTi, the measured rates seem to match within an order of magnitude with $\dot{x}_4$, $\dot{x}_6$, and $\dot{x}_8$, in different temperature ranges. The mechanism $\dot{x}_4$ (Fig. 4a) seems to be matching with 0.033 to 0.042 μm size range at lower temperature. The mechanism $\dot{x}_6$ (Fig. 4b) seems to be active in the very low temperature range only, and for the particles below ~022 μm size. Predicted rates for mechanism $\dot{x}_8$ (volume diffusion at stage-3) match well with external rates (Fig. 4c) for all size (0.003-0.045 μm) at higher temperatures.

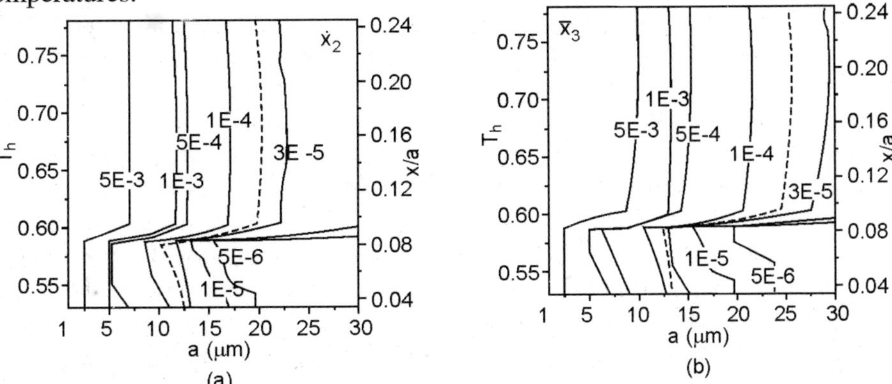

Fig. 3. Contour plots showing the predicted sintering rates for micron Ti system for two lattice diffusion mechanisms: (a) from surface source and (b) from grain boundary source. Experimentally obtained rated are shown by dotted lines.

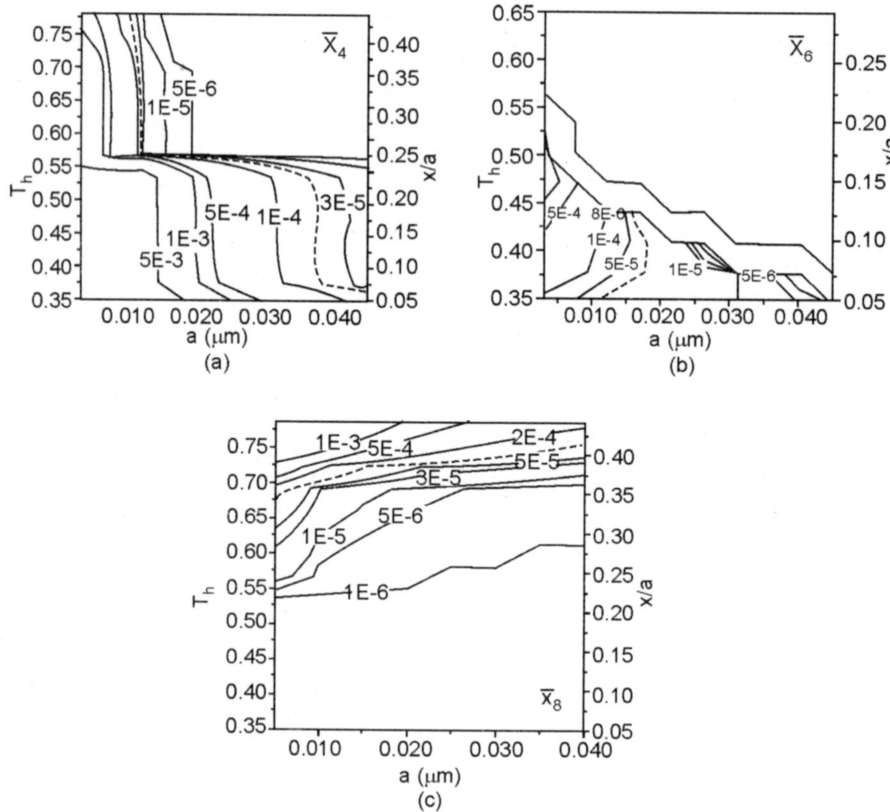

Fig. 4. Contour plots showing the predicted sintering rates for nano Ti system: (a) grain boundary diffusion (jc4), (b) lattice diffusion from dislocation sources (x_6) and (c) lattice diffusion (x_8). Experimentally obtained sintering rates are shown by dotted lines.

5. DISCUSSION

The evaporation and effusion of the vapour (stearic acid) tend to cause swelling in nTi samples (large expansion during initial period of beating compared to mTi system). The precipitation of impurities during cooling caused anomalous expansion during cooling. When a green compact (nTi) is heated to an elevated temperature, though the occluded gases tend to expand the compact, this expansion may partly be reduced if the gases effuse out causing particle rearrangement. Finally at the elevated temperatures various mass transport mechanisms could become active, which in turn would lead to sintering shrinkage. The observed volume shrinkage during sintering could thus be an arithmetical addition of these two processes (particle rearrangement and sintering). When shrinkage rates are plotted as a function of temperature a clear break is observed in the data set at about 700-800°C (Fig.2), which could be due to a drop in the phase transformation temperature in milled powder. This drop could be attributed to the presence of beta stabilizing impurities (*i.e.*, Fe, Ni, etc.). The apparent Q calculated for nTi for both the temperature ranges are much smaller than the self- diffusion data for titanium.

The reported values of activation energy for self-diffusion in α-Ti are in the range of 169 to 192 kJ/mol. It is possible that the apparent activation energy measured in mTi (α-range) refers to lattice self diffusion as the possible rate controlling mechanism of shrinkage. Interestingly there have been reports of *Q* for grain boundary self-diffusion in α-Ti also to be in the range of 187 kJ/mol and thus grain boundary diffusion, as a rate controlling step cannot also be ruled out in mTi. The *Q* reported for self-diffusion of β-Ti once again range from 131 to 328 kJ/mole and the *Q* increases with increasing temperature. The decrease in activation energy (92.5 kJ/mol) at very high temperature indicates the presence of particle rotation (creep like mechanism) due to very fine particle size.

It is necessary to make additional efforts to ascertain which of the mechanism (lattice or grain boundary or both) acts as the real rate controlling mechanism. One can get additional clues from the determination of theoretical rates of neck growth for the given system using the Ashby's sintering diagram approach. Such analysis in the current investigation indicates that in mTi system, $\dot{x}_2$ (lattice diffusion from surface source) and $\dot{x}_5$, (lattice diffusion from grain boundary source), appears to be the rate controlling mechanisms. In nTi system, lattice diffusions from dislocations as source and grain boundary diffusion from grain boundary sources, seem to have significant contribution to sintering. Similarly at the higher temperature (β phase), it is the stage 3, lattice diffusion seems to be the significant rate controlling mechanism.

The apparent difficulty in getting mirror finish polish on the sintered samples and the related grain pull out may be explained by this intra-agglomerate sintering. Possible contribution by mechanism, which can cause neck growth without shrinkage (such as surface diffusion) could also contribute to the above. The low *Q* values and shrinkage anisotropy at higher temperature could be attributed to the grain boundary sliding or grain boundary rotation.

The range of particle sizes, which have the values matching in order of experimental rates are slightly larger in size than the actual size of particles for both (mTi and nTi samples). This reveals an important fact that when particles are in ultrafine or nano size range, the size of agglomerates has strong influence on sintering. This effect is more pronounced in nano titanium, as the tendency of agglomeration is more in nano powders. In addition to that, the grain growth also takes place during sintering.

It is important to note that the milled powder picked up additional impurities. These impurities are expected to play a significant role in sintering as well as in phase transition temperature. These impurities, in present case turned out to be favourable for sintering, *i.e.,* increasing diffusion rate. Thus the increasing sintering rate in nTi system could be due to both the effects, decrease in particle size and impurities. Inspite of variation in purity level from starting powder to milled powder, this investigation is sufficient to give inside the complex sintering mechanisms and its variation through changing particle size.

6. CONCLUSIONS

Microcrystalline titanium powder shows isotropic shrinkage and sintering has been found to be controlled by lattice diffusion in the α range and combined lattice diffusion and grain boundary rotation at elevated temperatures in the beta region. The role of dislocation movements, particle rearrangements could be noticed at initial period of sintering. Further, the grain boundary diffusion at moderate temperatures and lattice diffusion at elevated temperatures and creep like behavior involving grain boundary rotation seem to be the rate controlling mechanisms of sintering.

REFERENCES

1. K. Akechi and Z. Hara, Increase of Sintering Rate of Titanium Powder during Cyclic Phase Transformation, Powder Metall., Vol. 24, pp 41-46 (1981).
2. R. Sundaresan, A. C. Raghuram, R. M. Mallya and K. I. Vasu, Sintering of B Titanium: Role of Dislocation Pipe Diffusion, Powder Metall., Vol. 39, pp. 138-142 (1996).
3. J. R. Groza and R. J. Dowding, Nanoparticulate Material Densification, Nanostruct. Mater., Vol. 7, pp. 749-768 (1996).
4. R. A. Andrievski, Compaction and Sintering of Ultrafine Powders, Int. J. Powder Metall., Vol. 30, pp. 56-59 (1994).
5. Moldovan, D. Wolf, S.R. Phillpot, A.J. Haslam, Role of Grain Rotation during Grain Growth in a Columnar Microstructure by Mesoscale Simulation, Acta Mater., Vol. 50, pp. 3397-3414 (2002).
6. J.G. Li and X. Sun, Synthesis and Sintering Behaviour of a Nanocrystalline α-Alumina Powder, Acta Mater., Vol. 48, pp. 3103-3112 (2000).
7. M. F. Ashby, A First Report on Sintering Diagrams, Acta Mater., Vol. 22, pp. 275-289 (1974).
8. M. German, Sintering Theory and Practice, John Wiley and Sons Inc., USA, (1996).

8

High Performance Cements and Advanced Ordinary Portland Cement Manufacturing by HEM-Refinement and Activation

H. Zoz
Zoz GmbH, D-57482 Wenden, Germany.
ESIQIE, National Polytechnic Institute, Mexico City, DF 07300, Mexico.
CISRI, Powder Metallurgy & Environmental Technology Div., Beijing 100081, P.R. China.

D. Jaramillo V
ESIQIE, National Polytechnic Institute, Mexico City, DF 07300, Mexico

S. Diaz de la Torre
Advanced Materials Research Center CIMAV S.C., Chihuahua CP 31109, Mexico

B. Trindade
FCTUC-University of Coimbra M P-3030 Coimbra, Portugal

H. Ren
Zoz GmbH, D-57482 Wenden, Germany

Z. Tian
CISRI, Powder Metallurgy & Environmental Technology Div., Beijing 100081, P.R. China

O. Chimal-V,
Advanced Materials Research Center CIMAV S.C., Chihuahua CP 31109, Mexico

ABSTRACT

Ordinary Portland Cement (OPC) is the material most widely used in construction industry all over the world and therefore consumed in super large volume. This causes general interest in improving this product with respect to materials properties. Compared to the raw material cost, the manufacturing cost is energy intensive and therefore cement industry has been and is interested in improvements in the efficiency of their milling operation and their rotary kilns.

If High Energy Milling (HEM) is applied for the grinding of cement, this can lead to substantial refinement (< 2 μm) and mechanically activation of the powder particles that in conventional material exhibit a particle size in the order of 50 mm.

Earlier work and preliminary studies have shown that this can result in a faster setting time, a faster curing time and in increased mechanical properties. Due to the far shorter grinding process at far higher energy efficiency, due to an expected reduced firing temperature in the processing and finally due to an expected significant saying in production space for the grinding step, an economically advanced manufacturing process of cement seems to be offered by HEM.

(*Contd...*)

Due to a tremendous refinement into submicron range at high active surface, the material can result into a super fast setting cement (SFC ≈ 3 min) where this feature being available for repair-work of concrete constructions would also lead to an advanced product application which would be the first appliance imagined for short term range. Later this maybe spread out by reduced construction dimensions and faster building due to increased mechanical properties.

The key question if HEM is available and can be scaled up for this kind of super large volume application can be carefully answered positively since the high kinetic semi-continuously processing route based on a carrier-gas in compression mode has been introduced.

The present paper reviews the preliminary studies, explains the novel technique and suggests the route into commercial application. Particular attention is paid to wear results with an applied Si_3N_4-grinding unit where no substantial wear was found after 4000 h of operation.

1. INTRODUCTION

Since ancient times, cement has been used as binder material to form concrete structures. Today, Ordinary Portland Cement (OPC) is a super large volume product with thousands of monthly produced tons all over the world. OPC has a particle size distribution (PSD) where 90% of total particles correspond to 50 mm, disclosing an onset setting time of 2 to 3 hours. Depending on its chemical composition OPC might attain 320 kgf/cm^2 of compressive strength after 28 days curing. The conventional firing temperature is today about 1450°C.

In this paper we suggest to apply High Energy Milling (HEM) as an innovative processing technique for the commercial production of superfine High Performance Portland Cement (HPPC) in short term range. As an almost natural consequence, then in long term range, this technique should be applied also for the manufacturing of OPC at lower energy consumption and then at finer structure as well.

HEM leads to a significant refinement of the cement particle size. The here reviewed preliminary studies [1-2] had lead to an average sub micron particle size which seems economically realistic. Compared to the conventional OPC which may have an average particle size in the order of 50 μm cement mortar made oil using refined cement can end up with larger strength, faster setting time and faster curing time.

Additionally, since HEM is based on the collision of grinding media rather than shear and friction effects of the same in low kinetic conventional ballmilling, this process results in a significantly higher efficiency related to the total energy consumption [3].

It is further shown, that the tremendously larger and high active surface of the HEM-refined powder which is also related to solid state reactions during HEM leads to significantly lower firing temperatures as well as to a super fast setting time (< 3 min) of the cement.

Previously reported work conducted on slightly refined cement has reveled that its compression strength might be increased by micron-size refining of cement particles [4]. In other work it was stated that refining cement leads to a number of drawbacks, such as strength retrogression, unusually fast setting features, and a large demand for water by the refined cement which dramatically reduces its compressive strength [5-6].

Based on the HEM-refined superfine cement and following the ASTM C109 norm as close as possible, a number of cement pastes have been produced using special sand as well as a water-reducing agent. Results verify that using proper additives and refined-clinker (RK) lead to a substantial increase on the material compressive strength, as compared to the typical strength of OPC. For an intermediate comparison to the state of the art, a mixture maintaining the total amount of each concrete ingredients as stated by the standards of RK, HPPC by HEM and other ingredients have been made and this material did undergo high early-strengths in the order of 320 kg/cm^2 (first day) and over 1100 kg/cm^2. after 28 days curing [2].

In result, next to a significant improvement of the mechanical properties of concrete, this technique also seems to offer a high potential in becoming a revolution for the conventional processing route of OPC since due to refinement and activation, setting time and firing temperature might substantially be improved. Thus, a promising safe of energy in the manufacturing process as well as cost in construction time and dimension supports the here given attempt, too.

Later this might even become interesting for tile-production where the firing temperature for the base- plate is equal to the one of cement but the melting temperature of the glaze-coating is significantly lower at around 450-600°C. Energy consumption in this industry is also a determining factor in the manufacturing costs [7]

The present paper reviews the preliminary work that has been done and proposes and explains the route to achieve commercial appliance in large volume.

2. HIGH ENERGY MILLING, TECHNIQUE AND APPLICATION

High Energy Milling (HEM) is one of the High Kinetic Processing techniques (HKP) which covers HEM, Mechanical Alloying (MA) and Reactive Milling (RM).

Mechanical Alloying (MA) is maybe the most historical one and has been described as a process where powder particles are treated by repeated deformation, fracture and cold welding by highly energetic collisions of grinding media in a milling process [8-9]. By this technique it is possible to synthesize new materials with new properties that cannot be created by conventional route e.g., due to a not present thermal equilibrium or immiscibility of their components. By structural design, important materials properties can be influenced (*e.g.*, nanocrystalline, amorphous).

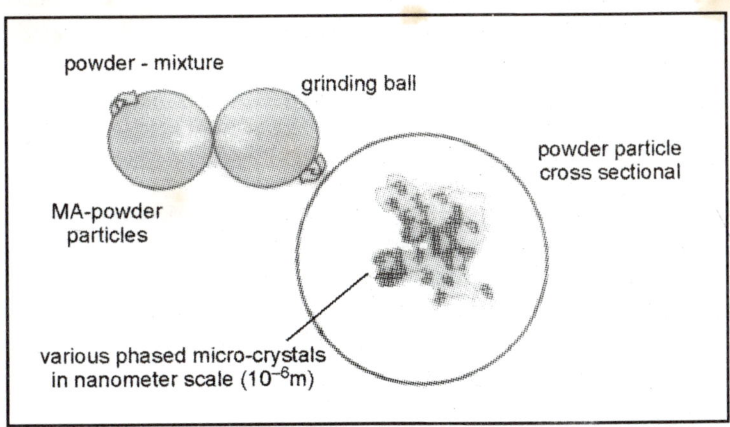

Fig. 1. Schematic of the collision as the main event of energy transfer.

High energy milling (HEM) and reactive milling (RM) are performed by the same processing principle where the variation is in general based on the target of the processing, the transformation effect by the kinetic energy and the starting materials.

The different procedures can be described as High Kinetic Processing (HKP) where the collision of the grinding media is the main event of kinetic energy transfer from the milling tools into the powder [8-9]. Fig. 1 shows the schematic of the collision. The basic equation (1) describes the relation between the kinetic energy (E_{kin}) and the velocity v of a single ball.

$$E_{kin} = 1/2 \ mv^2 \qquad ...(1)$$

It is clearly seen that the maximum relative velocity of the grinding media is the most determining factor contributing to the kiinetic energy.

A typical example of practical importance that demonstrates the effects that are inherent in Fig. 1 has been described in detail [10] in the system:

$$24 \ Ti + Al_{11}Nb \rightarrow Ti_{24}Al_{11}Nb$$

In this work the influence of the milling parameters have been evaluated and basically the processing route of Cycle Operation for CMB-materials [11] had been invented. By this technique, the problematic agglomeration and sticking behavior of the, in this case, Ti-Al-based material in a dry milling process could be significantly reduced which did lead to an increase of the powder yield from < 4% to 80%. Later on similar effects were found in the Ti-Ni-system [12] and in the processing of a number of ductile metal flakes [13-14]. In 1999, Kim et. al. investigated the difference of Cycle Operation and Constant Operation in the Fe-Co-system and found tremendous effects on particle and in particular crystallite-size reduction [15].

If the here described processing method is applied for particle and /or crystallite size reduction or particle deformation (*e.g.*, flakes) in single systems, this route is described as HEM. The goal can be to receive a special particle geometry *e.g.*, for rapid and large-scale production of ductile metal-flakes in dry process without solvents and less energy *e.g.*, for paint-pigments, conductive pastes and anti-corrosives [16-17]. On the contrary rapid particle size reduction of brittle solids like Enamels or Glass Fluxes has recently been introduced as a new application field [18-19].

The definition of RM is suitable if during milling a chemical reaction is wanted and observed. By this route, a dispersion of ultra-fine (nano-scaled) crystals and/or a homogeneous dispersion of transformed phases in a matrix can be achieved [20].

A typical example of practical importance here has been demonstrated in the Ag-Sn-system [21]:

$$Ag_3Sn + 2 \ Ag_2O \rightarrow 7 \ Ag + SnO_2$$

Due to the CMB-behavior of this ductile system, again the processing has to be performed by applying Cycle Operation. With respect to the starting materials (*e.g.*, Ag_2 instead of Ag), RM in this kind of principle can show economical advantage. Often these processes are environmentally benign as they avoid wastes *e.g.*, in solid state synthesis where organic solid-solid reactions and others can rapidly proceed without any wastes and in the absence of solvents [22-23].

3. EQUIPMENT

Industrial mills of different types are in practical use. Vibration-mills have to move the mass of the milling chamber. This limits their sizes and they are hard to run in normal environments which is also valid for simple (drum-) ball-mills with a rotating vessel. The efficiency related to the total energy impact of these types is known to be less than 5 % [24-25].

Jet-mills use large streams of energy intensive air or inert gas which very much limits their economical use and cannot supply solid state reactions for the cement application since the piece-mass of the single accelerated particles would be too small. Bead-mills, horizontally or vertically, do not exhibit a significant kinetic impact, planetary ball-mills [26] and shaker-mills [27] are limited to laboratory size. The most suitable choice are therefore high energy horizontal rotor-mills that can be operated in dry processing at high relative velocity of the grinding media (up to 14 ms^{-1}) that cannot be reached by the other types (up to 5 ms^{-1}) and which are mostly the referees in chapter 2 for process and application.

Furthermore these high energy mills can be operated in batch, continuous and semi-continuous operation mode using carrier gas in de- and compression mode where the last mentioned is expected to be of substantial importance for the large volume manufacturing of cement.

4. PRELIMINARY STUDIES (IN LABORATORY SCALE AND BATCH OPERATION)

4.1. Ordinary Portland Cement

For the preliminary studies, OPC supplied by Grupo Cementos de Chihuahua (GCC), Mexico was used as the starting material for the refinement tests by HEM. A typical raw-composition of GCC-cement, as released from the milling operation is given in Table 1. The typical particle size distribution (PSD) of the GCC-cement is given in Table 2.

Table 1. Average raw-composition of GCC-cement, as released from the milling operation

Average raw-composition of GCC-cement, as released from the milling operation									
Component	CaO	SiO_2	Al_2O_3	Fe_2O_3	MgO	K_2O	SO_3	Na_2O	other
Mass%	64.2	21.8	4.1	3.3	1.4	0.61	2.5	0.15	bal.

Table 2. Typical particle size distribution (PSD) analysis of GCC-cement

Typical particle size distribution (PSD) analysis of GCC-cement										
Fraction%	10	20	30	40	50	60	70	80	90	100
Size [μm]	2.5	5.1	7.2	10	15	20	28	34	49	85

4.2. High Energy Milling Process in Batch Operation (Preliminary Studies)

For the preliminary studies, a laboratory scale high energy mill (Simoloyer CM01-21) was used. In order to test the wear-resistance of recently available ceramic grinding units that are based on Si_3N_4, a grinding unit W01-21m-SiN has been used. where the vessel is lined inside with Si_3N_4-plates and the rotor is built up by restorable Si_3N_4-bulk-blades.

Fig. 2 shows a horizontal high energy mill with 2 L chamber volume, which can be operated on a table next to the process controlling computer. It runs with water cooling at rotation frequencies up to 1800 rpm.

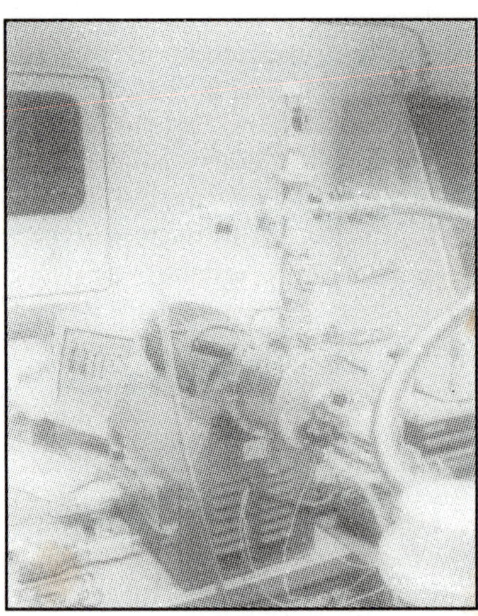

Fig. 2. 2-L horizontal high energy ball-mill (Simoloyer CM01-2l) with vacuum and inert-gas loading, operation and unloading.

Fig. 3a shows the disassembled grinding unit (flange, rotor, vessel and blind-lid) and Fig. 3b the draingratings and in the center the side-adapter.

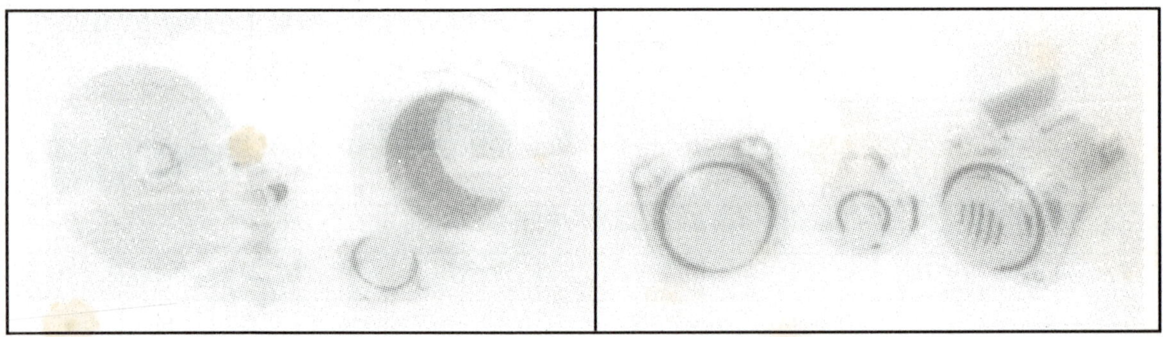

Fig. 3a. Disassembled grinding unit WO-2lm-SiN (flange, rotor, vessel and blind-lid).

Fig. 3b. Draingratings As-01-2l-SiN, As-G1-2l-SiN and side adapter RKDNO1-SIN (center).

With respect to the ceramic milling tools, the rotational speed of the mill was chosen not to exceed 900 rpm. It was further decided to operate the cooling system because of the temperature limitation of the bonding of the ceramic to the stainless steel base parts.

In the preliminary studies, 3 main parameters have been investigated. The processing time has been varied between 30 and 150 minutes at different powder/ball weight ratios from 1:20 to 1:40. This was additionally performed under varied rotational speed from 500 to 900 rpm and in another test-series, Cycle Operation had been applied which means during processing, the rotational speed of the rotor is changed in a special frequency in the range from here 500-900 rpm which has been proven to tremendously increase the powder yield of materials that tend to stick and agglomerate under MM to the milling tools and to each other [10-13].

Table 3. Varied parameters at the processing of GCC-cement powder

Analyzed parameters during the processing of GCC-cement powder			
Parameter	processing time [min]	powder/ball weight ratio	rotational speed (rpm)
Variation	30, 60, 90, 120 and 150	1:20, 1:30. 1:40	500, 750, 900 and 900/500*

The detailed parameter variation is given in Table 3 and the used Operation Cycle for the test-series at 900/500 rpm is given in Fig. 5 and the corresponding Discharging Cycle 500/900 in Fig. 6.

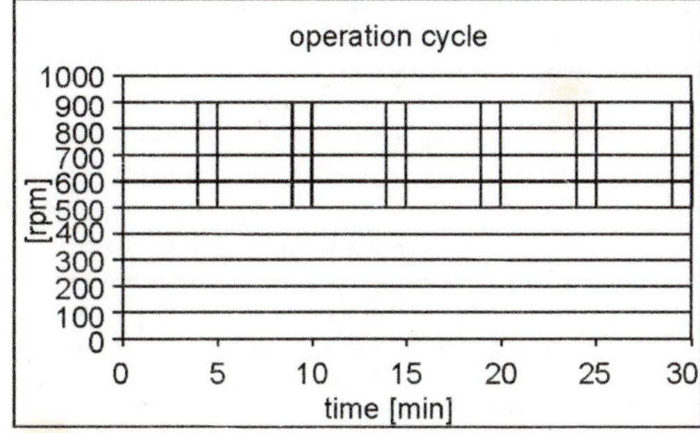

Fig. 5. Operation Cycle 900/500*

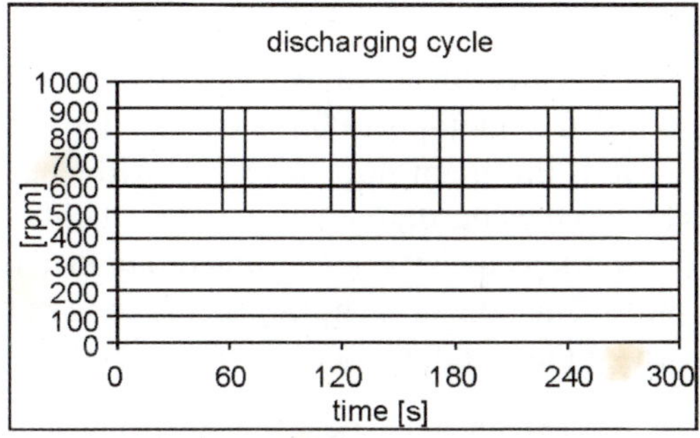

Fig. 6. Discharging Cycle 500/900**

The processing was done under air, which means no air-lock, no evacuation or flooding with inert-gas was used and the vessel was kept at RT by the cooling systems. The detailed base-parameters of the processing schedules are given in Table 4.

Table 4. Base Parameters for the processing of GCC-cement-powder.

Base parameters for the processing of GCC-cement powder	
Milling device	Simoloyer CM01, 2.7 kW
Operating software	Maltoz 3.1
Grinding unit	W01-2lm-SiN (2 liter, water-cooled, Si_3N_4 lining/bulk)
Grinding media	Zirconia, fully stabilized (YTZ), 5 mm, 2.3 kg
PCA (lubricant)	Drops of propylene-glycol
Starting powder	GCC-cement powder
Starting powder load	Varied at 57.5, 76.6 and 115 g
Powder/ball weight ratio	1:20, 1:30 and 1:40 varied
Atmosphere	No, operation, charging and discharging under air
HEM time I-V	30, 60, 90, 120 and 150 min varied
Operation Cycle	900/500 rpm -4/1 min
Discharging Cycle	500/900 rpm - 50/10 s
Average discharging time	5-10 min
Milling temperature	< 25°C (vessel inside by Maltoz)
Discharging/separation	Draingrating As 01-21 SiN
Average poweder yield	Close to 100%

For loading the starting powder for a single test, first the grinding unit is set into charging/ operation position which means the main-port P01 is on top. Then the powder is loaded through the open main- port. For operation, the main-port is closed with the blind-lid. After processing, for discharging, the blind-lid is replaced by the draingrating As-01-21-SiN with an adapted powder container. Then the grinding unit with the adapted container is turned into discharging position, which means the main- port is in bottom position. Then the discharging operation is started and the powder is transferred into the powder container where the grinding media remains in the grinding chamber.

4.3. Characterization and Specimen-Code (Preliminary Studies)

The as-received (initial) and the as milled powders were characterized by laser diffraction (PSD), XRD, BET and optical microscopy.

The particle size measurement was carried out with a Malvern Mastersize 2000 device and for X-ray diffraction we applied a Siemens Diffractometer D 5000 using monochromatic Cu-Kα radiation. The BET-surface area of the powders was analyzed with a sorption/desorption commercial Quantachrome device, and for optical microscopy an Olympus microscope AX70 model was used, optical pictures were taken with a JVC-TK 1270 video set. Compressive tests were carried out using a 50-C-21 H4 universal testing machine. The American standards for cements were used along this research work for ail samples, *i.e.*, using 5 × 5 × 5 cm large specimens (ASTM 109/C).

For the definition of the quite large number of up to 60 different powder-charges, where some have been disregarded corresponding to received results, a code as given in Table 5 was used to determine each powder charge:

Table 5. Specimen-code for the different powder charges of processed GCC-cement powder

Specimen-code for determining the different powder charges of processed GCC-cement powder			
Parameter	*processing time [min]*	*powder/ball weight ratio*	*rotational speed (rpm)*
Example	60	1:40	500
Code	m60	40	500
Full code	m60-40-500		

The wear of the milling tools was determined by visual observation and measurement of the blade diameter.

4.4. Results (of Preliminary Studies)

Figure 7 gives the X-ray diffraction patterns of the as received (OPC) cement and of the refined cement (Milled HPPC) where the investigated powder refers to the charge m60-40-500.

After the HEM-process, most of the peaks appear very much shortened and broadened verifying structure refinement.

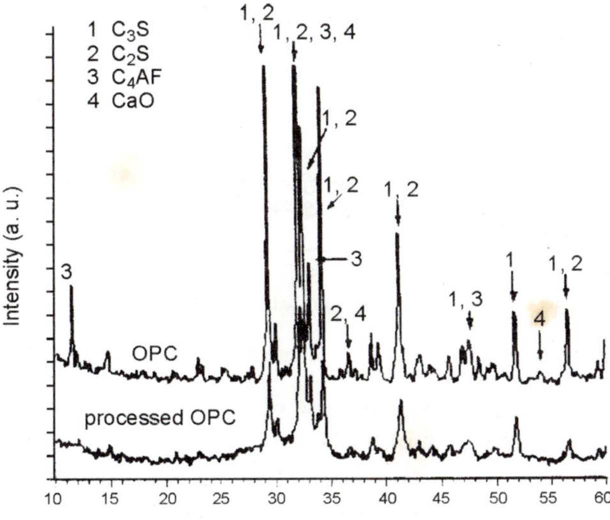

Fig. 7. XRD-patterns of GCC-cement poweder (OPC) and of processed powder m60-40-500.

Most interesting is here, that next to refinement a number of solid state reactions [28] seem to proceed which is caused by the kinetic energy inpact. E.g. in case of the quartz (SiO_2) given at 21.8

wt% and forming an infinite covalent crystal lattice, fragmentation of Si-0 bonds leads to completely unsaturated freshly cleaved faces with free Si and O-Si surface radicals and other highly reactive species. They represent a local plasma which tends to be saturated immediately and leads to tribochemistry effects [22-23].

Figure 8: gives the particle size distribution of selected results. The finest powder at most economical parameters is represented by the test m30-20-500. This is in so far surprising, as this powder is achieved after the shortest applied processing time at lowest applied rotational speed and at lowest applied powder/ball weight ratio but does belong to the finest achieved powders in total. In other words, here the lowest applied kinetic results into one of the finest achieved powders. This can only mean, that the general attempt of the processing schedule was to long and gives the expectation, that here a rapid particle size reduction of the naturally brittle solids leads to very fine powder after much shorter time and that this fine powder leads to a dumping effect for the kinetic impact (acts like a liquid in the dry process) [18-19] which is a general problem in the batch process. Further processing time of further increase of kinetic by increased velocity (rpm) or by increases ball/powder weight ratio does then not lead to further significant particle size reduction.

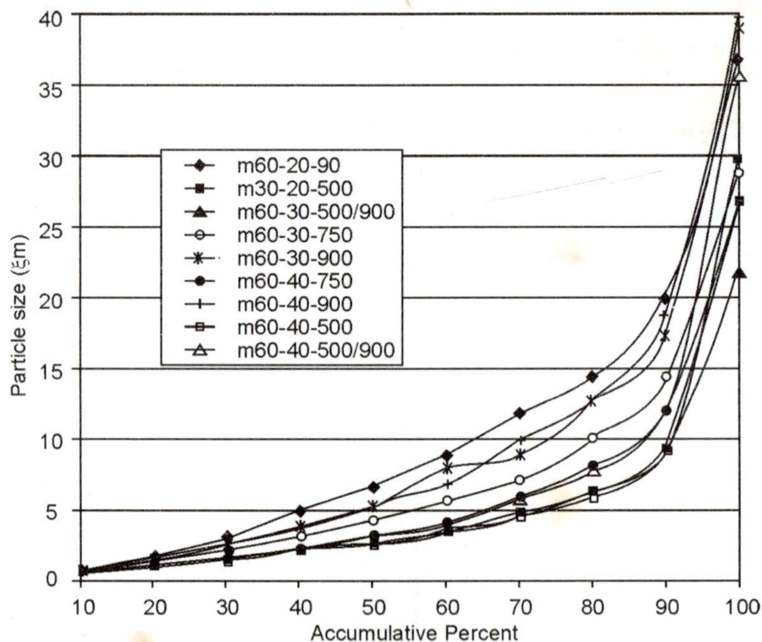

Fig. 8. Particle size distribution of processed GCC-cement powders.

As far as we can observe here a slight increase of the particle size with the processing time, this leads to the assumption of agglomeration of the fine particles in an "over-milled" stage.

What is further interesting, is the observation, that the finest powder in total, which means the finest 100% value, is achieved when Cycle Operation is applied (test m60-30-900/500). This might be explained by the known effect of this processing route to avoid dead-layers in the process, where particles could he "hidden" resulting into less or non-processed particles after discharging. A further hint to this can be seen in figure 9b where among the very fine processed powder a few (< 10 μm) mediums sized particles and one large particle is observed.

In general Figs. 9a and b show optical micrographs of the as received and of the processed powder with the code m60-40-500. Impressively the difference in microstructure and morphology is presented here. The non-refined cement appears as a very heterogeneous material with a huge number of large sized particles in the order of 30-60 μm and some very large particles in the order of 100-120 μm.

On the country, the refined cement shows a much more homogeneous microstructure at very small particle size (D50 is given at about 2 μm by PSD) with a few (<10 μm) mediums sized particles below 20 μm size and one large particle in the order of 100 μm. The BET-surface area was found to be $15 m^2/g$.

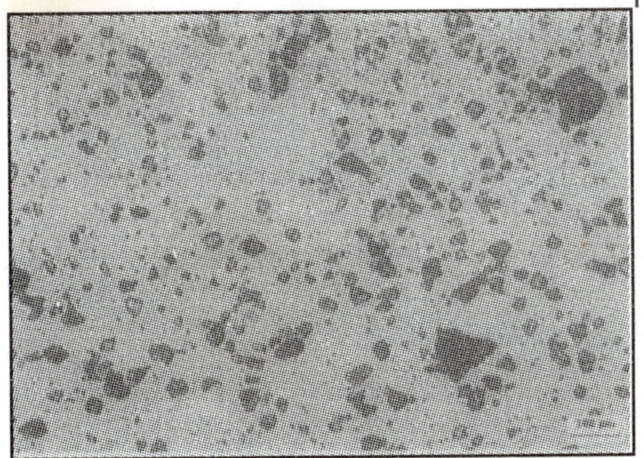

Fig. 9a. Optical micrograph of GCC-OPC cement powder (as received).

Fig. 9b. Optical micrograph of GCC-HPPC cement powder (processed under the code m60-50-500).

As a demonstration of the chemical and physical behavior of studied cements, using both materials shown in Fig. 9a-b, very fluid mortar-cement pastes were carefully prepared and the two little sculptures, as given in Figure 10 were molded applying no pressure.

Here, impressively the filling capability of finer grade cement for the expansion into fine and complex structures compared to the one of conventional material is shown. The setting time of the fine graded high performance material was 2-3 minutes only.

In particular these properties can be of major importance for the future advanced repair of cracks and hollow spheres of concrete constructions like bridges or dams.

In order to compare the strength of the High Performance Portland Cement (HPPC) by HEM-refinement, samples were tested using a universal testing machine 50-C-21 H4 shown in Fig. 11 at a compressive rate of 900 N/s. Tests were applied after 1, 3, 7 and 28 days, results are given in Fig. 12:

The compressive strength results of conventional OPC from around 94 kgf/cm^2 after the first day to 320 kgf/cm^2 after 28 days curing do agree well with typical expected values as reported by the C-109 ASTM standards.

The HPPC-material achieves a compressive strength of 550 kgf/cm^2 after the first day and of 1018 kgf/cm^2 after 28 days.

In comparison, the refined and activated HPPC achieves a very high early strength where already after 1 day the strength is about 70 % higher than the strength of the conventional material after 28 days. After 28 days, the strength of the HPPC is more than 3 times higher than the corresponding value of the conventional material.

Fig. 10. Sculptures made by conventional OPC (left) and by refined HPPC (right).

Fig. 11. Universal testing machine for HPPC specimens strength evaluation.

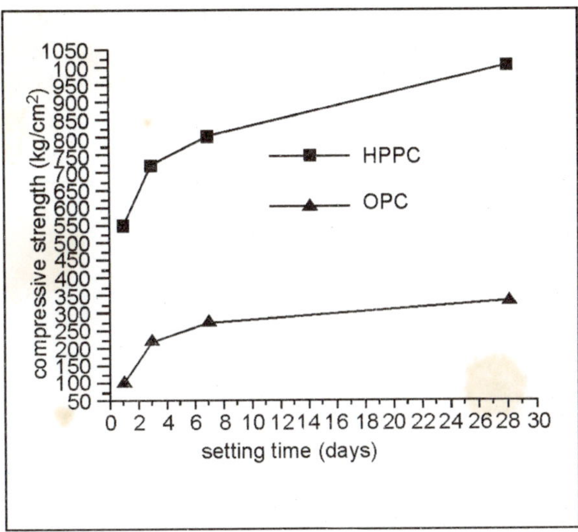

Fig. 12. Compressive strength results of HPPC as compared to OPC.

Finally we checked the wear of the milling tools which is important to be estimated in early stage since in case of success, extremely huge production numbers must be considered. The results are in so far surprising and impressive at the same time.

Figs. 13-16 shows pictures taken in CIMAV from the disassembled grinding unit after in total 4000 hours of operation with a grinding media load of 2300 g and an OPC-powder load varying between 115-230 g at rotational speeds varying between 500 and 900 rpm.

In detail Fig. 13 shows the inner Si_3N_4-lining of the vessel where in total (all components) 2 cracked plates have been found. One of these cracks is in the area of the given picture and is marked with an arrow. In both cases, no fracture of the plates did fall off which means the cracks are not considered as being problematic, in particular not after that long milling operation at relatively high kinetic.

Fig. 13. One of two cracks in the lining plates, both appeared in cylinder-lining.

Fig. 14 shows the inner view of the main-port POI with the assembled draingrating As-01-21-SiN where it can be seen, that the border of the vessel-lining causes in so far problems, as the not well defined circle-edge can act as a dead-zone in the process. This should be improved by using a 2- or 4-piece form-part for the lining, where then the circle-adge would be preformed when producing the Si_3N_4-lining plates.

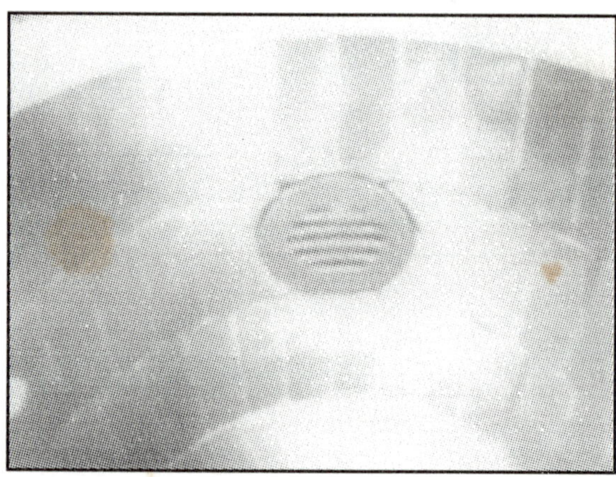

Fig. 14. Circle-edge of P01 acts as a dead zone which should be improved.

Fig. 15 shows the draingrating As-01-21-SiN. One of the center-edges was broken and had to be repaired. The edges are considered to be too weak for the kinetic impact during discharging, in particular when Cycle Operation is used since then at least for short time periods, high kinetic impact is exposed. This matter should be subject of dimensional improvement (thickness of lining or width of center-edges).

Fig. 15. Broken center-edge of draingrating, dimensional improvement is recommended.

Figure 16 shows flange and rotor. No cracks or damage could be noticed in the flange-lining and most surprisingly, no damage or cracks could be found at the Si_3N_4-bulk rotor-blades as well.

The diameter of the blades has been measured and a diameter-loss of less than 2 mm has been determined. This is a surprisingly very good result and it should be pointed out, that the given total processing time of 4000 hours would refer *e.g.*, to a 24-h-operation for a period of more than 5 months.

The YTZ grinding media did not show significant diameter-loss as well.

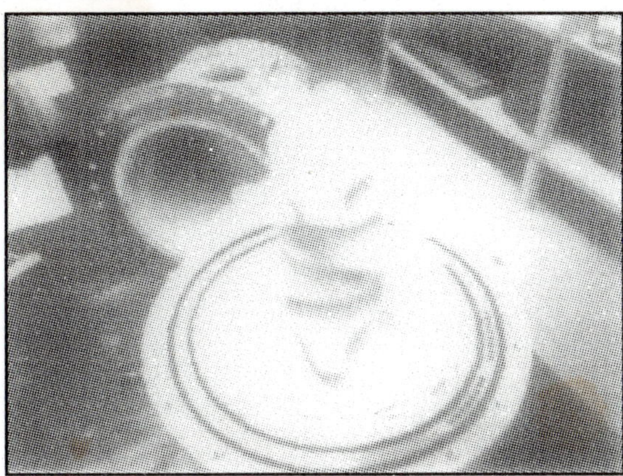

Fig. 16. No damage on the rotor-blades, wear surprisingly low, diameter loss less than 2 mm (after 4000 h!).

4.5. Conclusions (of Preliminary Studies)

In conclusion, the preliminary studies do show, that HEN/RM is capable to refine and activate OPC which then results into a High Performance Cement with very interesting properties such as high early strength higher than the final strength of conventional material and 3 times higher strengths after 28 days.

There seems to be a good chance to produce far better material at lower cost.

As far as the refinement-operation by HEM is concerned, the results lead obviously to the assumption, that the set processing times have been too long and that the batch operation procedure might be suitable to show up attempts in laboratory scale, but not in large scale production and not for optimum processing results.

Here, the semi-continuously route using a carrier-gas in compression mode has to be considered because simultaneously to the appearance of fine fraction, this fine fraction is extracted from the grinding process and therefore cannot any longer dump the kinetic of the impact during HEM. This technique will be discussed in the following chapter.

5. MAJOR GOAL: CONTINUOUS HEM-PROCESS

Since OPC is a super large volume product based on relatively cheap raw materials, the manufacturing costs play the major role in the total product costs. The main operation steps in the manufacturing process are grinding of the raw-materials and the firing of the compound.

The grinding process today is done in huge drum(ball)mills in a low- kinetic batch process. Fig. 17 shows such a huge mill that can exceed the dimensions of a 2- floor building easily. The firing step is done in even bigger rotary kilns at temperatures around 1150°C. In fact in the literature we found a statement, that most of the total electric energy used in the world is consumed by milling devices producing cement [29] which probably should include the kilns.

If an advanced processing/grinding technique could reduce both, the energy consumption in the milling operation as well as the one during the heat treatment (firing) due to reduced temperature caused by high active large surface after finer milling, this might lead to a significant cost reduction of the entire manufacturing process.

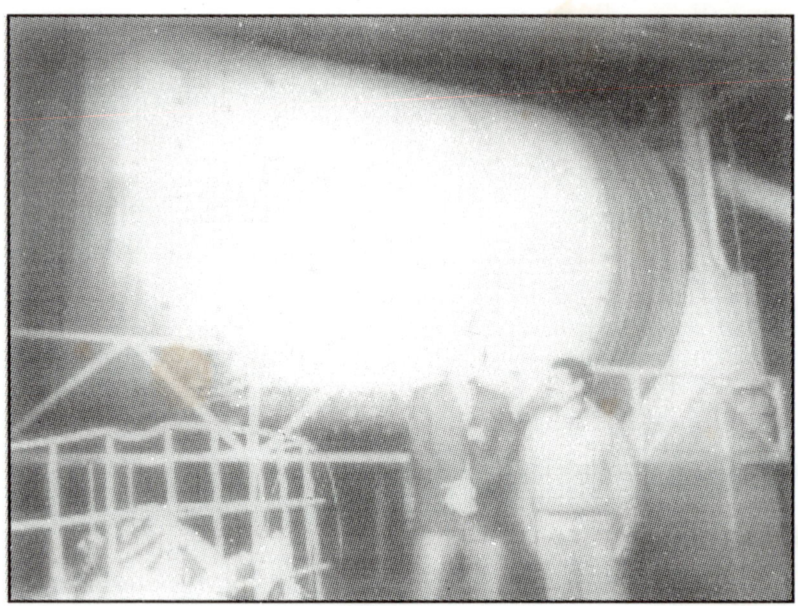

Fig. 17. Huge drum (ball) mill at Interceramic in Chihuahua, Mexico.

Since the preliminary studies also end with the conclusion, that the used batch operation should not be considered for mass production, another technique shall be considered, which here most obviously must be the continuous process, first because of naturally larger throughput (if applicable) and second in this case because of expected far better parameters (faster process, finer particles) where the last expectation is based on experience in rapid particle size of enamels and glass-fluxes [18-19]

The process principle and a mass-production attempt for cement are explained in the following chapters.

5.1. Semi-Continuously Processing

The semi-continuous operation procedure can be divided into two principles, the one using depression and the one using compression. The difference describes a different use of carrier-gas where in depression mode the gas system is open and suction is used to carry powder particles out of the grinding chamber into a cyclone. In compression mode, the carrier-gas is cycled in a closed gas-circuit which may also be divided into a primary and secondary circuit. In both cases, the simoloyer is continuously fed with starting material and continuously processed material is discharged. Important issues here are the control of the amount of powder in the grinding chamber and the separation res. classification of the product out of the gas-flow.

5.2. Semi-Continuously Processing in Depression Mode

Experience in depression mode is already available in industrial production of ductile metal flakes [1 6- 17]. Fig. 18 shows a picture of the HEM-devise in such a process where the cyclones are located in the second floor and can therefore not be seen on this picture.

However, it must be noticed that the production rates in this application cannot be compared to the numbers that would be applicable in the cement-world.

The entire unit is based on a load cell which is connected to a control unit that also controls the automatic feeder for the starting powder and in this case another one for a PCA. A fan is initiating depression through 2 cyclones that are connected with the grinding chamber where the first one separates to heavy particles and returns the separated fraction into the Simoloyer (high energy mill), the second one then separates product and gas.

- oversized powder from cyclone (>8μm)
- flakes from cyclone operation
- starting powder
- processed powder to cyclone

Fig. 18. Simoloyer CM100-s1 in semi-continuous configuration in depression mode (Picture: Fukuda, Japan).

5.3 Semi-Continuously Processing in Compression Mode

For the here discussed cement-application, for a number of reasons, the second mentioned principle, the semi-continuous processing based on compression mode is considered. Experience in this is based long term project work in rapid particle size reduction of enamels [18-19, 30] where one of the pilot-plants is shown in figure 19 which is based on the laboratory scale Simoloyer CM01.

Fig. 20 shows the corresponding scale-up to a CM 100-Simoloyer in compression mode.

This unit is suggested to be used for the continuously refinement of OPC in large volume and is based on the experience of the earlier and smaller pilot-plants VS01 a and VS20a based on CM01 and CM20 scale respectively [18-19].

Fig. 19. VS01a (Simoloyer CM01-2ls1, pilot plant in compression mode, Picture: Zoz, Germany).

In general, the Simoloyer is equipped with a closed gas circuit where the side channel turbine on top blows the gas at high velocity through a heat-exchanger, then the carrier-gas picks up the starting material at the injection unit and goes into the grinding chamber at 2 tangentional ports. On the other side of the unit, the multiphase flow leaves the chamber via 2 more tangentional ports and goes into the first cyclone which has a return path into the chamber. In line after the first cyclone the classified multiphase-flow is separated in the second cyclone and the final material is collected in a container that is adapted with and automatic air-lock. Before the carrier-gas is returned into the turbine, it has to pass a filter-unit that automatically returns filtered dust into the second cyclone. In order to classify the multiphase-flow at the first cyclone, the gas system is equipped with various by-pass units which allow to adjust different velocities in each of the cyclones, the grinding chamber and the piping.

With respect to application, the main criteria is the remaining time. Up to now, this principle can only be realized for those processes that require in batch operation a processing time in the range of several minutes *e.g.*, rapid particle size reduction of enamels [18-19]. Due to the effect, that the fine fraction of particles is continuously, which means here also immediately discharged out of the vessel, the dumping effect of this fraction (like liquid) is terminated and consequently the kinetic of the system is increased tremendously. Additionally it is important to understand, that in case of treating large brittle particles in up to several mm-scale by collisions of grinding media at high kinetic, these particles will collapse not just into two pieces but into a number of pieces of different size and finally with a size distribution. This means that also a very fine fraction is present right away. For the today's understanding, this all together leads then to the found extremely short processing times in continuously mode in the range of several seconds for these materials-systems which is expected to be also valid for the cement-system.

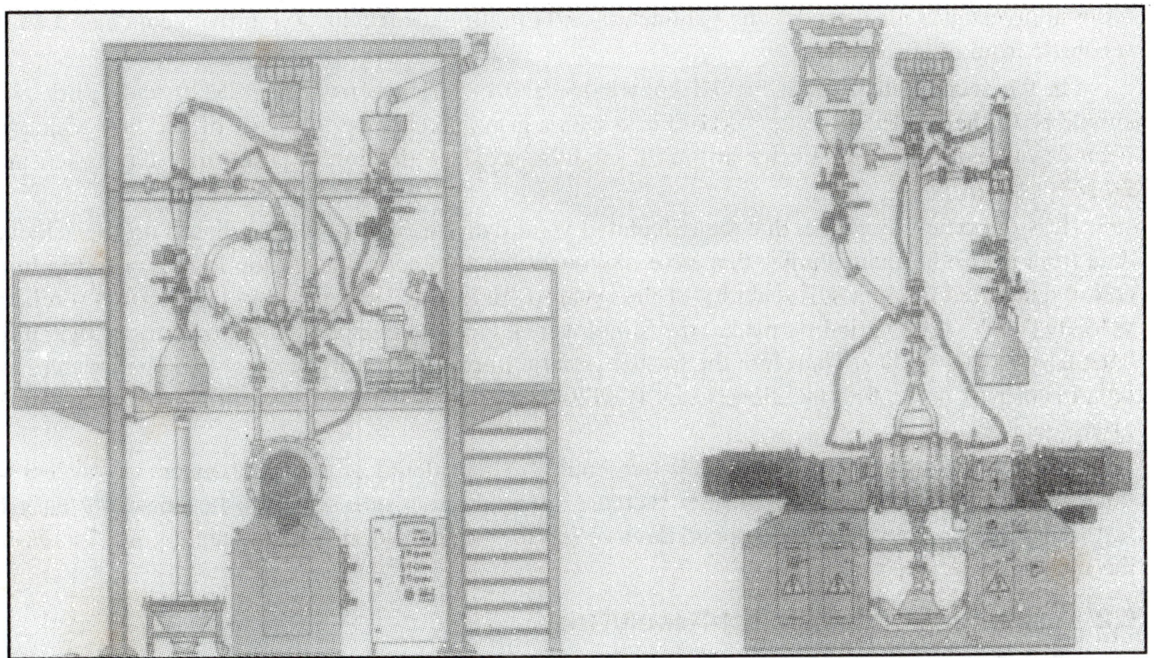

Fig. 20. Simoloyer CM100-s1 in semi-continuous configuration (compression mode).

For the refinement of OPC, this suggests additionally, that the starting material should probably not be the already ground cement powder in a range of 50 μm. Economically it appears more interesting to use much coarser material in the order of 4-10 mm which then maybe produced only by crushing and not by grinding which finally means the today existing milling step in the previously mentioned huge drum mills could be skipped.

Since the preliminary studies did show a potential, that achievable processing times will be below 30 min and since the experience in the glass-systems did show a cut of processing time of factor 10-50, this seems to open a good chance to achieve final processing times in the range of 1-3 minutes which could then be estimated to a production rate of several tons per day with a single CM100-system as given in Fig. 19. If we then take into account, that Simoloyer-systems are available up to 900 liters size with the potential of further enlargement, then this technique seems to offer the application for the here discussed super large volume product. In any case and from this point of view it will be far capable to produce smaller volumes of High Performance Cement in the range of 10-100 to/day.

6. FUTURE WORK AND TARGET

6.1 Next Steps

In the next step it will be important, to prove that the here described laboratory scale knowledge and potential can be scaled up to an industrial scale. For this it will first be necessary to set up the semi-continuously operation in compression mode where for this we suggest a CM20-scale (VS20*) which is a device with a grinding chamber capacity of 20 liters and has been calculated to be suitable for an initial small-scale production of > 720 kg/day based on a 20 h/day [31]. Thus being obeyed as a not able production rate already, the CM20-plant could still be operated in a laboratory which is necessary in order to investigate ideal processing parameters and to introduce technical improvements based on the feedback from online testing.

In the second step, the achieved knowledge and experience from the CM20-scale pilot plant should be further scaled up to the CM100-size with a grinding chamber capacity of 100 liter which has been calculated to be suitable for an initial medium-scale production of > 3.6 to 1 day based on a 20 h/day [31].

It is important to notice, that the calculated yield data for both systems is based on the achieved data from the preliminary studies that have obviously been processed at far too long processing times and at quite medium kinetic capability of the systems. In fact and compared to the maximum relative velocity (MRV) of the grinding media, the Simoloyer-device can be operated up to 4x faster than it has been used in this studies. Therefore the further general improvement of the processing parameters will be a major part in the first mentioned step (CM20-base) which then can give a different view for the given yield data.

Finally, the scale up to the CM100-size could be considered as the achievement of success for this project goal since then any further increase of production rate seems to be relatively easy by further increasing the size of the applied devices (up to 900 liter volume today) and by just increasing the number of devices.

6.2 Application of the Project-Result/Goal

The short-term application of the project-goal is to produce HPC and take advantage of it's accelerated curing time and at the same time substantially increased mechanical properties. This would he *e.g.*, an ideal application in repairing bridges and dams by injection technique where in particular under-water repair conditions would tremendously benefit from the new materials' properties.

For these particular applications, today some high-performance materials usually based on polymer and polymer water management are available at costs from $US 700,00-1.000,00/ton. A careful full-cost calculation in the production order of 10 tons per day leads to a value of around $US 600,00/to for the HPPC [31] which promises even if disregarding superior properties an immediate acceptance on the market.

Here it must be pointed out, that the calculated cost level for the HPPC is based on the operation of pilot plants, which means after the achieved project goal, these cost are expected to be substantially decreased.

In mid-term application, the increased mechanical properties shall potentially be used for key-units in construction and building in order to safe limited and cost-intensive space. This can *e.g.*, be pillows of shifted mass-transportation like highway or monorail. Any successful introduction here will directly depend on decreased product cost by increased production rates, where this seems quite possible since the today's standard cost of OPC can be estimated to about $US 60.00/to and a good share of this is related to the energy cost of the heat treatment step [31].

In long-term range, it might even be possible to replace entirely the conventional manufacturing technique of OPC and gradually reduce the firing temperature by refinement with higher activation and then meet the key-points at maximum mechanical properties, accelerated curing time and producing cost reduction by energy saving. In this case the mark-line of product market cost based on the average $US 60,00 per ton of OPC today will be about $US 100,00/ton which would mean that an expected 100 % increase of compressive strength (to be on the very safe side) might lead to a general material volume saying of at least 30 % equivalent to about $US 20,00/ton already and another $US 20,00 /ton would be to be benefited by reduced labor- and construction cost and building time at least.

CONCLUSIONS

It has been proved, that HEM of OPC can lead to substantial property improvements. It has further been proved, that the semi-continuously HEM-route using carrier-gas in compression mode can offer a manufacturing process in super-large volume capability for brittle solids.

These two general findings must be combined in a project with cement industry to introduce HEM as an innovative processing technique for the commercial production of superfine High Performance Portland Cement (HPPC) as well as later for the manufacturing of OPC at lower energy consumption and then at finer microstructure as well.

This is finally the suggestion of this work in order to achieve:

(*a*) a better product

- by reducing the particle size and activating the surface of OPC which can lead to:
- a faster setting time
- a faster curing time and
- substantially increased mechanical properties

(*b*) an economically advanced process

- by a far shorter grinding process at far higher energy efficiency
- by reduced firing temperature in the processing
- by reducing the production space for the grinding step

(*c*) an advanced product application

- by availability of super fast setting cement for repair-work
- by reduced construction dimensions and faster building

ACKNOWLEDGEMENT

The authors appreciate and acknowledge important contribution of CIMAY technicians: MS. Enrique Torres-Moye, Eng. J. Pablo-Mundloz-Mendoza and M. S. H. de la Garza Gutierrez. Part of this work was supported by Grupo Cementos de Chihuahua GCC, Mexico and SNI-Conancyt, Mexico.

The authors further appreciate and acknowledge the important contribution of

The State of Nordrhein-Westfalen,

Germany

for the main funding of TPW/TIP-projects:

(a)

development of a Ceramic-Simoloyer, a high kinetic processing device to perform Mechanical Alloying, High Energy Milling and Reactive Milling at high kinetic energy impact

applying ceramic processing tools, project no. TPW 9403–858-02/02-642-214

(b)

continuously and monitored flake-production by HEM with aerodynamic separation & classification and rapid particle size reduction of brittle solids (aux. target, enamel) project no. TIP 001-0102-0016

REFERENCES

1. H. de la Garza-Gutierrez, J.P. munoz-MEndozal, O. A. Chimal-Valencial, R. Martinez-Sanchez-S. D. De la Torre, A. Garcia-Luna and P. Garcia-Casillas: Optimizing Milling Parameters for Refining Portland Cenment, Journal of Metastable and Nanocrystelline Materials, 15-16 (2003) pp 395-400
2. S. D. De la Torre, et al., in preparation.
3. H. Zoz, H. Ren: Energy Balance during Mechanical Alloying, Measurement and Calculation Method supported by the Maltoz-software, PM_2Tech 1999, proceedings MPIF (1999)
4. Z. Sekulic, M. Stefanovic, B. Zivanovic: Mechanical activation of OPC and cement with additives. Proc. Int. 10th Congr. Chem. Cem., (1997) 65-IBAI-3
5. L. Shondeep Sarkar and J. Wheeler: Important properties of an ultra fine cement-Part 1. Cement and Concrete Research 31 (2001) pp. A 119-123.
6. L. Shondeep Sarkar and J. Wheeler: Microstructural development of an ultra fine Cement - Part 1. Cement and Concrete Research 31 (2001) pp. 125-128.
7. Authors discussion with A. Garcia, Interceramic, Chihuahua March 2003.
8. J.S. Benjamim, T.E. Volin, Metall. Trans. Vol. 5, 1929 (1974).
9. C. Suryanarayana, Guo-Hao Chen, F.H. Froes, Milling Maps for Phase Identification during Mechanical Alloying, Scripta Metallurgica et Materialia, Vol. 26, 1727 (1992).

10. H. Zoz, D. Ernst, H. Weiss, M. Magini, C. Powell, C. Suryanarayana, F.H. Froes, Mechanical Alloying of Ti- 24A1-11Nb (at%) using the Simoloyer Metall Vol. 50, 09/96, pp. 575-579,1996.

11. H. Zoz, D. Ernst, Mechanical Alloying using Cycle Operation— A New Way to Synthesize CMB-Materials, 5th International Conference on Advanced particulate Materials and Processes, proceedings MPIF (1997).

12. H. Zoz, D. Ernst, I. S. Ahn, W.H. Kwon: Mechanical Alloying of Ti-Ni-based Materials using the Simoloyer, TMS Annual Meeting 1997, eds. C.M. Ward-Close, F.H. Froes, S.S. Cho, D.J. Chellman: Synthesis/Processing of lightweight Metallic Materials, proceedings (1997).

13. H. Zoz, H. Ren, H.U. Benz, Ductile Metal Flakes based on [Au], [Ag], [Cu], [Ti], [Al], [Ni] and [Fe] by High Energy Milling (HEM) Part I PM2Tech 1999, MPIF, proceeding (1999).

14. R. Savin, H.U. Benz and H. Zoz, Zn and Zn-Ni-flakes for anti-.corrosives, unpublished exploitation report 05.02.-08.02.2000, Palm Springs, CA.

15. J.Y. Chung, J. Kim and Y.D. Kim: Formation of Nanocrystalline Fe-Co Powders Produced by Mechanical Alloying, Dept. of Metallurgy and Materials Science, Hanyang University, Ansan, Korea, 1999.

16. H. Zoz, D. Erst, T. Mizutani, H. Okouchi, Simoloyer CM100s, semi-continuously Mechanical Alloying in a production scale using Cycle Operation-Part 1, Metall Vol. 51, 10/97, pp. 568-572, 1997

17. H. Zoz, D. Ernnst, Simoloyer CM 100s, semi-continuously Mechanical Alloying in a production scale using Cycle Operation-Part II, Metal Vol. 51, 09/98, pp. 521-527,1998.

18. H. Zoz, H.U. Benz, 6. Schafer, M. Dannehl, J. Krull, F. Kaup, H. Ren and D. Jaramillo V.: High Kinetic Processing of Enamel, p. 1-a, INTERCERAM Intern. Ceramic Review, Vol. 50 (2001) [5] pp. 388-395.

19. H. Zoz, H.U. Benz, G. Schafer, M. Dannhl, J. Krull, F. Kaup, H. Ren and D. Jarainillo V.: High Kinetic Processing of Enamel, p. I-b, INTERCERAM lntern. Ceramic Review, Vol. 50 (2001) [6] pp. 470-477.

20. T. Nowicki, C. Carbonnaux, Precious-metal-base advanced materials, Journal de Physique IV, Colloque C 7, Supplement au Journal de Physique III, Vol. 3, Nov. 1993, pp. 509-518.

21. H. Zoz, H. Ren, N. Späth: Improved Ag-SnO_2 Electrical Contact Material Produced by Mechanical Alloying, Metal Vol. 53, 07-08/99, pp. 423-428, 1999.

22. G. Kaupp, J. Schmeyers, M. R. Naimi-Jamal, H. Ren, H. Zoz : Reactive milling with the Simoloyer: environmentally benign quantitative reactions without solvents and wastes, Elsevier, Chemical Engineering Science 57 (2002) 763-765.

23. G. Kaupp, M. R. Naimi-Jamal, H. Ren and H. Zoz: Environmentally Protecting Reactive Milling, CHEMIE TECHNIK, 31, Vol. 6 (2002) p. 58-60.

24. H. Rumpf, K. Schönert : Zerkleinern, Ullmanns Enzyklopädie der Technischen Chemie, Weinheim 1972.

25. H. Zoz, J.S. Kim, Anwendung and Auslegung von Trommeluhlen, Keramische Zeitschrift, Vol. 53 (2001)[5] pp. 384-392.

26. K. Ameyama, S. Umekawa, H. Ren, D. Jaramillo V., H. Zoz : Alternative Mechanical Milling routes for grain-refinement of conventional High-speed Steel powder for later consolidation by SPS, PM2TEC 2003, MPIF (2003).

27. H. Zoz, S. Morales and D. Jaramillo V. : Comparative routes of solid-solution-formation by MM of Ag-70Cu (at%), PM2TEC 2003, MPIF (2003).

28. S. D. De la Torre, K. N. Ishihara, P. H. Shingu, D. Rios-J., and H. Miyamoto: Mechanochemical Synthesis of Refined Ag- and Zn- Composites Starting from Oxides. Innovative Processing/ Syntheis: Ceramics, Glasses, Composites II. Ceramic Transactions, Am. Ceram. Soc. 94 (199) pp. 287-294.

29. P. C. Hewlett: Lea's Chemistry of Cement and Concrete, Young Wiley & Sons. Inc., Fourth Ed. 1997.

30. H. Zoz, H. U. Benz, T. Suzuki, H. Ikehata and T. Saito, Zoz GmbH (Germany), Toyota Central R&D LABS., INC. (Japan), ESIQIE (Mexico) : HKP using Carrier-gas Assisted Discharging, PM2TEC 2003, MPIF (2003).

31. Authors discussion with P. Burciaga, Grupo Cementos, Chihuahua, March 2003.

Index

D

E

F

G

H

I

L

M